TEETH: Form, Function, and Evolution

TEETH: *Form, Function, and Evolution*

Edited by Björn Kurtén

Columbia University Press
New York *1982*

Library of Congress Cataloging in Publication Data
Main entry under title:

Teeth—form, function, and evolution.

Papers presented at the 5th International Symposium on Dental Morphology, held Aug. 13–15, 1979, in Turku, Finland.
Includes bibliographies and index.
1. Teeth—Congresses. 2. Teeth—Evolution—Congresses. 3. Dental anthropology—Congresses. 4. Anatomy, Comparative—Congresses. I. Kurten, Bjorn. II. International Symposium on Dental Morphology (5th : 1979 : Turku, Finland) [DNLM: 1. Evolution—Congresses. 2. Odontogenesis—Congresses. 3. Tooth—Anatomy—Histology—Congresses. 4. Tooth—Physiology—Congresses. W3 IN917D 5th 1979t / WU 101 T258 1979]
QL858.T35 599.04'3 81-10210
ISBN 0-231-05202-2 AACR2

Columbia University Press
New York Guildford, Surrey

Printed in the United States of America

Clothbound editions of Columbia University Press books are Smyth-sewn and printed on permanent and durable acid-free paper.

Contents

Contributors

Lassi Alvesalo
Deparment of Dentistry, University of Kuopio, PL 138, 70101 Kuopio 10, Finland

Gudjón Axelsson
University of Iceland, Faculty of Odontology, Landspitalínn, Reykjavik, Iceland

Patricia L. Belier
Department of Anthropology, University of Toronto, Toronto, Ontario, Canada

Percy M. Butler
Department of Zoology, Royal Holloway College, Alderhurst, Bakeham Lane, Englefield Green, Surrey TW20 9TY, England

J. G. Clement
Department of Dental Anatomy, The London Hospital Medical College, Turner Street, London E1 2AD, England

Albert A. Dahlberg
Department of Anthropology, University of Chicago, Chicago, Illinois 60637

Thelma Dahlberg
Department of Anthropology, University of Chicago, Chicago, Illinois 60637

E. Dykes
Department of Dental Anatomy, The London Hospital Medical College, Turner Street, London E1 2AD, England

Mikael Fortelius
Department of Geology, University of Helsinki, Snellmaninkatu 5, 00170 Helsinki 17, Finland

David G. Gantt
Institute of Dental Research, School of Dentistry, University of Alabama in Birmingham, University Station, Birmingham, Alabama 35294

Denys H. Goose
School of Dental Surgery, University of Liverpool, Pembroke Place, P.O. Box 147, Liverpool L69 3BX, England

Pentti Kirveskari
Institute of Dentistry, University of Turku, 20520 Turku 52, Finland

Wighart von Koenigswald
Hessisches Landesmuseum, 6100 Darmstadt, Friedensplatz 1, Federal Republic of Germany

Edward J. Kollar
Department of Oral Biology, School of Dental Medicine, University of Connecticut, Health Center, Farmington, Connecticut 06032

Cornelis A. W. Korenhof
Tandheelkundig instituut, Rijksuniversiteit Utrecht, Sorbonnelaan 16, 3508 TB Utrecht, Netherlands

P. W. Lucas
Unit of Anatomy in Relation to Dentistry, Department of Anatomy, Guy's Hospital Medical School, London SE1 9RT, England

A. G. S. Lumsden
Department of Anatomy, Guy's Hospital Medical School, London SE1 9RT, England

Fumihiro Masaki
Child Study Clinic, School of Dentistry, University of Oregon Health Sciences Center, 611 SW Campus Drive, Portland, Oregon 97201

John T. Mayhall
Faculty of Dentistry, University of Toronto, Toronto, Ontario, Canada

Melvin L. Moss
Department of Anatomy, College of Physicians & Surgeons of Columbia University, 630 West 168th Street, New York, N.Y. 10032

Letty Moss-Salentijn
Department of Anatomy, College of Physicians & Surgeons of Columbia University, 630 West 168th Street, New York, N.Y. 10032

G. Gisela Nass
Overseas Division, University of Maryland [PSC Box 100, APO San Francisco, California 96328

Susan Nielsen
Child Study Center, School of Dentistry, University of Oregon Health Sciences Center, 611 SW Campus Drive, Portland, Oregon 97201

J. F. van Reenen
School of Dentistry, University of the Witwatersrand, Johannesburg 2001, South Africa

John M. Rensberger
Burke Museum DB-10, University of Washington, Seattle, Washington 98195

Eric E. Roberts
School of Dental Surgery, University of Liverpool, Pembroke Place, P.O. Box 147, Liverpool L69 3BX, England

Takuro Sakai
Department of Anatomy, School of Dentistry, Aichi-Gakuin University, Chikusa-ku, Nagoya, Japan

Shelley R. Saunders
Department of Anthropology, University of Toronto, Toronto, Ontario, Canada

Bhim Sen Savara
Child Study Clinic, School of Dentistry, University of Oregon Health Sciences Center, 611 SW Campus Drive, Portland, Oregon 97201

Jeffrey H. Schwartz
Department of Anthropology, University of Pittsburgh, Pittsburgh, Pennsylvania 15260

G. Richard Scott
Anthropology Program, University of Alaska, Fairbanks, Alaska 99701

Richard James Shadel
Child Study Clinic, School of Dentistry, University of Oregon Health Sciences Center, 611 SW Campus Drive, Portland, Oregon 97201

Patricia Smith
Department of Anatomy, The Hebrew University-Hadassah Medical School, P.O. Box 1172, Jerusalem, Israel

R. M. S. Taylor
Department of Anatomy, School of Medicine, University of Auckland, Auckland, New Zealand

TEETH: Form, Function, and Evolution

INTRODUCTION

Björn Kurtén

The Fifth International Symposium on Dental Morphology was held August 13–15, 1979, in Turku, Finland—more precisely, on the island of Ruissalo, just outside the city. The history of these symposia goes back to 1965, when the first one was held in Fredensborg, Denmark; its proceedings were published in 1967 in the *Journal of Dental Research* (vol. 46). The second symposium, at the Royal Holloway College, London, was held in 1968, and published as *Dental Morphology and Evolution* (University of Chicago Press, 1971), edited by A. A. Dahlberg. The third symposium, in Brussels in 1971, has not been published. The fourth was in Cambridge, England, in 1974, and was published as *Development, Function, and Evolution of Teeth* (Academic Press, 1978), edited by P. M. Butler and K. A. Joysey.

The present volume constitutes the proceedings of the Turku symposium. Most of the papers (except numbers 6 and 20) were read at the symposium.

The papers contained herein represent a broad survey of the present-day frontiers of our subject. In topic they range widely: from experimental work to speculative theorizing, from classical studies of dental traits in human populations to mathematical and computer-based analyses of form and function, from investigation of enamel microstructure to paleontologic documentation of evolution; and we need them all. Every student of dental morphology will find here contributions of special interest to himself. For my part, as an evolutionist, I have found many papers bearing, sometimes unexpectedly, on that subject. Take just one example—Kollar's amazing discovery of the interaction between avian epithelium and dental mesenchyme, despite the fact that birds have been toothless for more than 70 million years (by no means the sole *Aha-Erlebnis* that came to me while reading the contributions). It would, however, be unfair to enumerate "favorites" from such a one-sided viewpoint.

I have attempted to collect the papers into four sections, according to their main theme: papers on ontogeny; papers on the structure, shape, and function of teeth; papers on populations; and papers on evolution. The distinction is artificial in many instances. A population study, for example, may have a direct bearing on evolution. The arrangement, however, gives the volume the backbone of a coherent story. We begin with the early development of the dental lamina, proceed through ontogeny, then look at its result—the dentition in action—and proceed to explore its variation and differentiation in living populations, finally to bring in the phylogeny that lies behind it all.

The papers in the section on ontogeny range from reports of advanced experimental work to speculative approaches to new areas of study. Letty Moss-Salentijn's study of the dental lamina in relation to positional landmarks suggests that the initial invasion of the lamina may be a passive event, in contrast with the morphogenetic activity which follows in the enamel organs. Edward J. Kollar shows that avian epithelium may be made experimentally to interact with dental mesenchyme and partake in odontogenesis, in spite of the edentulous nature of bird jaws. The role of innervation in the initiation of odontogenesis is explored by A.G.S. Lumsden, who shows that a temporal relationship exists. *Post hoc, propter hoc*?

The initiation phase of dental ontogeny is followed by that of morphogenesis, in which the cusp pattern develops. P.M. Butler's speculative paper attempts to pose the questions to be asked in studying these events, at present poorly understood. A still later phase of ontogeny is treated by Bhim Sen Savara et al., who are developing methods for the simultaneous study of the growth of the head, face, and dentition, as recorded for individual subjects over a number of years. R.M.S. Taylor, treating the final stage of dental ontogeny—the development of the third molar—frames a hypothesis to account for deformation of M^3 which is likely to be argued pro and con.

In the final paper of this section, Melvin L. Moss speculates upon the role of allometry and the logarithmic spiral as expressions of thermodynamics in the genesis of enamel structure and tooth crown size and shape.

Enamel structure and the form and function of teeth are the central themes of the papers in part 2. David D. Gantt, studying hominid teeth, shows that hominids and pongids differ in enamel thickness and prism pattern, and that this difference may be comparatively ancient. Wighart von Koenigswald elucidates the relationships between enamel prism pattern, tooth shape, and function in the molars of arvicolid rodents. Both studies have important ramifications in the areas of functional morphology and phylogeny. Jeffrey H. Schwartz's broad discussion of heterodonty and homology in mammalian dentitions ranges from ontogeny to phylogeny, and takes a critical look at some of our ingrained assumptions.

Next follow two papers seeking new methods for the characterization of tooth design. A new, three-dimensional recording technique for serial sectioning is introduced by J. G. Clement and E. Dykes, and applied to the study of dental asymmetry in man, while P. W. Lucas develops a generalized model for the study of tooth function in mastication: the interaction between tooth shape and the food to be comminuted.

The functional aspect is also stressed in the two last papers in this section. Dealing with rhinoceros dentitions, Mikael Fortelius uses functional analysis to reach paleobiological conclusions, while J. F. van Reenen studies the effect of attrition on dental dimensions in a human population.

This brings us to the theme of part 3: populations, G. Gisela Nass' analysis of dental asymmetry in a free-ranging macaque population establishes important differences between the developmental canalizations in deciduous and permanent teeth, and also, to some extent, between the sexes. Aspects of dental morphology in human populations form the main theme in the four contributions that follow. Denys M. Goose and Eric E. Roberts chart the dentitions in a sharply localized sample from North Wales; Gudjón Axelsson and Pentti Kirveskari chart that of Icelanders; and J. T. Mayhall, S. R. Saunders, and P. L. Belier, chart that of the whites of North America—represented, to be sure, by a local population from Ontario, yet suggesting a revised "Caucasoid Dental Complex." G. Richard Scott and Albert A. Dahlberg analyze differentiation in Indians of the North American Southwest, and its possible relation to linguistic grouping. These papers typify the mainstream of anthropological odontology, and contribute to the build-up of a firm foundation of knowledge in this field.

The final two papers in this section deal with the genetics of dental patterns, using opposite and complementary approaches. In a short, abstract article, Albert A. Dahlberg, Pentti Kirveskari, and Thelma Dahlberg study pedigrees for dental traits, thus using the phenotype as their starting point. On the other hand, Pentti Kirveskari and Lassi Alvesalo, starting from the genotype, study the teeth of a chromosomal-aberration group (45, X females) in relation to their normal siblings.

While many contributions to this volume have important evolutionary connotations, those collected in the fourth and final section are concerned with evolution as their main theme, at various levels, from the deme to the order. The origin of molar cusps and tubercles in the primate dentition is traced by Takuro Sakai in a comparative study. John M. Rensberger analyses dental change in two exceptionally well-documented lineages of aplodontid rodents from the Miocenc of Oregon, evaluating their functional significance. C. A. W. Korenhof argues that the dentine-enamel partition, as reflected in endocasts of enamel caps of deciduous molars, enables us to take a look at a more original condition of dental traits than the occlusal surface. Patricia Smith studies dental reduction in human populations and considers the ecological factors that may be conducive to such trends.

So the circle is closed. It remains for the editor to express his gratitude to the contributors, to the publishers, and last but not least, to the Secretary of the Symposium, Pentti Kirveskari, upon whom fell the brunt of the organizational work, and whose assistance has been invaluable at all stages of the editing.

I

ONTOGENY

1. Morphological Aspects of the Growth Behavior of the Early Dental Lamina in the Cat and Rat

LETTY MOSS-SALENTIJN

ABSTRACT

The initial development of the dental lamina (up to the cap stage of dental development) was studied in serial, coronal sections of cat and rat embryos. Tracings of these coronal sections, as well as horizontal and sagittal graphical projections of these tracings, were prepared.

Relative to the principal trigeminal nerve bundles of the face, the basal portion of the dental lamina remained approximately stationary. The buccolingual widths of the attachments of the dental lamina to the oral epithelium became thinner. There were distinct differences in the shapes of the dental laminae in upper and lower jaws. On the basis of these findings, the suggestion is supported that the initial "invasion" of the dental lamina, in contrast with the later morphogenetic events in the enamel organs, is a passive event, determined by the growth patterns of the surrounding tissues.

The ontogenesis of vertebrate teeth is characterized by a series of interdependent epithelio-mesenchymal interactions, which produce a distinct pattern of morpho- and cytodifferentiations essential for tooth formation. In recent years attention has been focused largely on those epithelio-mesenchymal interactions, which result in specific cytodifferentiations within the tooth germ (see Slavkin 1978; Thesleff 1978 for recent reviews).

Despite the impressive advances made by such studies, relatively little is known about the positional relationships that exist between a developing mammalian tooth and the structures surrounding it during its early growth and morphodifferentiation.

Descriptions in most textbooks of mammalian dental histology suggest that the dental lamina, and subsequently the enamel organs, progressively invade the subjacent ectomesenchyme in the early stages of odontogenesis. Some authors have proposed an alternative explanation, namely that this *apparent* invasion may be a passive event, resulting from the proliferation of ectomesenchymal cells around the relatively stationary enamel organs (Blechschmidt 1974; Blechschmidt and Gasser 1978; Osborn and Ten Cate 1976; Schroeder 1976).

In this paper some observations will be presented dealing with the morphology and the growth behavior of dental lamina and enamel organs in cat and rat embryos and fetuses. Based on these observations the alternative hypothesis, that the apparent invagination of epithelial components during early tooth morphogenesis is a passive event, will be supported.

Materials and Methods

Serial histological sections, cut parallel to the coronal plane, of 20 cat embryos and fetuses (from 13 to 26 mm crown-rump length [CRL]) and of 10 rat embryos and fetuses (14.5 to 17.5 days post conception) were studied. The specimens had been selected from the Detwiler collection, presently housed at Columbia University.

The rat embryos had been fixed in 10 percent neutral formalin. The fixation of the cat embryos is unknown, but based on the presence of crystals precipitated in the sections, a sublimate-containing fixative was probably used (Romeis 1968). All specimens had been embedded in celloidin.

In both groups of embryos the youngest specimens were characterized by the presence of thickened epithelial zones in the future tooth-bearing regions of the jaws, while in the oldest specimens the earliest developing enamel organs had reached the cap stage.

The sections of the cat and rat embryos and fetuses were displayed on the large projection screen of a Zeiss Ultraphot II microscope. With the aid of acetate tracing paper, tracings were prepared at a magnification of 40 X of every fourth section through the oral region. Included in the tracings were: the outline of the oral cavity; the transversely sectioned infraorbital and inferior alveolar nerves; and the interfaces between oral epithelium and jaw ectomesenchyme in the future tooth-bearing regions (fig. 1.1a).

On each tracing one reference line was drawn running through the centers of the outlines of both inferior alveolar nerves and another running through the centers of the outlines of both infraorbital nerves. The midpoint between the nerves was marked on these lines. From these tracings sagittal and horizontal projections were constructed graphically.

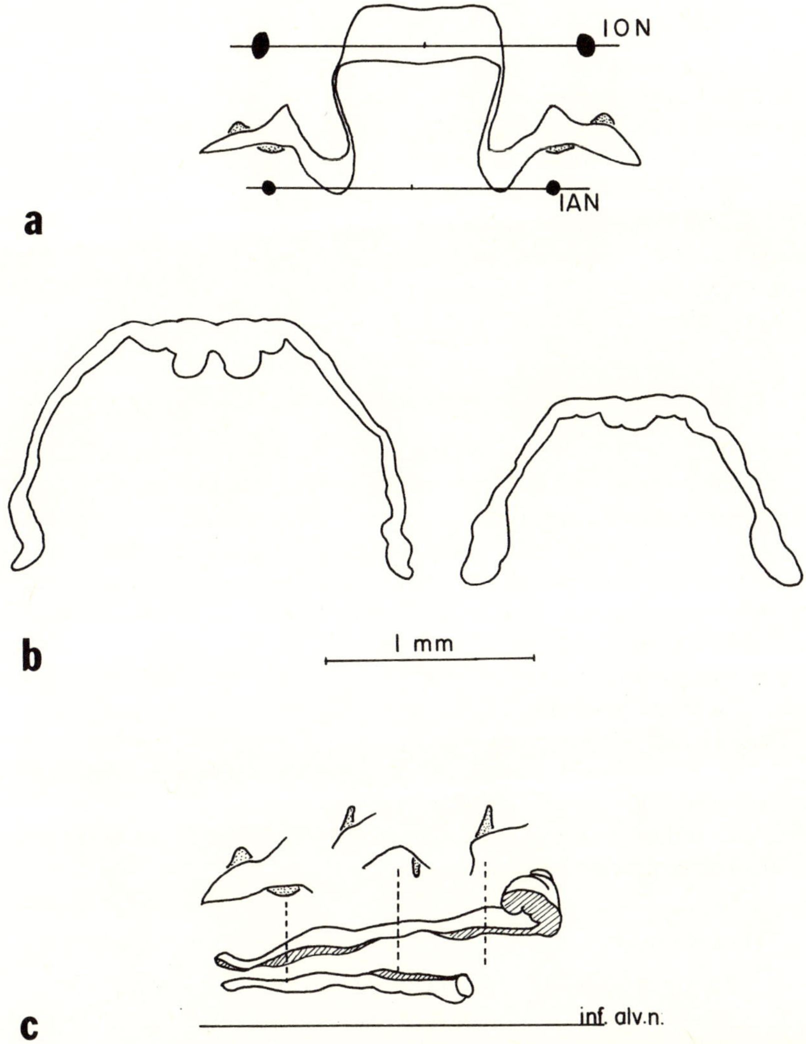

Figure 1.1. *a.* Tracing of a coronal section through the oronasal region of a 15 mm CRL cat embryo. In addition to the outline of the oronasal cavity, the outlines of the cross-sectioned infraorbital nerves (ION) and inferior alveolar nerves (IAN) are shown in black. Reference lines are drawn through the centers of these nerves and midpoints are indicated. The dental laminae in upper and lower jaws are shaded lightly.

b. Horizontal projections of upper (left) and lower (right) dental laminae of a 15 mm CRL cat embryo. The buccolingual widths of the future odontogenic areas exceed those of the non-odontogenic areas. Note the zig-zag arrangement, especially in the molar region of the upper lamina.

c. Sagittal projections of left upper and lower dental laminae of a 15 mm CRL cat embryo, relative to the plane of the inferior alveolar nerve (inf. alv. n). The oral surfaces are shaded. The morphology of the laminae in coronal section at three selected sites is shown.

All diagrams in this figure are shown at the same magnification.

SAGITTAL PROJECTIONS

A straight, horizontal line, representing the reference plane through the inferior alveolar nerves in sagittal view, was drawn. Subdivisions were marked on this line, corresponding accurately to the combined thickness of four histologic sections multiplied by a magnification factor of 40 X. Relative to this line at each subdivision, the vertical distances were marked to the oral surface of the dental lamina and to its deepest point of invagination (as measured in the corresponding tracings). When similar points were connected, proportionately accurate sagittal projections of the dental lamina were obtained (fig. 1.1c).

HORIZONTAL PROJECTIONS

A line was drawn with subdivisions, the width of each corresponding to the combined thickness of 4 histological sections, multiplied by 40 X. This line represented the midline between either the two inferior alveolar nerves or the two infraorbital nerves. Relative to this line, at each subdivision, the horizontal distances were marked to the most buccal and lingual points of the epithelial thickenings, the dental lamina and the enamel organs (as measured in the corresponding tracings). When similar points were connected, proportionately accurate horizontal projections of the dental lamina were obtained (fig. 1.1b).

The original sections, the tracings, and the projections were further analyzed. Sagittal and horizontal projections of the cat tracings were superimposed at specific registration points, such as greater palatine foramina and mental foramina, to compare the development of the dental laminae at successive stages.

Observations

In the youngest rat and cat embryos presently studied, the oral surfaces of the upper and lower jaws were covered with a thickened epithelial layer. In the rat embryos separate epithelial thickenings were found in the incisor and molar regions.

In the cat embryos continuous bands of thickened epithelium were present on the oral surfaces of upper and lower jaws (fig. 1.2). The buccolingual widths of these epithelial thickenings, measured on selected histological sections, was approximately 175 μm in the rat embryos. The thickened epithelium in the cat embryos had similar buccolingual widths at the presumptive odontogenic sites, but only one-third of this width at non-odontogenic sites.

In 15 mm CRL cat embryos and 15 days post conception rat embryos,

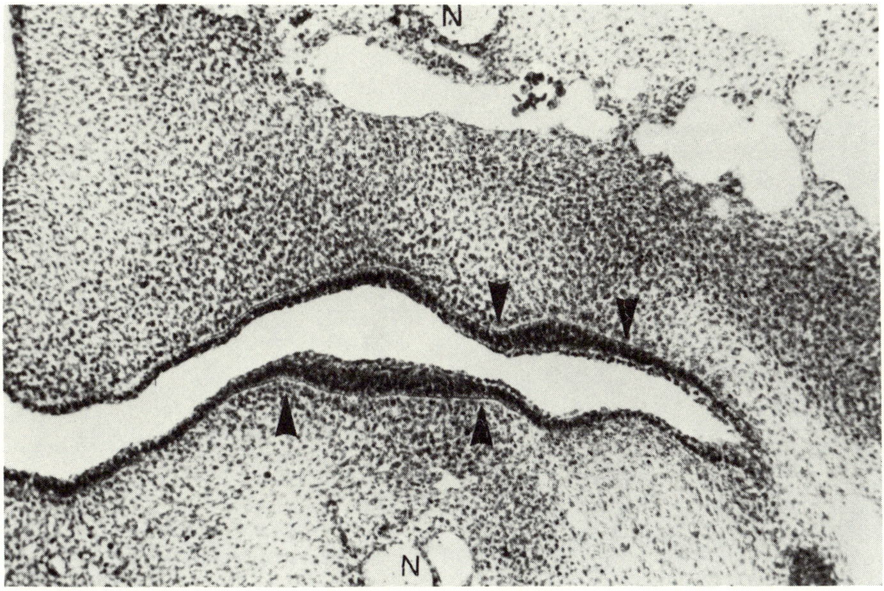

Figure 1.2. Epithelial thickenings (between arrowheads) in the future molar regions of upper and lower jaw in a 13 mm CRL cat embryo. Sections through inferior alveolar and infraorbital nerves (N, N) are visible. Magnification 135 times.

the dental lamina extended to some depth into the jaw mesenchyme. This was most notable at sites of future teeth, where the dental lamina extended into a dense mass of well-vascularized mesenchyme. In the cat embryos bulbous epithelial buds had become visible at the sites of the future deciduous incisors and canines.

In the older rat embryos (16.5–17.5 days post conception) one bulbous epithelial bud was found in the incisor region and one in the molar region of each jaw quadrant. In the older cat embryos (between 21 and 26 mm CRL) the epithelial buds in incisor, canine, and molar regions had developed into distinct caps (three deciduous incisors, one deciduous canine, and one deciduous molar (Pd_3) in each jaw quadrant). An additional, small epithelial bud (Pd_4) had appeared posterior to the molar cap in each quadrant.

In the oldest cat (26 mm CRL) and rat (17.5 days post conception) specimens the greatest buccolingual widths of the molar enamel organs in cats and rats were approximately 135 μm. The widths of the dental lamina attachment to the oral epithelium were approximately 30 μm in both species (fig. 1.3).

Of particular interest was an observation made on the horizontal projections of the dental laminae. In both species, but especially in the cat

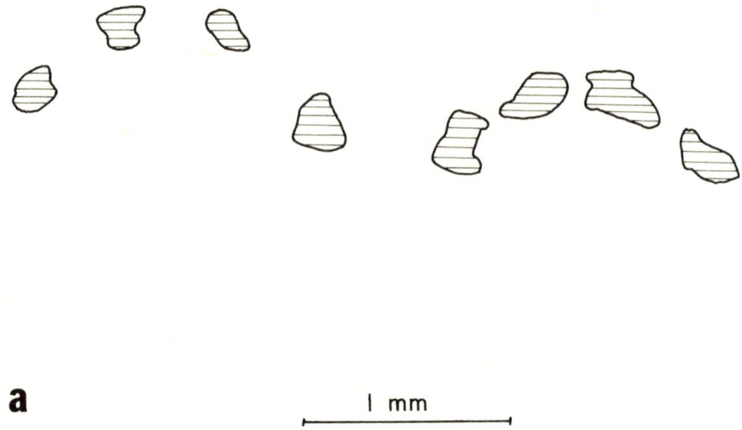

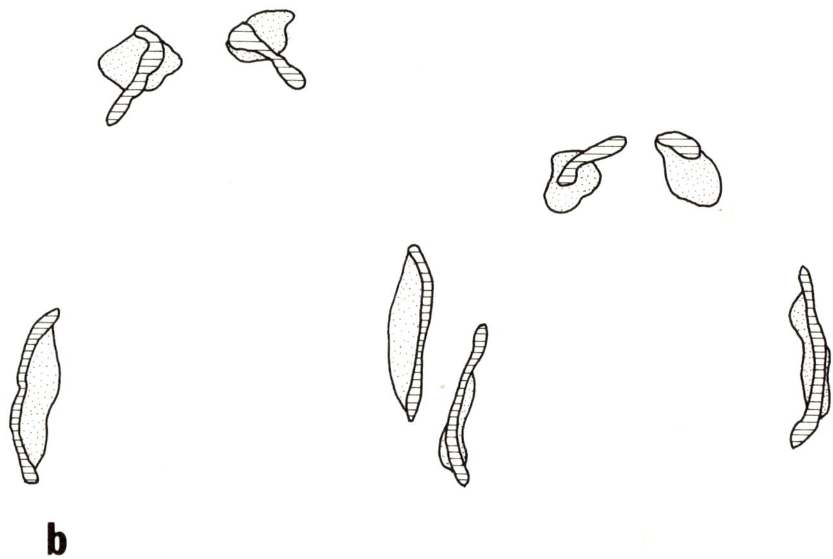

Figure 1.3. Horizontal projections of dental laminae in a rat embryo and fetus.

a. Upper (left) and lower (right) dental laminae of an embryo (14.5 days post conception). Separate epithelial thickenings exist in incisor and molar regions.

b. Upper (left) and lower (right) dental laminae of a fetus (17.5 days post conception). The attachment of the dental lamina to the oral epithelium is shown as horizontally hatched zones, while the bulbous thickenings, protruding into the mesenchyme, are indicated as stippled zones.

Both sets of projections are shown at the same magnification.

embryos, the horizontal projections of the dental laminae exhibited a zig-zag pattern (figs. 1.1 and 1.5). A similar but somewhat less obvious pattern was noted in sagittal projections.

In the older cat embryos a marked difference could be observed in the appearance of the dental laminae in the non-odontogenic regions of upper and lower jaws. In coronal section the dental lamina of the lower jaw was triangular, with its base at the oral surface and its apex protruding over a short distance near-vertically into the mesenchyme (fig. 1.4a). In the upper jaw the rod-shaped dental lamina protruded deeply into the mesenchyme, slanting strongly with its deepest tip pointing in palatal direction (fig. 1.4b).

For the purpose of comparing specimens of various ages, superimpositions were made of the horizontal projections of the cat tracings. The projections were superimposed on the midline, with the sites of the mental foramina as registration points for the lower jaw, and the sites of the greater palatine foramina as registration points for the upper jaw.

Relative to these registration points the horizontally projected molar primordia remained relatively stationary. The dental laminae extended further anteriorly largely by increases in antero-posterior lengths of the non-odontogenic areas and posteriorly by addition of a new deciduous molar primordium.

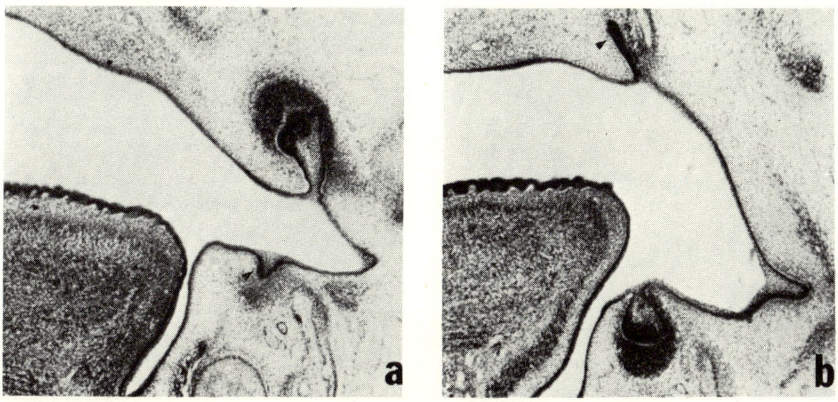

Figure 1.4. Coronal sections through the molar region of a 26 mm CRL cat fetus show the characteristic appearances of the non-odontogenic parts of the dental laminae in upper and lower jaws.

a. Non-odontogenic part of the dental lamina in the lower jaw (arrow). In the upper jaw Pd_3 is visible in the cap stage of development.

b. Non-odontogenic part of the dental lamina in the upper jaw (arrow). In the lower jaw Pd_3 is visible in the cap stage of development. In both photographs the tongue is in the left half of the picture. Magnification 55.75 times.

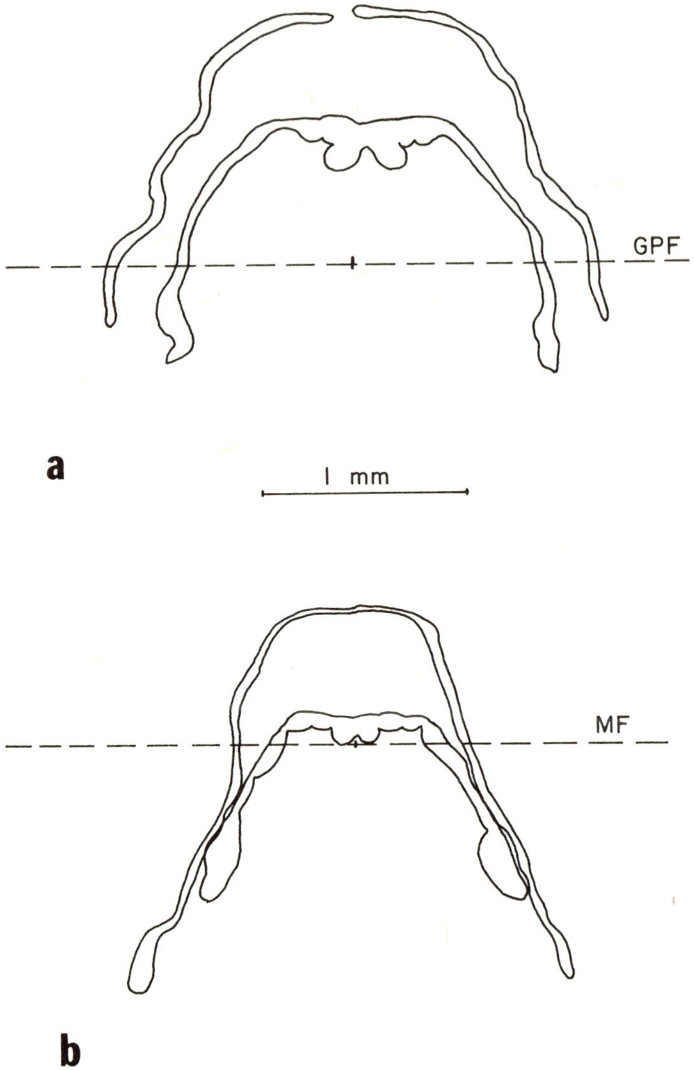

Figure 1.5. Superimpositions of horizontally projected dental laminae of cat embryo (15 mm CRL) and cat fetus (26 mm CRL). For clarity, only the zones of the lamina attachment to the oral epithelium are indicated.

a. superimposition of horizontally projected upper laminae on the midline, with registration on the greater palatine foramina (GPF).

b. Superimposition of horizontally projected lower laminae on the midline, with registration on the mental foramina (MF).

In the lower jaw almost all changes in dimensions occurred in antero-posterior direction, with very little increase in width in the molar region. In the upper jaw there was some gradual increase in width of the entire dental arch, along with an increase in antero-posterior dimensions (fig. 1.5).

At the time of secondary palate closure (in 21.5–22 mm CRL specimens) the intermolar width decreased somewhat, but afterwards it gradually increased again.

Superimposition of individual coronal tracings, as well as sagittal projections on the inferior alveolar nerve reference line for the lower jaw and on the infraorbital nerve reference line for the upper jaw, showed that the basal cells of the thickened epithelium in early embryos were located at the same distance or nearer to these reference lines than the basal cells of the enamel organs in cap stage of the older specimens (fig. 1.6).

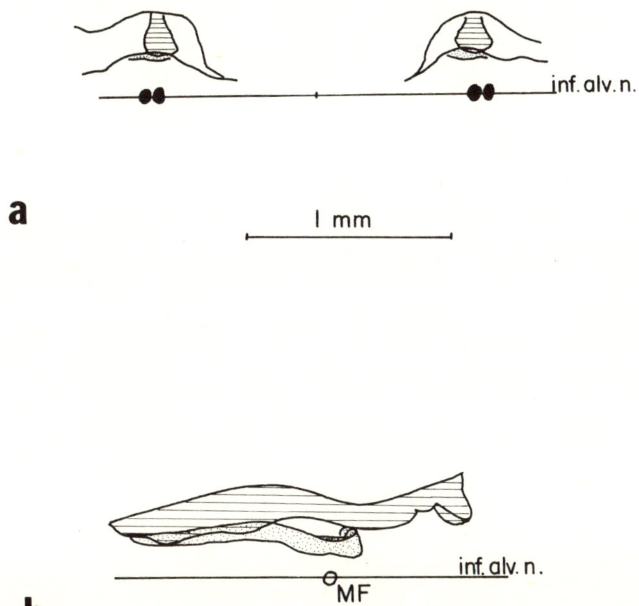

Figure 1.6. *a.* Superimposition of comparable coronal tracings through the lower molar region of a 13 mm CRL cat embryo (stippled) and a 26 mm CRL cat fetus (horizontally hatched), registered on the reference lines through the inferior alveolar nerves (inf. alv. n.). Relative to the nerves, the basal part of the dental epithelium remains relatively stationary.

b. Superimpositions of sagittal projections of lower dental laminae of a 13 mm CRL cat embryo (stippled) and a 26 mm CRL cat fetus (horizontally hatched). The projections are superimposed on the inferior alveolar nerve plane (inf. alv. n.) and registered on the mental foramina (MF).

Discussion

The methods of graphically producing horizontal and sagittal projections, based on the measurements of traced coronal sections, has been described in detail by Gaunt and Gaunt (1978). In existing, serially sectioned embryologic specimens, such as those in the Detwiler collection, no external registration landmarks are present. Therefore, the principal nerves in the facial region were chosen as landmarks for registration since they were the structures whose *relative* locations in the growing face were most constant.

This was based on the generally accepted premise that the "neural skeleton" of the face, and indeed of the rest of the body, has a location that is relatively stable, especially at the sites of future foramina. If the relative position of a foramen were to change considerably during growth, the enclosed nerve and accompanying blood vessels might be compromised, and there is no biological evidence that this normally occurs.

The use of specific foramina as additional registration points, when the need arose for comparison between embryos of different ages, was based on findings of earlier studies (Moss-Salentijn 1981a; 1981b) that had demonstrated the biological appropriateness of those points.

The use of trigeminal nerve roots as registration landmarks introduces constraints on the interpretation of the graphic projections and superimpositions depicted in this study. Registration on the infraorbital and the inferior alveolar nerves permits only the interpretation of *local* growth behavior of structures in their neighborhood, and this for only the period of time of the present study. However, within these constraints, the relational changes between the nerves and their adnexa may be recorded with confidence.

Since the pioneering work on dental laminae and enamel organs by Röse (1891) many investigators have published three-dimensional reconstructions, illustrating the changes in shape and dimensions of the early dental lamina and enamel organs in human embryos and fetuses (Burdi 1968; Burdi and Lillie 1966; Nery et al. 1970; Ooë 1956, 1957, 1959), cats (Gaunt 1959; Scheidt 1894), dogs (Williams and Evans 1978), rats and mice (Gerber et al. 1974; Nagai et al. 1976; Peterková 1974; Strassburg et al. 1971). All these studies have dealt largely with isolated dental laminae without reference to their positional changes relative to surrounding structures.

Even those authors who proposed the alternative explanation, that the apparent invasion of the dental lamina may be a passive event (Blechschmidt 1974; Blechschmidt and Gasser 1978; Osborn and Ten Cate 1976; Schroeder 1976), did so indirectly, on the basis of either the growth patterns of lips or the patterns of disruption of the dental lamina in later stages of development.

As was shown in the present study, the distance between the plane of the related nerve (inferior alveolar or infraorbital nerve) and the basal portion of the dental lamina did not decrease, but either remained approximately the same or increased to some extent during the early period of development of the dental lamina. At the same time there was a distinct increase in distance between the nerve plane and the oral epithelium covering the surfaces of the jaws.

The buccolingual width of the dental lamina at the level of the basal layer of the oral epithelium decreased substantially in this same period (an observation also noted by Gaunt 1959), while the bulbous epithelial thickenings in the presumptive odontogenic areas of the dental lamina had buccolingual widths approximately equal to the buccolingual widths of the initial epithelial thickenings.

Finally, a distinct difference was noted in the outlines of the coronally sectioned, non-odontogenic parts of the dental laminae in upper and lower jaws of the cat embryos. The differences in shape of the dental lamina outlines may reflect the different directions of the growth vectors that are active in the tissues of upper and lower jaws. In the lower jaw a pronounced horizontal growth in antero-posterior direction is found. In the upper jaw there is some antero-posterior growth but less than in the lower jaw, while in addition an asymmetric vertical downward growth of the palatal processes is taking place in this period.

The different growth patterns in upper and lower jaws also govern the different increases in distance between presumptive odontogenic sites. Thus, the lengths of the non-odontogenic parts of the dental lamina appear to be passively increasing.

These three sets of observations, while still not conclusive, provide some additional support for the notion that the initial "invasion" of the dental lamina may be a passive event, distinct from the subsequent "active" morphogenesis of the enamel organs in the presumptive odontogenic locations. This conclusion agrees well with the findings of Osman and Ruch (1975, 1976), who failed to note any difference in mitotic activity in the odontogenic epithelium and mesenchyme of mouse embryos relative to the surrounding tissues in the early stages of dental lamina development.

Two additional observations were made in this study, not directly related to the posed problem, but worth mentioning. The decrease in intermolar width at the time of secondary palate formation has also been noted by Diewert (1973) in rat embryos. The secondary palate is formed when the two initially vertical palatal processes become horizontal and fuse. This change in orientation is predominantly due to a rotation of the palatal processes around an antero-posterior axis (Moss-Salentijn 1981b). The dental lamina is involved passively in this rotation, and as a result, the intermolar distance decreases somewhat, temporarily.

Of additional interest is the zig-zag (or serrate) arrangement of the dental laminae, especially in cat embryos. This arrangement was seen in both horizontal and sagittal projections. A similar pattern has been illustrated by Burdi (1968) in graphic horizontal projections of human dental laminae. One interpretation could be that this pattern is the two-dimensional projection of a (logarithmic?) spiral. If this interpretation is correct, it would have certain implications for the existence of an allometric growth pattern of the dental lamina (Moss, this volume).

The suggestion, supported in this paper, that the apparent initial invasion of the dental lamina may be a passive event, is not an isolated one in the developmental literature. There is a growing awareness among embryologists that several instances of apparent cell migrations (Gasser 1979) or invaginations (Jacobson et al. 1979) may have alternative explanations. These become evident only when the relative growth behavior of a growing system is studied as a proper adjunct to the studies of isolated growing organ systems.

References

Blechschmidt, E. 1974. *Humanembryologie. Prinzipien und Grundbegriffe.* Stuttgart: Hippokrates.

Blechschmidt, E. and R. F. Gasser. 1978. *Biokinetics and Biodynamics of Human Differentiations: Principles and Applications*. Springfield, Ill.: C. C. Thomas.

Burdi, A. R. 1968. Morphogenesis of mandibular dental arch shape in human embryos. *J. Dent. Res.* 47:50–58.

Burdi, A. R. and J. H. Lillie. 1966. A catenary analysis of the maxillary dental arch during human embryogenesis. *Anat. Rec.* 154:13–20.

Diewert, V. 1973. The course of the palatine arteries during secondary palate development in the rat. *J. Dent. Res.* 52:1273–1280.

Gasser, R. F. 1979. Evidence that sclerotomal cells do not migrate medially during normal embryonic development of the rat. *Am. J. Anat.* 154:509–524.

Gaunt, W. A. 1959. The vascular supply to the dental lamina during early development. *Acta Anat.* 37:232–252.

Gaunt, W. A. and P. N. Gaunt. 1978. *Three-Dimensional Reconstruction in Biology.* Tunbridge Wells: Pitman Medical Publishers.

Gerber, R., V. Karcher-Djuricic, and J. V. Ruch. 1974. Matériel presomptif odontogène et topographie de l'epithelium dentaire dans la mâchoire inférieure de l'embryon de souris. *J. Biol. Buccale* 2:249–258.

Jacobson, A. G., D. M. Miyamoto, and S.-H. Mai. 1979. Rathke's pouch morphogenesis in the chick embryo. *J. Exp. Zool.* 207:351–366.

Moss, M. L. The logarithmic spiral and allometry as expressions of odontogenic thermodynamics. This volume.

Moss-Salentijn, L. 1981a. Reattachment of the mammalian mylohyoid muscle during late embryonic development. In *Monograph 10*, D. S. Carlson, ed. Craniofacial Growth Series, Center for Human Growth and Development.

——1981b. The vasculature of the palatal processes during secondary palate development in the cat. In B. Kehrer, T. Slongo, B. Graf, and M. Bettex, eds., *Long-Term Treatment in Cleft Lip and Palate with Coordinated Team Approach,* pp. 58–72. Bern: Hans Huber.

Nagai, H., K. Kambara, H. Sudo, A. Hiura, S. Yokoyama, N. Nagai. 1976. Development and growth of the tooth germs of rats. *Okaj. Folia Anat. Jap.* 53:279–290.

Nery, E. B., B. S. Kraus and M. Croup. 1970. Timing and topography of early human tooth development. *Arch. Oral Biol.* 15:1315–1326.

Ooe, T. 1956. On the development of position of the tooth germs in the human deciduous front teeth. *Okaj. Folia Anat. Jap.* 28:317–340.

Ooe, T. 1957. On the early development of human dental lamina. *Okaj. Folia Anat. Jap.* 30:197–210.

Ooë, T. 1959. On the development of the position of the tooth germs in the human deciduous molar teeth. *Okaj. Folia Anat. Jap.* 32:97–108.

Osborn, J. W. and A. R. Ten Cate. 1976. *Advanced Dental Histology*, 3rd ed. Bristol: Wright.

Osman, A., and J. V. Ruch. 1975. Répartition topographique des mitoses dans les territoires odontogènes de la mâchoire inférieure de l'embryon de souris. *J. Biol. Biuccale* 3:117–132.

Osman, A., and J. V. Ruch. 1976. Répartition topographique des mitoses dans l'incisive et la lère molaire inférieure de l'embryon de souris. *J. Biol. Buccale* 4:331–348.

Peterková, R. 1974. Growth of tissue components in early development of tooth primordia in mice. *Folia Morph.* 22:296–304.

Romeis, B. 1968. *Mikroskopische Technik.* München-Wien; R. Oldenbourg Verlag.

Röse, C. 1891. Ueber die Entwicklung der Zähne des Menschen. *Arch. f. mikr. Anat.* 38:447–491.

Scheidt, P. 1894. Morphologie und Ontogenie des Gebisses der Hauskatze. *Morph. Jb.* 21:425–462.

Schroeder, H. E. 1976. *Orale Strukturbiologie.* Entwicklungsgeschichte, Struktur und Funktion normaler Hart-und Weichgewebe der Mundhöhle. Stuttgart: Georg Thieme.

Slavkin, H. C. 1978. The nature and nurture of epithelial-mesenchymal interactions during tooth morphogenesis. *J. Biol. Buccale* 6:189–203.

Strassburg, M., S. Peters and H. Eitel. 1971. Zur Morphogenese der Zahnleiste. II. Histologische Untersuchungen über die frühesten Differenzierungsphasen der Zahnleiste bei der Maus. *Dtsch. Zahnärztl. Z.* 26:52–57.

Thesleff, I. 1978. Role of the basement membrane in odontoblast differentiation. *J. Biol. Buccale* 6:241–249.

Williams, R. C. and H. E. Evans. 1978. Prenatal dental development in the dog, *canis familiaris*: chronology of tooth germ formation and calcification of deciduous teeth. *Zbl. Vet. Med. C. Anat. Histol. Embryol.* 7:152–163.

2. Interspecific Epithelio-Mesenchymal Interactions *Avian Epithelium Participation in Odontogenesis*

EDWARD J. KOLLAR

The driving force of morphogenesis is embryonic induction, which operates between interacting tissues and supports specific histogenesis and cytodifferentiation in the interacting tissues. Embryonic induction has been shown to operate normally between tissues of broadly divergent phylogenetic origins. Both primary induction of the nervous system (Waddington 1934) and inductive tissue interactions between the ectoderm and its supporting mesenchyme (Spemann and Schotte 1932; Briggaman and Wheeler 1968; Garber et al. 1968; Sawyer et al. 1972; Dhouailly 1973; Propper and Gomot 1973) have been examined in xenoplastic (interspecific) tissue combinations. From these studies the conclusion has been drawn that the inductive signal or principle is general in character and can be interpreted by the interacting tissues despite genetic and phylogenetic disparities. In addition, once the inductive influence has been exerted, the responding tissues must express a mode of differentiation compatible with the genetic history of the responding tissue. For example, in a most convincing experiment by the Coulombres (1971), undifferentiated corneal epithelium in chick embryos could respond to hair-inducing dermis

My thanks are due to C. Fisher who participated in the most recent series of experiments, to S. Pearson for typing the manuscript, and to the University of Connecticut Research Foundation for financial support.

But, most of all I wish to thank S. Glasstone, to whom I dedicate this paper. Dr. Glasstone's pioneering work in odontogenesis links this paper to a series of informative and enduring studies of interspecific inductive tissue interactions. If it had not been for Glasstone's personal intervention some years ago in my floundering attempts to culture teeth, this paper as well as many others would not have been written.

isolated from embryonic mice by forming feathers. The conclusion is clear: competent tissues will respond to foreign inductive stimuli and will express a mode of differentiation compatible with the genetic repertoire expected to reside in the responding tissue.

There have been four reports of experimental combinations of avian epithelium with mouse dental tissues (Kollar 1972; Karcher-Djuricic and Ruch 1973; Hata and Slavkin 1978; Kollar and Fisher 1980). In several of these reports some unexpected and novel results were obtained.

Preliminary data obtained from chorioallantoic-membrane (CAM) grafts of 5-day old chick first pharyngeal arch epithelium which was recombined with 15-day old embryonic mouse molar dental papillae (Kollar 1972) indicated that the chick epithelium could act in an unusual fashion in the presence of mouse molar dental papilla, a mesenchyme with tooth-inducing capabilities (Kollar and Baird 1970). After one week's growth on the CAM, the chick epithelium in these combinations displayed an enamel organ-like configuration and in the dental mesenchyme, differentiated odontoblasts had begun to secrete dentin. This observation was unexpected because dental mesenchyme must participate in an obligatory interaction with an enamel organ before dentin synthesis can occur (Thesleff 1977). Thus, these data suggested that the chick epithelium had been influenced by the mouse mesenchyme to form an enamel organ which then provided the permissive conditions necessary for dentin synthesis. These grafts were maintained on the CAM for only one week. Despite this suggestive result, it was not possible, because of the limited growth period, to determine whether a true enamel organ capable of enamel protein secretion could develop in avian epithelium.

Then in 1973, Karcher-Djuricic and Ruch reported a variation of this experiment. They tested combinations of 5-day-old chick epithelium from the first pharyngeal arch and mandibular mesenchyme from 11-day-old mouse embryos. These combinations were grown as *in vitro* organ cultures. Their results were negative. The chick epithelium did not display any mammalian features and the mouse mesenchyme did not differentiate into odontoblasts. They did report, however, that the chick epithelium could differentiate into mucus-secreting cells (see below) as well as a stratified surface epithelium. In addition, mouse mesenchyme from these early embryos, in the presence of chick epithelium, developed cartilage nodules and muscle fibers. However, the young age of the mouse mesenchyme used by Karcher-Djuricic and Ruch drastically reduced chances of constructing combinations with inductive mesenchyme, and the lack of a blood supply severely hampered the possibility of achieving a positive result in their study.

Similarly, dental structures were absent in another study in which mouse dental mesenchyme was associated with chick epithelium. Hata and

Slavkin (1978) recombined 17-day-old mouse dental papilla with chick limb bud epithelium from stage 22–23 chick embryos and grew them on either Millipore-raft assemblies *in vitro* or as grafts to the CAM. It is important to note that in the Hata and Slavkin study, chick limb bud (i.e., non-oral) epithelium was used. Dental structures were not recovered, the epithelium did not differentiate, and the dental mesenchyme produced "molariform" cartilagenous nodules.

There are some surprising aspects to these data. First, the isolated limb bud epithelium that presumably contained the inductively active apical ectodermal ridge at the time of isolation cannot function in the *absence* of limb mesoderm or its maintenance factor (Zwilling and Hansborough 1956; Calandra and MacCabe 1978). Thus, the appearance of cartilage in the older mouse mesenchyme is a surprising finding, since the chick tissue rapidly loses its inductive capability in the absence of its homologous mesenchyme. Second, the "molariform" shape attributed to the cartilaginous nodules in the mouse dental mesenchyme is inconsistent with the inductive capabilities of the avian apical ridge epithelium. If the avian epithelium were acting as expected, the cartilaginous nodules should have been *limb-like* since the avian epithelium controls the distal outgrowth of *limb* skeletal elements. Thus, either the mouse dental papilla is acting as surrogate limb mesoderm and thereby supports the isolated apical ectodermal ridge, or the isolated ectodermal ridge is operating in a totally new and unprecedented manner in this combination.

Furthermore, the absence of odontogenesis in these combinations is complicated by a number of considerations. The disparity in the ages of the tissues used should be examined more closely. There is usually a narrow range of ages that permits successful inductive interactions (Kollar and Baird 1970; Kollar 1972) especially when non-oral epithelium is involved. This aspect of the experimental design certainly contributed to the results reported, and a broader range of ages should be tested. In addition, the uncertainty as to the actual development status of the isolated chick epithelium is a serious problem. If on the one hand, the epithelium itself is acting inductively, then the experimental combination is certainly less than optimal to permit an influence from the dental mesenchyme to be felt. Alternatively, if the chick limb epithelium is indeed undergoing regression, the interaction is hampered as well. The Hata and Slavkin work raises a number of questions concerning the range of possibilities that derive from experimental combinations of avian limb and mouse dental tissues. But, morphological expressions arising from the interactions of cranial and limb components offer, perhaps, too divergent and complex a structural disparity to permit reasonable interpretations from their data at the present time. In view of these ambiguous data, the preliminary results of Kollar (1972) needed to be repeated.

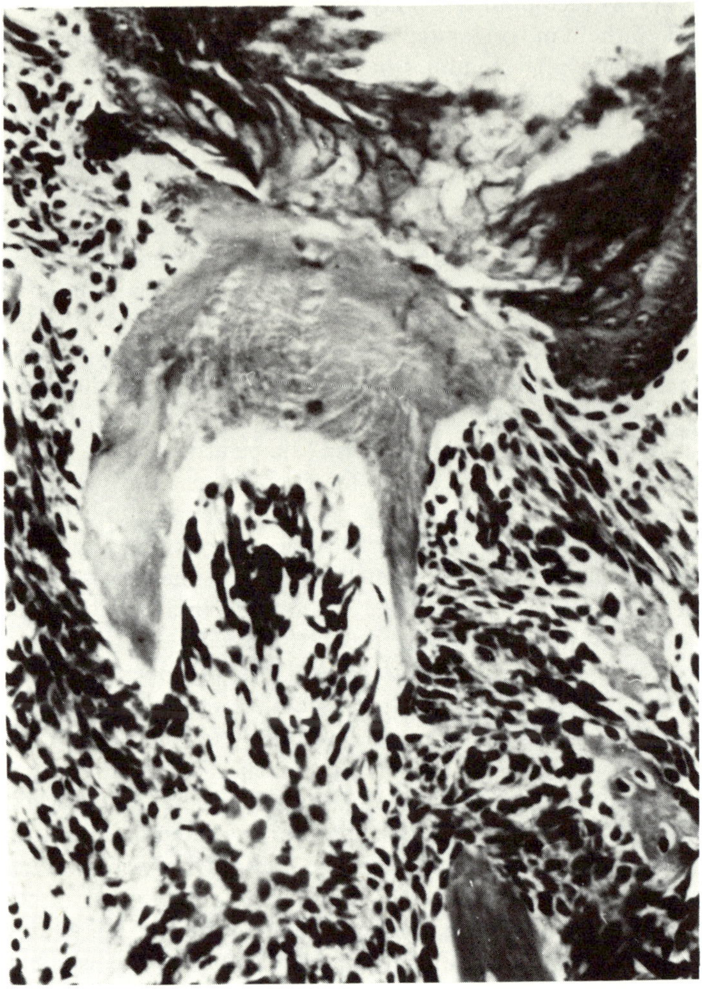

Figure 2.1. A tooth-like configuration of dentin formed in close apposition to the chick epithelium. The mesenchyme had differentiated into true odontoblasts. The chick epithelium displayed unusual cytologial features distinct from the contiguous stratified, keratinizing epithelium.

The limitations of short growth periods for the experiments reported earlier that did produce encouraging results (Kollar, 1972) have been overcome. The availability of an athymic nude mouse strain has permitted long-term grafts to be grown in these animals without eliciting tissue rejection of the foreign tissues. Thus, the earlier preliminary experiments have been repeated and the combinations grafted for as long as one month in the anterior chamber of the eyes of adult nude mice.

A report of these findings (Kollar and Fisher 1980) demonstrated that combinations of the first and second pharyngeal arch epithelium from 5-day-old chick embryos, when combined with embryonic mouse molar dental papilla, can participate in tooth morphogenesis.

A variety of dental structures was seen in successful grafts. In addition to the enamel organ-like invasiveness of the epithelium similar to that seen in our earlier experiments, the increased explantation time also permitted significant deposition of hard-tissue matrices. The configuration of the dentin was often displayed in aberrant tooth-like patterns in close association with the chick epithelium (fig. 2.1). But these aberrant structures were identified as dentin on the basis of dentinal tubules and the exclusion of cell bodies from the matrix. In contrast, spongy bone that was found in all grafts and especially in control grafts of isolated dental papillae was clearly identifiable as a distinct and characteristic mode of development.

Although the dentin deposits were molariform, they were often devoid of discernable enamel matrix. While sometimes at the surface of the dentin

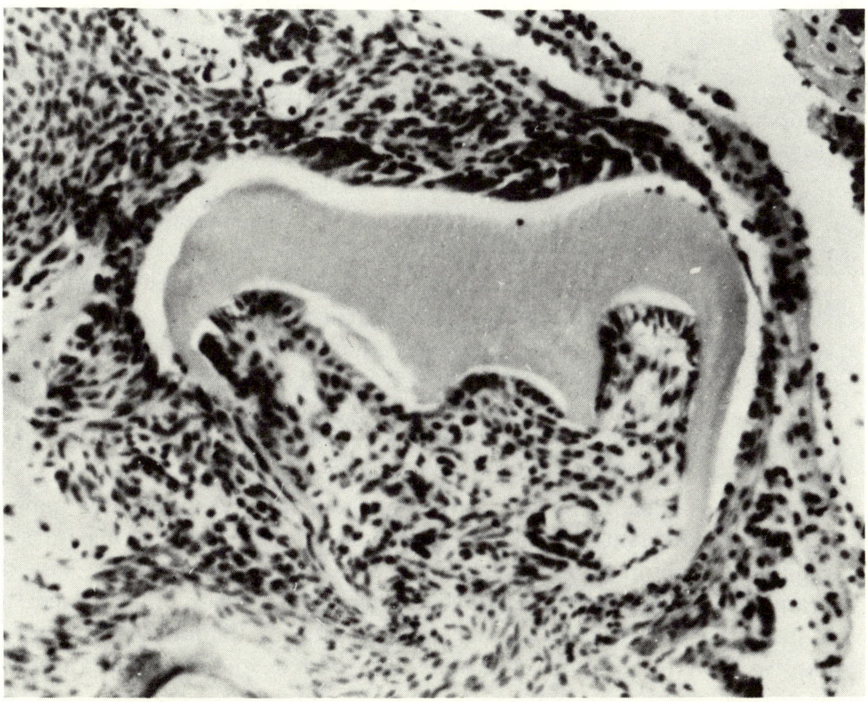

Figure 2.2a. A molariform configuration of dentin. The mesenchyme had differentiated into true odontoblasts.

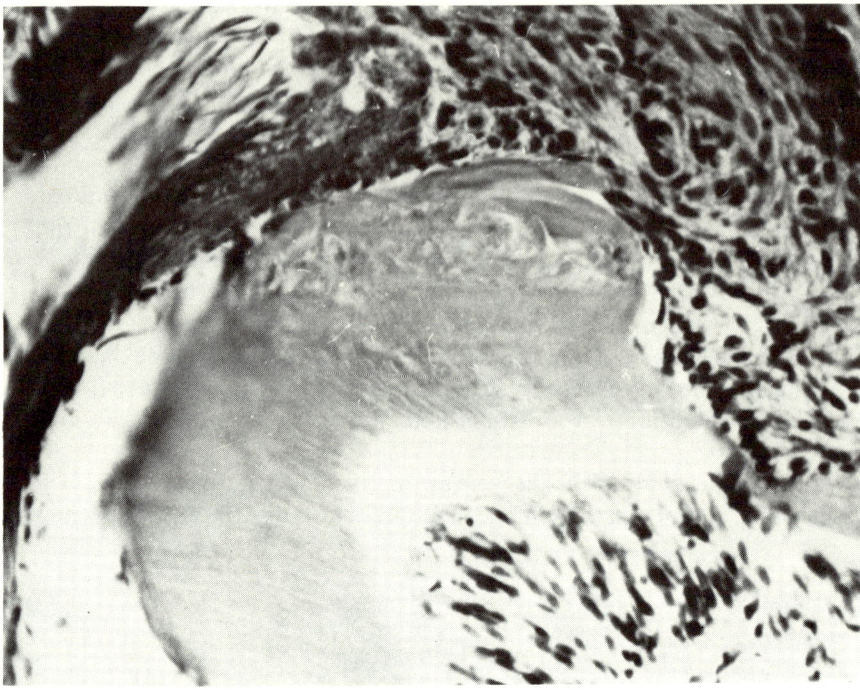

Figure 2.2b. In addition to true dentin deposition, an aberrant formation of hard tissue was seen on the surface of the true dentine layer.

some unusual deposits of hard tissue did occur, the staining reaction was not that of enamel (fig. 2.2).

The chick epithelium was isolated from the ectodermal and endodermal surfaces of the first and second pharyngeal arches. Consequently, in some of the grafts, cysts of ciliated mucus glands (Ivey and Edgar 1952) were found in association with cysts of keratinizing epithelium (fig. 2.3). The presence of these ciliated mucus-secreting glands is corroborated by the findings of Karcher-Djuricic and Ruch (1973) mentioned earlier. In addition, keratinizing epithelium devoid of keratohyaline granules was observed. This characteristic feature of the avian epithelium was maintained in the explantation site. It is important to emphasize that in contrast mouse epithelium produces keratohyaline granules in a well defined granular layer in control grafts. These are some of the criteria used to identify the source of the epithelium. When the epithelium was found to express both the keratinizing surface epithelium devoid of granules as well as ciliated esophageal glands, the dental differentiation was remarkably normal (fig. 2.4). However, the crown morphology of the teeth was often grossly disturbed and in the best cases took on a smooth, peg-like appearance, without distinct cusps.

The enamel space on the crowns contained remnants of enamel matrix protein, and the inner enamel organ epithelium was typical of a resorbing or reduced enamel epithelium (fig. 2.5). The nuclear staining and morphology agreed with the corresponding nuclei in the ciliated mucus glands. The glands could not develop from contaminating mouse dental epithelium. In addition, the epithelium of the enamel organ was contiguous with this epithelium as well as the keratinizing cysts devoid of a granular layer. Thus, we are confident that the well developed teeth seen in these grafts were induced from the chick epithelium.

The range of dental structures that were recovered suggests that the chick epithelium expresses normal and abnormal responses to the inductive stimulus. Apparently, chick epithelium may participate in early tooth formation by supporting dentin deposition but then does not continue its own differentiation with the deposition of observable amounts of enamel matrix proteins in response to the dentin layer. These data support a similar idea of Moss (1960) who suggested that the variously incomplete odontoblastic differentiations seen in phylogeny might be the result of several

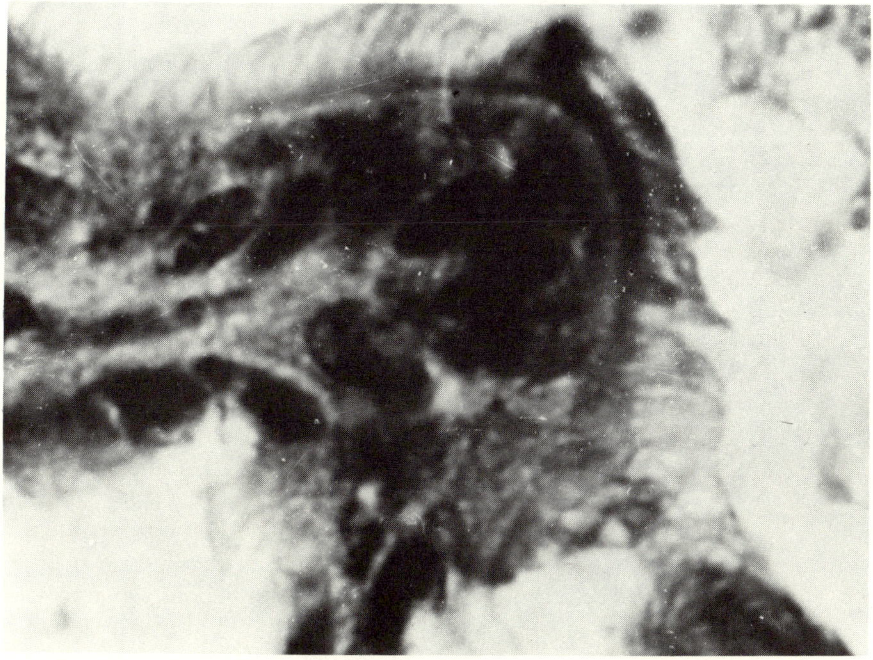

Figure 2.3. A detail of ciliated mucus gland.

causes; namely, variations in the strength of the inducing stimulus or the competency of the responding tissues, or variations in the duration of the inductive interactions or variations in the developmental stages at which the inductions occurred. I think the data from these experimental tissue interactions confirm this general idea, even if we still can't determine which of the variations in the interactions is the major cause; all are possible.

These data suggest further that in the developing chick the appropriate inductive mesenchyme does not appear or act at the appropriate time

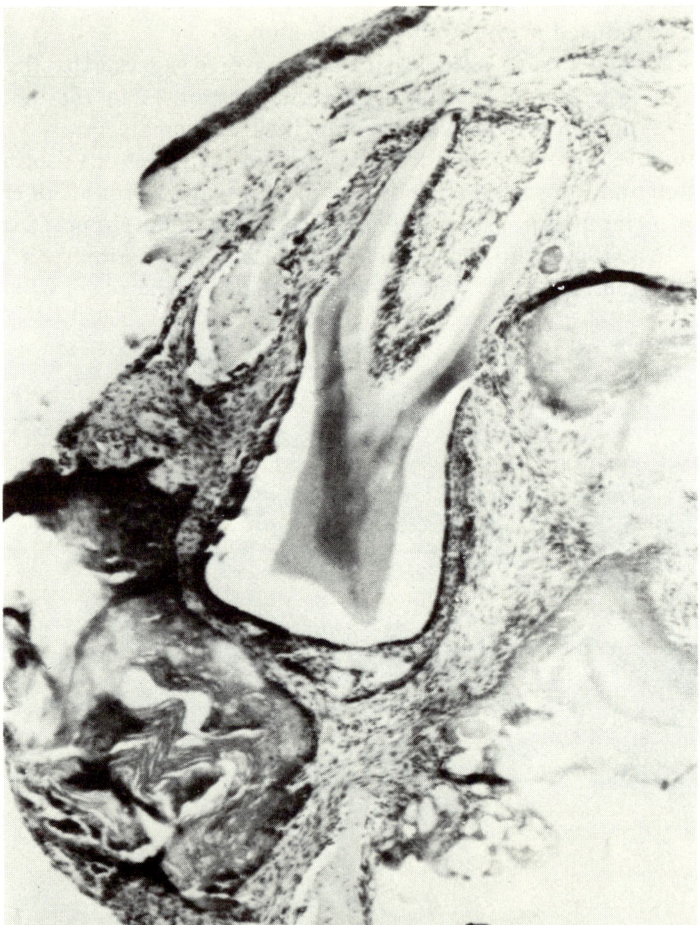

Figure 2.4. A well-formed tooth developing in association with a cyst of keratinizing epithelium, and an adjacent cyst formed by a ciliated mucus-secreting epithelium.

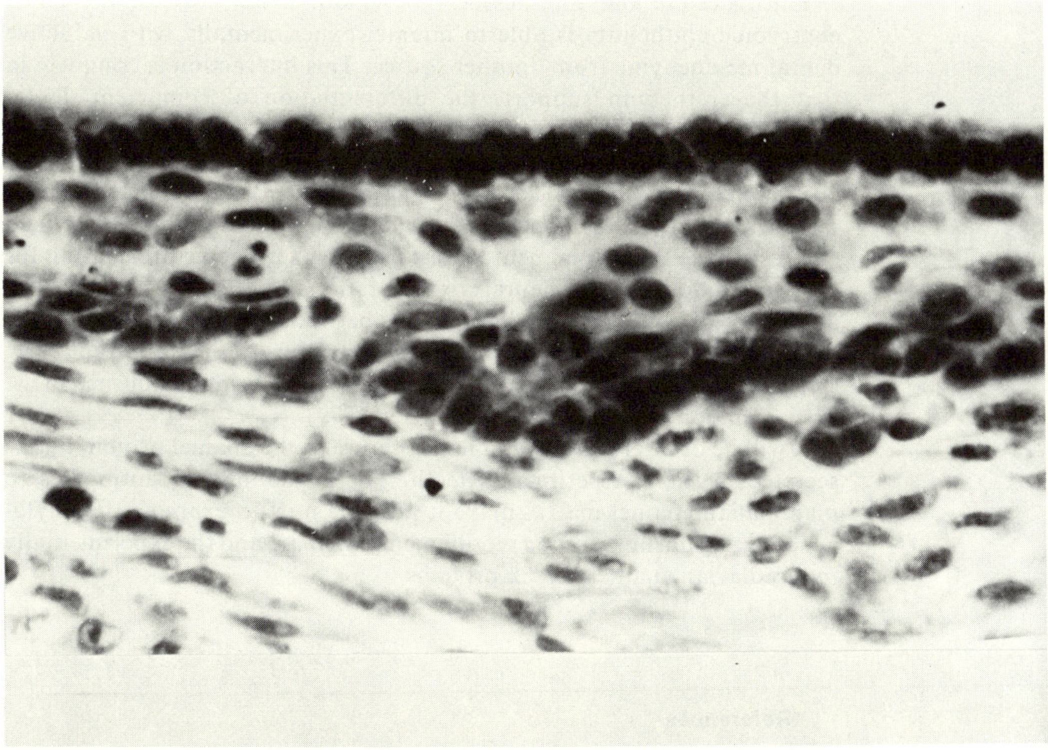

Figure 2.5. A detail of the enamel organ formed by the chick epithelium. The ameloblasts were reduced in height. Remnants of the enamel matrix proteins were seen in the enamel space.

during ontogeny, and thus the sequence of dental development is interrupted. Jacobson and Fell (1941) identified the beginnings of a dental lamina in chick embryos between the fifth and sixth days of incubations, but this primitive dental lamina regresses. Presumably the absence of the appropriate reciprocal interaction between this epithelial ridge and the developing dental mesenchyme is interrupted. The genetic information for enamel synthesis and enamel organ formation remains quiescent in the genome of embryos and is not permitted to be expressed.

Jacob (1977) in a recent discussion of the mechanisms of evolution suggests that normal development and evolution depend in part on such tissue-level mechanisms to provide diversity of and alteration of form. A recent observation (Lemanski et al. 1979) that normal endoderm corrects abnormal heart development in cardiac-less mutant salamanders provides an additional example in which restoration of the appropriate tissue interaction similarly restores gene function.

Thus, despite the toothlessness of extant avian species, the chick embryonic epithelium is able to interact experimentally with inductive dental mesenchyme from another source. This interaction is complete in that the epithelium supports the differentiation of competent dental mesenchyme into odontoblasts with the secretion of dentin matrix, as well as responding to the dentin by undergoing ameloblast differentiation and enamel matrix synthesis. The data suggest further that competent dental mesenchyme does not appear in the developing mandible of the chick, either because the cranial neural crest has lost the genetic information for tooth induction; because neural crest migrations into that region have been altered, or possibly because interactions between the neural crest and the dental epithelium are not permitted to occur during the important early stages of odontogenesis when dental competence is being established in the tissue components (Kollar and Lumsden 1979).

It will be of interest to examine the nature of the enamel protein that is secreted in these experimental teeth. The availability of antibodies to mammalian enamel matrix make it possible now to examine the phylogenetic relationships among reptilian, mammalian, and the experimentally induced avian enamel matrix proteins.

References

Briggaman, R. A. and C. E. Wheeler. 1968. Epidermal-dermal interactions in adult human skin: role of dermis in epidermal maintenance. *J. Invest. Dermatol.* 51:54–65.

Calandra, A. J. and J. A. MacCabe. 1978. The *in vitro* maintenance of the limb-bud apical ridge by cell-free preparations. *Devel. Biol.* 62:258–269.

Coulombre, J. L. and A. J. Coulombre. 1971. Metaplastic induction of scales and features in the corneal anterior epithelium of the chick embryo. *Devel. Biol.* 25:464–478.

Dhouailly, D. 1973. Dermo-epidermal interactions between birds and mammals: Differentiation of cutaneous appendages. *J. Embryol. Exp. Morph.* 30:587–603.

Garber, B., E. J. Kollar, and A. A. Moscona. 1968. Aggregation *in vivo* of dissociated cells. II. Effect of state of differentiation of cells on feather development in hybrid aggregates of embryonic mouse and chick cells. *J. Exp. Zool.* 168:455–472.

Hata, R.-I. and H. C. Slavkin. 1978. *De novo* induction of a gene product during heterologous epithelial-mesenchymal interactions *in vitro*. *Proc. Nat. Acad. Sci.* 75:2790–2794.

Ivey, W. D. and S. A. Edgar, 1952. The histogenesis of the esophagus and crop of the chicken, turkey, guinea fowl, and pigeon with special reference to ciliated epithelium. *Anat. Rec.* 114:189–211.

Jacob, F. 1977. Evolution and tinkering. *Science* 196:1161–1166.

Jacobson, W. and H. B. Fell. 1941. The developmental mechanics and potencies of the undifferentiated mesenchyme of the mandible. *Quart. J. Micros. Sci. N. S.* 82:563–586.

Karcher-Djuricic, V. and J. V. Ruch. 1973. Cultures d'associations interspecifiques de constituants mandibulaire d'embryons de souris et de poulet. *C. R. Soc. Biol.* 167:6–7.

Kollar, E. J. 1972. The development of the integument: spatial, temporal, and phylogenetic factors. *Am. Zool.* 12:125–135.

Kollar, E. J. and G. R. Baird. 1970. Tissue interactions in embryonic tooth germs. II. The inductive role of the dental papilla. *J. Embryol. Exp. Morph.* 24:173–186.

Kollar, E. J. and A. G. S. Lumsden. 1979. Tooth morphogenesis: The role of the innervation during induction and pattern formation. *J. Biol. Buccale* 7:49–60.

Kollar, E. J. and C. Fisher. 1980. Tooth induction in chick epithelium: Expression of quiescent avian genes for enamel synthesis. *Science* 207:993–995.

Lemanski, L. F., D. J. Paulson and C. J. Hill. 1979. Normal anterior endoderm corrects the heart defect in cardiac mutant salamanders (*Amblystoma mexicanum*). *Science* 204:860–862.

Moss, M. L. 1960. Experimental induction of osteogenesis. In *Calcification in Biological Systems*, pp. 323–348. Washington, D. C.: American Association for the Advancement of Science.

Propper, A. and L. Gomot. 1973. Control of chick epidermis differentiation by rabbit mammary mesenchyme. *Experientia* 29:1534–1544.

Sawyer, R. H., U. K. Abbot, and J. D. Trelford. 1972. Inductive interactions between human dermis and chick chorionic epithelium. *Science* 175:527–529.

Spemann, H. and O. Schotte. 1932. Ueber xenoplastische transplantation als Mittel zur analyze der embryonalen induktion. *Naturwissenschaften* 20:463–467.

Thesleff, I. 1974. Tissue interaction in tooth development *in vitro*. In M. Karkinen-Jaashelainen, L. Saxen, L. Weiss, eds., *Cell Interactions in Differentiation*, pp. 191–207. New York: Academic Press.

Waddington, C. H. 1934. Experiments on embryonic induction. III. A note on inductions by chick primitive streak transplanted to the rabbit embryo. *J. Exp. Biol.* 11:224–226.

Zwilling, E. and L. A. Hansborough. 1956. Interaction between limb bud ectoderm and mesoderm in the chick embryo. III. Experiments with polydactylous limbs. *J. Exp. Zool.* 132:219–239.

3. The Developing Innervation of the Lower Jaw and Its Relation to the Formation of Tooth Germs in Mouse Embryos

A. G. S. Lumsden

The mammalian tooth germ provides a unique model for the study of developmental processes. Not only does the individual tooth germ exhibit a complex of both permissive and instructive tissue interactions analogous to primary embryonic induction (Kollar 1979), but the developing dentition as a whole provides us with an excellent model system in which to investigate another developmental process, the spatial organization of cytodifferentiation, or pattern formation. For example, by what mechanism do potential tooth forming cells become precisely distributed in the developing jaws so as to form a series of discrete primordia? Or is the potential for odontogenesis imposed in an ordered way on indifferent cells?

What little experimental work has been directed at this question indicates that, within a tooth class (e.g., molar), the first formed primordium contains all the developmental information needed for generation of the complete sequence of tooth germs (Lumsden 1979; Osman et al. 1979). Sequentially later formed primordia appear to be generated autonomously by growth of the first primordium, and since this process will continue in isolation from the jaws, it can be presumed that the patterning mechanism does not rely on extrinsic developmental cues. Both tooth number within a class and quantitative changes of form along the tooth row may be reasonably accounted for by the autonomous, time related changes of positional value in the cells generated by a "progress zone" (Summerbell et al. 1973; Wolpert et al. 1975) at the proliferating margin of the progenitor tissue. Osborn (1973, 1978) has proposed an alternative pattern-generating mechanism which is compatible with available experimental data but does not require the assumptions of either extrinsic cues or special controlling regions.

Early in mammalian development, three tooth primordia differentiate in each jaw process. According to Osborn's theory (1978) each of these primordia is a progenitor from which will subsequently develop all the elements of the tooth class. Thus the most anterior primordium forms the deciduous and permanent incisors, the middle primordium forms the canines, and the posterior primordium forms the postcanines. The progenitor tooth germs appear at about the same time in development of the embryo (Gaunt 1964; Oöe 1957); further primordia then appear sequentially within each region by budding in anterior, posterior or lingual directions. In most mammals, budding from the progenitors has been repressed; later formed primordia fail to differentiate, resulting in the "loss" of teeth from the ends of series. In the mouse embryo one of the progenitors (probably the canine) itself fails to develop. What mechanism controls the number of progenitors and their positions in the developing jaws? If a similar mechanism to "within-class" patterning is proposed, it is difficult to see how these regions could develop almost simultaneously, and even more difficult to explain how, in the mouse and in other mammals which lack canines, the middle progenitor of the series could be "lost." Alternatively the progenitors may be specified by quite a different mechanism.

Recently, Pearson (1977) described early stages of tooth formation in 5–6 week human embryos, and noted a close relationship between the sites of primary epithelial thickening and the presence in subjacent ectomesenchyme of fine nerve branches. Although he presented no data for earlier embryos (from which temporal priority could be established) he suggested that the innervation may play an inductive role in the initiation of odontogenesis. This possibility has prompted investigation of earlier stages of development.

In serially sectioned mouse embryos, impregnated by a recently refined silver gelatin method, fine nerve fibers of less than 1 μm diameter were found beneath the oral epithelium in two regions of each jaw quadrant at between 10.5 and 11 days of development; that is, prior to any histological evidence of tooth initiation (Kollar and Lumsden 1979; Al-Takriti and Atkinson 1980).

The present paper is concerned with the pattern of the developing innervation of the lower jaw, as it is revealed in silver impregnated whole embryos, particularly in relation to the positioning of initial odontogenic events.

Materials and Methods

Inbred CD1 albino mice with a gestation period of 20 days were used. Embryos were aborted at 12-hour intervals between 9.5 and 13 days following the appearance of a vaginal plug (= day 0) and were staged on the basis of external features according to the normal tables of Theiler (1972).

Transparent whole mounts of embryos from each stage of development were prepared using Lewis' (1978) copper-protargol technique. The following modifications were made to Lewis' original procedure to suit the size and nature of the material. Following fixation for 48 hours in Bodian's fluid, the embryos were transferred to 100 ml of 1 percent aqueous Protargol (strong silver protein, Roques), to which was added simultaneously 0.83g of 0.91 mm diameter copper wire. The length of impregnation time was varied; a period of 12 hours at 37° C gave good superficial staining whereas 15 hours at the same temperature gave improved staining of deeper fibers. This procedure yielded embryos with their entire peripheral innervation silver-stained dark red or black against a pale gold transparent background (fig. 3.1).

In its ability to reveal the complex three-dimensional pattern of the developing innervation, this technique offers considerable advantages over the conventional methods of graphic or solid-reconstruction, most especially in topographical precision. Although at first sight the procedure may appear crude and inadequate for revealing subtle patterns of innervation, in practice it is equally capable of disclosing developing neuroanatomy and revealing fine details. The great transparency of the specimens permitted observation under the light microscope at up to 250× magnification, while the use of objectives with high numerical aperture and correspondingly shallow depth of focus allowed reasonably precise depth measurement. For clarity of illustration drawings have been prepared by camera lucida.

A fate map of the developing mandibular arch showing the regions of odontogenic potential was prepared by the following method. Lower jaws were dissected from the embryos under cold Tyrode's solution enriched with 20 percent fetal calf serum. Each jaw was then divided into sectors approximating to 15° of arc (centered on the pharynx), and each sector was then grafted to the anterior eye chamber of an homologous adult host (Lumsden 1979). After three to four weeks *in vivo* culture, the grafts were recovered, stained whole with Alizarin red S to reveal calcified tissue, and cleared in 50 percent aqueous glycerin containing 2 percent potassium hydroxide.

Results

Between 8.5 and 9.5 days of development the trigeminal ganglion appears, and potential tooth-forming ectomesenchyme has migrated from

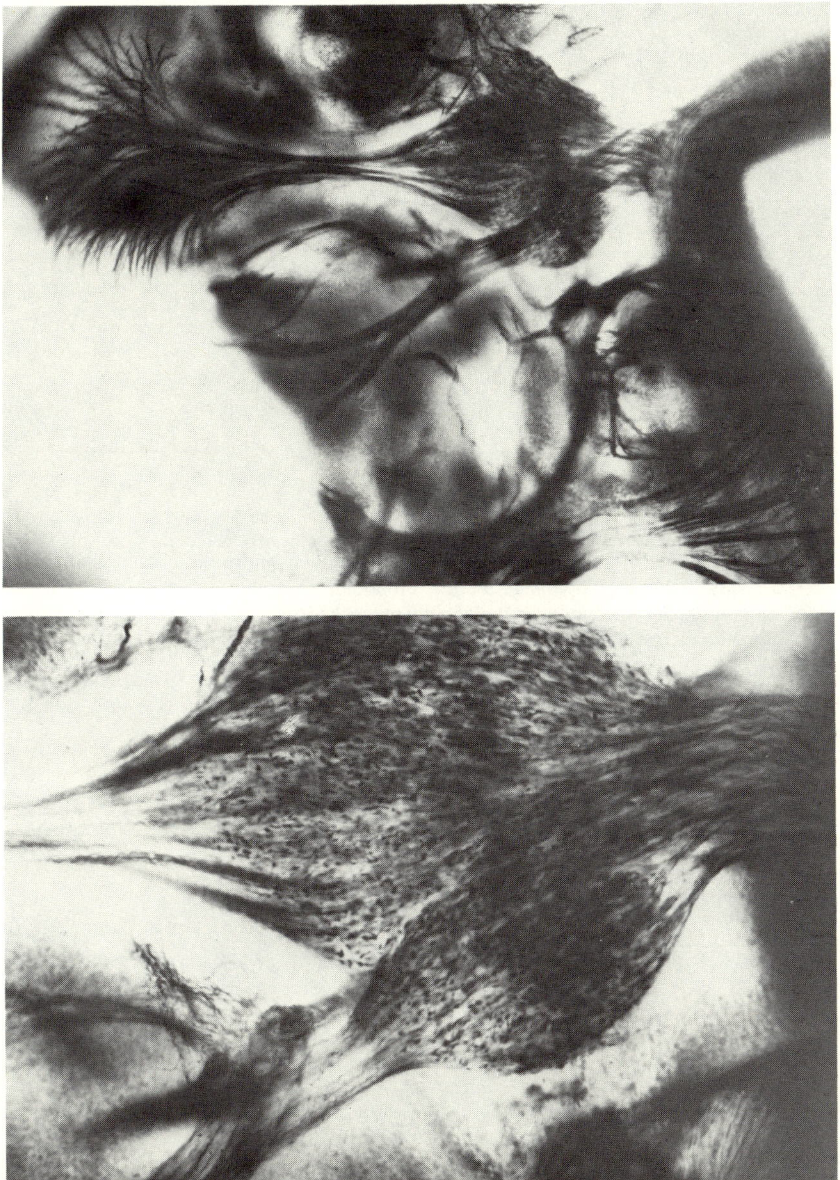

Figure 3.1. Silver stained and cleared 11.5 day mouse embryo, transmitted light. *Above:* lateral aspect of head showing the courses of the maxillary and mandibular divisions of the trigeminal nerve. *Below:* higher power micrograph of the trigeminal ganglion.

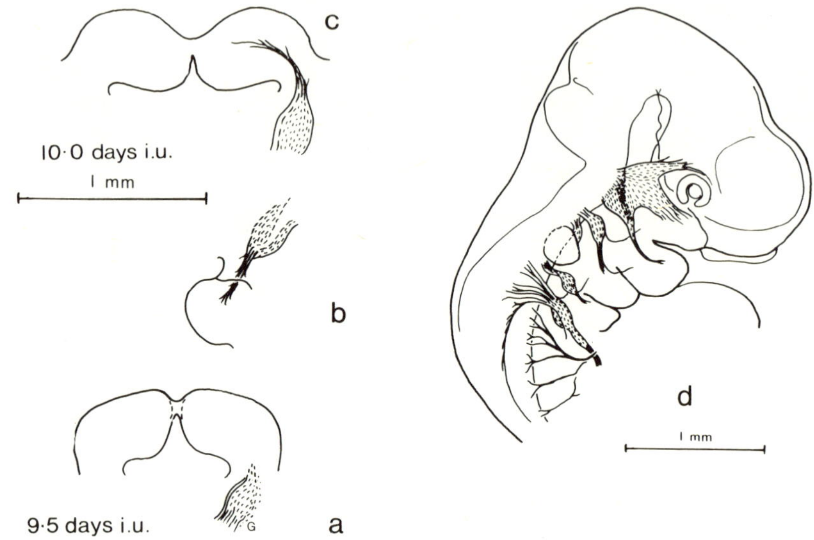

Figure 3.2. Camera lucida drawing showing the developing pattern of innervation in 9.5–10.5 day mouse embryos: *a.* 9.5 day mandibular arch, oral aspect; *b.* 10 day mandibular arch, hemisectioned in the sagittal plane, lingual aspect; *c.* 10 day mandibular arch, oral aspect; *d.* 10 day whole head, lateral aspect. Key: see figure 3.7.

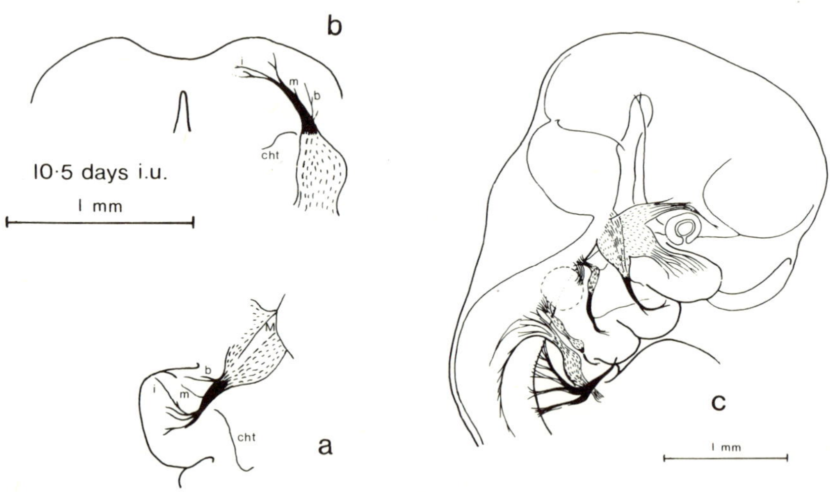

Figure 3.3. Camera lucida drawing showing the developing pattern of innervation in 9.5–10.5 day mouse embryos: *a.* 10.5 day mandibular arch, hemisectioned, lingual aspect. *b.* 10.5 day mandibular arch, oral view. **c.** 10.5 day whole head, lateral aspect. Key: see figure 3.7.

the rhombencephalic neural crests into the expanding mandibular processes by way of cleavage planes beneath the epithelia (Milaire 1959; Johnston and Listgarten 1972; Noden 1973).

In the stage 15 (9.5 day) embryo (Fig. 3.2a) axons can be seen emerging from both the neurons in the trigeminal ganglion and the neurons in the more medially situated motor nucleus. By stage 16 (10 days, fig. 3.2d) the mandibular nerve, consisting of both motor and sensory axons, has grown out into the mandibular process in which it occupies a more or less central position (fig. 3.2b,c). Along much of its length the nerve consists of terminal growth cones and more mature fibers, while lateral branches are found at many positions along its course. None of these branches approaches the oral epithelium.

By stage 17 (10.5 days, fig. 3.3) the mandibular nerve has diverged into a number of distinct branches, four of which approach the epithelium. The buccal nerve leaves the main trunk anteriorly close to its emergence from the trigeminal ganglion, and at the distal extremity of the mandibular nerve a number of fine fibers represent the earliest appearance of the mental nerve. Between these two there are two branches which leave the main trunk of the inferior alveolar nerve in a superficial direction and reach within 10–20 μm of the oral epithelium. As will be demonstrated, these nerves terminate in the presumptive incisor and molar regions and are the same nerves which previously were observed in serially sectioned material (Kollar and Lumsden 1979).

At stage 18 (11 days, fig. 4) the lingual nerve grows out and joins the chorda tympani branch of the facial nerve which had entered mandibular arch territory at stage 16. The motor branches of the trigeminal are now growing out into the developing muscle masses. The mental nerve branches fan out to come into close relation with the skin. The "incisor" and "molar" nerves, though still prominent, are now some 50–70 μm beneath the surface.

At stage 19 (11.5 days, fig. 3.5) all the branches of the mandibular nerve are present, but whereas well defined mental, buccal and auriculotemporal branches are seen running to cutaneous regions, the "incisor" and "molar" nerves are very small, and in the latter case especially, indistinct. Both nerves appear to terminate more than 100 μm beneath the epithelial thickenings which mark the appearance of the tooth primordia that were visible in the whole mounts, on account of their refractility.

At stage 20 (12 days, fig. 3.6) a re-innervation of the tooth germs begins. Especially prominent in the incisor region is a developing plexus of fine fibers which appears to envelop the mesenchyme that will later contribute to the tooth follicle. Some fibers of this incisor plexus were seen to cross the midline and terminate close to the contralateral incisor bud. Beneath the molar germ there are a few fine twigs.

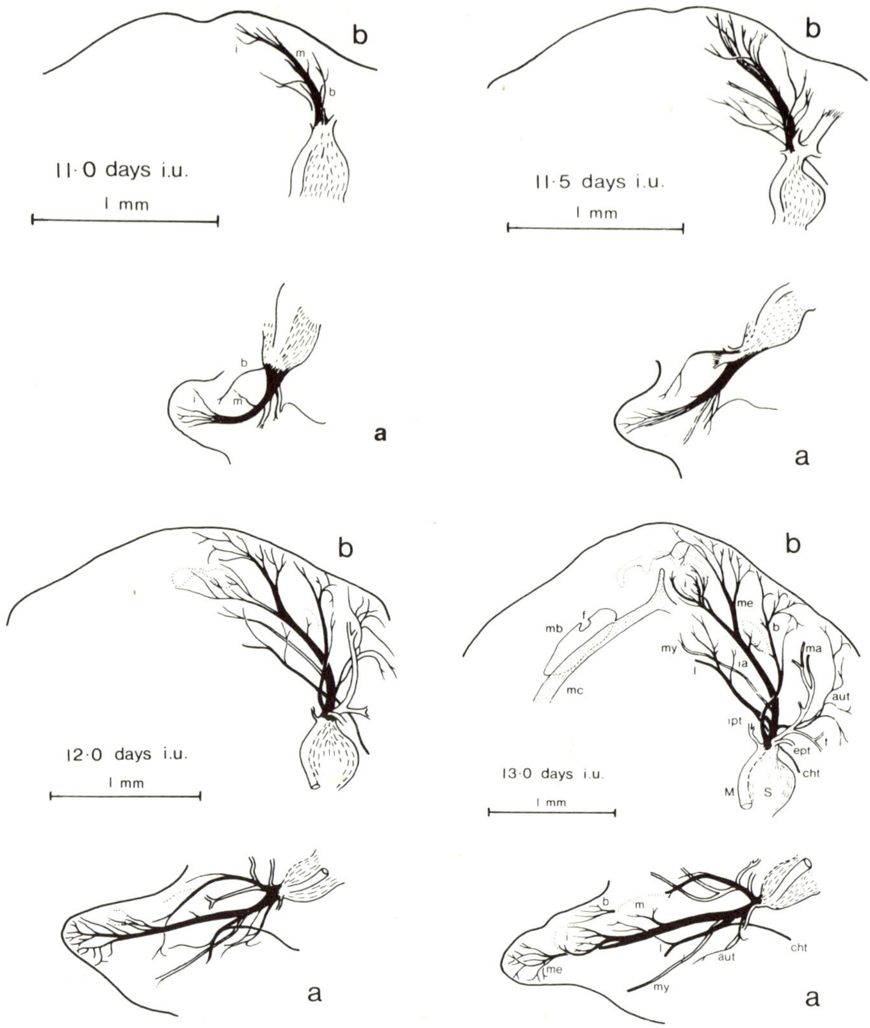

Figures 3.4, 3.5, 3.6, 3.7. Camera lucida drawings showing the developing pattern of innervation in 11–13 day mouse embryo mandibular arches from lingual aspect (*a*) and oral aspect (*b*). Sensory nerves, black; motor nerves, white. Dotted lines indicate the positions of tooth primordia.

Key to Figures 3.2–3.7: aut: auriculotemporal nerve. b: buccal nerve. cht: chorda tympani. ept: external pterygoid nerve. f: mental foramen. G: trigeminal ganglion. i: incisor nerve. ia: inferior alveolar nerve. ipt: internal pterygoid nerve. l: lingual nerve. M: motor root of trigeminal nerve. m: molar nerve. ma: masseteric nerve. mb: mandibular bone. mc: Meckel's cartilage. me: mental nerve. my: mylohyoid nerve. S: sensory root of trigeminal nerve. t: nerve branches to temporalis muscles.

By stage 22–23 (day 13, fig. 3.7) Meckel's cartilage has chondrified, and a thin plate of woven bone has appeared on its lateral side with upgrowths around the root of the mental nerve, marking the formation of the mental foramen of the young mandible. The buccal nerve passes forward in the developing cheek lateral to the molar dental lamina. Each tooth germ is now supplied by a plexus of nerves, but with no nerve fibers close to the epithelium: particularly significant is the absence of nerves from the posterior end of the M_1 lamina, the region which will form a primordium for M_2 at 14 days and later, at 23 days, a primordium for M_3 (Hay 1961; Lumsden 1979).

The results of grafting isolated sectors of 10.5 day mandibles are shown in figure 3.8. Identifiable teeth with normal or nearly normal crown shapes were formed in culture. Well-formed incisor teeth were formed by sector 15°–45° and less well formed teeth by subsectors 15°–30° and 30°–45°. Since the more anterior of these subsectors produce a higher yield than the posterior, it can be presumed that while the ontogenic potential is spread over a wider area, the focus or center of incisor-forming activity lies between 15° and 30° from the midline.

Well formed molar teeth were formed by grafts of 45°–75° and also, in decreased incidence, by 45°–60°. In a few instances two teeth, recognizable as M_1 and M_2, were formed by the larger sector. Since 60°–75° produced only rudimentary calcified nubbins, it can be presumed that the presump-

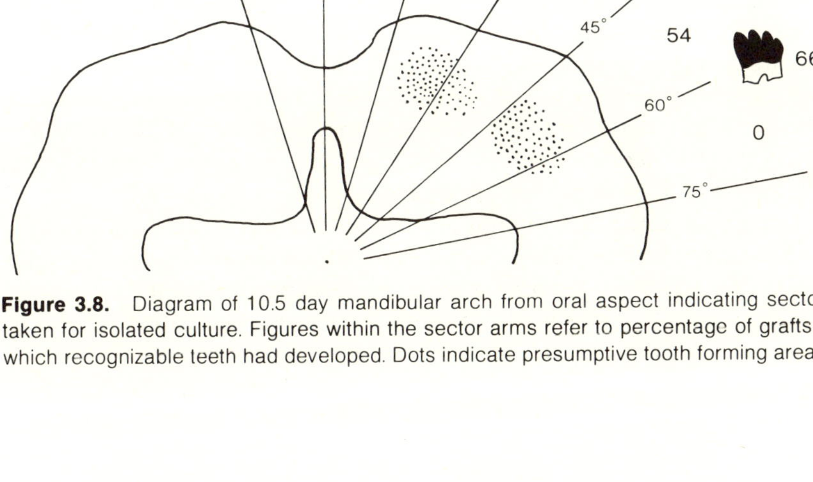

Figure 3.8. Diagram of 10.5 day mandibular arch from oral aspect indicating sectors taken for isolated culture. Figures within the sector arms refer to percentage of grafts in which recognizable teeth had developed. Dots indicate presumptive tooth forming areas.

tive molar district is centered in sector 45°–60° but spreads into the posteriorly adjoining territory.

The diastema region, if present at all, is clearly of small dimensions at this early stage of development.

Discussion

As far as a relationship between innervation and the inception of odontogenesis is concerned, the critical phase of development appears to be between 10 and 10.5 days. At 10.5 days, the nerves which lie close to the epithelium in the presumptive tooth forming areas appear to be those which were noted in serial sections to reach within one cell diameter of the basement membrane and possibly, as unstained terminal fibers, into the epithelium (Kollar and Lumsden 1979). Whole mount preparations reveal that the nerves leave the main trunk as a bundle of fibers which fan out toward the surface to innervate a small area of about 6–10 cells in diameter. Only a few of these 1–2 μm near terminal fibers reach to within 10–15 μm of the epithelium. Between 10.5 days and 11 days this distance increases considerably.

Subsequent events, the innervation of the peripheral prefollicular layer of the papilla at 13–14 days and of the papilla itself coinciding with the commencement of dentinogenesis at 19–20 days (Atkinson and Al-Takriti 1980), appear to be related to a marked increase in the vascular supply to those areas. Atkinson noted that most, if not all, of the second phase "pioneer" fibers of the cap stage follicle are adrenergic; possibly they are involved initially with angiogenesis (Smith and Wolpert 1975) and later with control of the vascular supply to the rapidly growing tooth germs. However, since the superior cervical ganglion is not clearly organized prior to 13 days (Coughlin et al 1977) the early innervation of odontogenic areas would not involve fibers of the sympathetic system.

What is the relationship between innervation and tooth development? Regarding correspondence of the regions in the 10.5 day mandible—those which are innervated and those with prospective tooth forming potential—in each case the innervation supplies a region of the mandible which, when explanted into culture, produces the highest yield of teeth. The innervation is therefore present in the appropriate location before there is any histological evidence for commencement of odontogenesis.

The retreat, either real or apparent, of these nerve fibers from proximity to the oral epithelium during the early stages of odontogenesis would imply that in the early phase the innervation has some causal significance in tooth initiation, and that this is of a transitory nature. Once accomplished the nerve relationship is lost. The only experimental evidence for a causal relation is that while isolated regions of 10–11 day mandibles will form teeth in culture (Glasstone 1967; Lumsden 1979), isolated regions of 9–10 day mandibular arches do not (Kollar, pers. comm; Lumsden, unpublished). Of significance in this regard are the findings of Karcher-Djuricic and Ruch (1970) and Gerber et al. (1972) who have obtained molar rudiments from isolated whole mandibles of 9–10 day embryos. Whereas the former were hesitant in interpreting the structures found in their 9-day grafts as tooth germs, Gerber et al. found that sometimes quite well developed teeth, in which both dentine and enamel were being deposited, formed in their intratesticular grafts of 9-day arches. They concluded that the 9-day (± 12 hours) mandibular arch already contains incisor and molar progenitor tissues in a determined state. However, since they explanted the entire arch, there is a possibility that cell bodies of neurons in the trigeminal ganglion were also included in the grafts. Kollar (1976) has found that the incidence of tooth formation by isolated parts of the early mandible can be increased by including ganglion tissue with the grafts. Since contrary data is inconclusive, it would appear that the tooth forming potential of the mandibular arch is not determined until the 11th day, a time which coincides with the appearance of nerve fibers in the prospective odontogenic regions. The nervous system is known to be involved with the differentiation of several epithelial organ systems. For example, Kollar (unpublished) has observed a prepatterning influence of infraorbital nerve branches on the positioning of vibrissae rudiments. Farbman (1972) has clearly shown that specific innervation is a prerequisite for the development and regeneration of taste buds. If specialized sensory organ systems in general are shown to have an ontogenetic dependence on their innervation, then the same would be expected for teeth which, although now modified for ingestion and mastication of food, originally evolved as specialized exteroceptors (Halstead-Tarlo 1965). The requirement of innervation for tooth initiation may be an essential primitive developmental mechanism which has been retained by the embryo long after the primary sensory role of the adult structure had become redundant.

References

Al-Takriti, S. and M. E. Atkinson. 1980. The innervation of developing mouse teeth. *J. Dent. Res.* 59 (Special issue D): 1806.

Atkinson, M. E. and S. Al-Takriti. 1980. A histochemical study of the innervation of developing teeth. *J. Dent. Res.* 59 (Special issue D): 1807. Supplement. (British Division I. A. D. R. Abstract).

Coughlin, M. D., D. M. Boyer, and I. B. Black. 1977. Embryologic development of a mouse sympathetic ganglion *in vivo* and *in vitro*. *Proc. Nat. Acad. Sci.* 74:3438–3442.

Farbman, A. I. 1972. The taste bud: a model system for developmental studies. In H. C. Slavkin and L. A. Bavetta, eds. *Developmental Aspects of Oral Biology*, pp. 109–123. New York: Academic Press.

Gaunt, W. A. 1964. The development of the teeth and jaws of the albino mouse. *Acta Anat.* 57:115–151.

Gerber, R., V. Karcher-Djuricic, and J-V. Ruch. 1974. Matériel présomptif odontogène et topographie de l'epithelium dentaire dans la mâchoire inférieure de l'embryon de Souris. *J. Biol. Buccale* 2:249–258.

Glasstone, S. 1967. Morphodifferentiation of teeth in embryonic mandibular segments in tissue culture. *J. Dent. Res.* 46:611–614.

Halstead-Tarlo, B. 1965. The sensitivity of dentine. *Roy. Dent. Hosp. Mag.* 1:11–13.

Hay, M. F. 1961. The development *in vivo* and *in vitro* of the lower incisor and molars of the mouse. *Archs. Oral Biol.* 3:86–109.

Johnston, M. C. and M. A. Listgarten. 1972. Observations on the migration, interactions and early differentiation of orofacial tissues. In H. C. Slavkin and L. A. Bavetta, eds. *Developmental Aspects of Oral Biology*. pp. 53–80. New York: Academic Press.

Karcher-Djuricic, V., and J. V. Ruch. 1970. Différenciation d'ébauches dentaire jeunes et présomptifs d'embryon de Souris. *C. R. Soc. Biol.* 164:1347–1349.

Kollar, E. J. 1976. The use of organ cultures of embryonic teeth for teratological studies. In M. Marois, ed. *Tests of Teratogenicity in vitro*, p. 303–304. Amsterdam: North Holland.

——. 1979. Epithelial-mesenchymal interactions in the control of tooth size and shape. In M. Mebrick, E. D. Shields, and N. J. Burzynski, eds., Clinical dysmorphology of oral-facial structures: genetics, embryology, diagnosis, and management. Acton, Mass: Sciences Publishing Group.

Kollar, E. J., and A. G. S. Lumsden. 1979. Tooth morphogenesis: the role of the innervation during induction and pattern formation. *J. Biol. Buccale* 7:49–60.

Lewis, J. H. 1978. Pathways of axons in the developing chick wing: evidence against chemo-specific guidance. *Zoon* 6:175–179.

Lumsden, A. G. S. 1979. Pattern formation in the molar dentition of the mouse. *J. Biol. Buccale*, 7:77–103.

Milaire, J. 1959. Prédifferentiation cytochimique de diverses ébauches céphaliques chez l'embryon de Souris. *Archs. Biol.*, Liège, 70:587–730.

Noden, D. M. 1973. The migratory behaviour of neural crest cells. In J. F. Bosma, ed. *Oral Sensation and Perception; Development in the Fetus and Infant*, pp. 9–36. Bethesda: U.S. Deparment of Health, Education and Welfare. Publication No. 73–546.

Oöe, T. 1957. On the early development of human dental lamina. *Okajimas Folia Anat. Jap.* 30:197–210.

Osborn, J. W. 1973. The evolution of dentitions. *Am. Sci.* 61:548–559.

Osborn, J. W. 1978. Morphogenetic Gradients: Fields versus Clones. In P. M. Butler and K. A. Joysey, eds. *Development, Function and Evolution of Teeth*, pp. 171–201. London: Academic Press.

Osman, M., V. Karcher-Djuricic, and J.V. Ruch. 1979. Differenciation *in vitro* du matérial molaire présomptif de l'embryon de Souris. *C. R. Acad. Sci. Paris*, 289:149–151.

Pearson, A.A. 1977. The early innervation of the developing deciduous teeth. *J. Anat.* 123:563–577.

Smith, A.R., and L. Wolpert. 1975. Nerves and angiogenesis in amphibian limb regeneration. *Nature* 257:224–225.

Summerbell, D., J.H. Lewis, and L. Wolpert. 1973. Positional information in chick limb morphogenesis. *Nature* 244:492–496.

Theiler, K. 1972. *The House Mouse* (Development and normal stages from fertilization to 4 weeks). Berlin, Heidelberg, New York: Springer.

Wolpert, L., J.H. Lewis, and D. Summerbell. 1975. Morphogenesis of the vertebrate limb. In Cell patterning. *Ciba Found. Symp.* 29:95–119.

4. Some Problems of the Ontogeny of Tooth Patterns

PERCY M. BUTLER

Paleontologists and anthropologists are accustomed to examining teeth in minute detail. The differences they observe between species and between populations have led to the belief that dental characters are genetically controlled to a high degree. This is confirmed by twin studies in man (e.g., Kraus and Jordan 1965), and by the existence of differences in molar pattern between genetic strains of mice (Grüneberg 1965).

Examination of a large sample of any species of mammal will reveal variation in the teeth (e.g., Green 1977). Most of this variation affects small details of the pattern, such as the presence or absence of minor cusps and the development of crests and cingula, as well as the shape and size of whole teeth. Much of it is homeotic: variant features may occur on more than one tooth, or features characteristic of one tooth may appear as variants on adjacent teeth.

In addition, there are major departures from the norm, such as dwarfing or absence of teeth, twinning, the fusion of teeth, supernumeraries, and unusual cusp patterns. Though rarer than the minor fluctuations, they are more obvious and have consequently been recorded in a wide diversity of mammals (Colyer 1936). *Tabby* and similar mutants in the mouse have been the subject of genetic and embryological studies (Grüneberg et al. 1972; Sofaer 1969a, 1969b, 1977; Miller 1978). There is no evidence that such major abnormalities have evolutionary significance. The paleontological record indicates that dental evolution proceeds by the selection of minor variatons. Presumably, large departures from the normal pattern would be functionally deleterious. Genetic experiments on "normal" variations of teeth have not, as far as I know, been made.

The link between gene and character is ontogeny. This is a complex sequence of processes, involving physical and chemical interactions between cells. Operating at different times and in different sorts of cells, the genes determine the proteins that are synthesized at various stages. Kollar and Lumsden (1979) divide dental ontogeny into three main phases, characterized respectively by initiation, morphogenesis, and cytodifferentiation. In the initiation phase the general pattern of the dentition is laid down. Mesenchyme cells migrating from the neural crest arrange themselves in contact with the oral epithelium on the maxillary and mandibular processes. The dental lamina appears, and mesenchyme cells accumulate at positions where tooth buds begin to form. During morphogenesis the cusp patterns develop by folding of the surfaces of contact between enamel organs, derived from the oral ectoderm, and papillae derived from the mesenchyme. There follows the differentiation of ameloblasts and odontoblasts, and deposition of the hard tissues of the tooth. The phases overlap in time: some teeth develop later than others, and morphogenesis may continue at the base of a tooth germ after apposition has begun at the tips of the cusps.

Despite the considerable advance made in recent years, there is still much about dental ontogeny that we do not understand. My purpose here is not to contribute any new information, but to ask some questions.

I have long been interested in the mammalian dentition as a unitary structure, to be studied as a whole rather than as an assortment of individual teeth. To explain variations and evolutionary changes in the dentition, I proposed what has become known as the Field Theory (Butler 1939). I was struck by the fact that tooth form is independent from numerical position: for example, M^2 of one species can resemble M^3 of another if both are last molars. I postulated a series of tooth-forming locations along the jaw, produced by a process analogous with the segmentation of the somitic mesoderm. Tooth germs forming at these locations were acted upon by what would nowadays be called morphogens, corresponding to the incisor, canine and molar districts. The highest concentration of the molar morphogen was conceived as being normally in the region of dP 4 or M 1, falling off in either direction. Its distribution was primarily related to the head rather than to the dentition, for the position of maximum cheek-tooth development changes when in the course of evolution the teeth are displaced forward or backward in the skull (Butler 1941). To explain why anterior premolars did not resemble posterior molars, though both were subjected to similarly reduced concentrations of morphogen, it was necessary to conceive of the field as polarized: tooth germs developing at different levels along the anterior-posterior axis of the jaw react differently to the same morphogen.

This theory did not take into account the considerable range of time over which the dentition develops. It seems very improbable that a morphogenetic field persists up to the time of formation of the buds of the last molars and the replacing premolars. Lumsden (1979) has recently shown that in the mouse all three molars are determined at a very early stage. Prospective molar-forming tissue, removed from the jaw of a 10-day embryo in which the enamel organ of M_1 has not yet appeared, and grown in the anterior chamber of the eye, produced M_2 as well as M_1; when the experiment was carried out at 12 days a complete molar series could be obtained. If a morphogen is operating it must do so before the appearance of enamel organs, perhaps not later than the first appearance of the lamina.

It is now known (Kollar and Baird 1969, 1970) that in the 13-day mouse, determination of tooth form resides in the papilla, which can induce the formation of molariform or incisiform enamel organs in accordance with its origin. Differences between the teeth of a dentition therefore reflect differences in mesenchyme. Two questions arise. First, what determines the differences in the mesenchyme? Is it possible that at some time between the 8th day, when it enters the jaws, and the 13th day, the mesenchyme undergoes an induction from the primitive oral ectoderm which is already regionally differentiated? Or is the mesenchyme already determined when it enters the jaws, perhaps due to contact with endoderm or other tissues during its passage from the neural crest? The second question is, are there as many sorts of mesenchyme as there are teeth? Or are there only three sorts, corresponding to the incisor, canine and molar tooth districts, the gradation within each district being due to some other factor, such as the reactivity of the oral ectoderm? Osborn (1978a, 1978b), who believes that there are three sorts of mesenchyme, suggests that competence of the ectoderm to react may play a part, but he ascribes gradations of pattern mainly to an aging process in the mesenchyme. As it grows along the jaw, backwards and forwards from a "stem-precursor," the mesenchyme loses its tooth-forming potency with repeated cell division. This loss of "morphogenetic vigor' (Osborn 1978b), though consistent with the progressive simplification of mouse molars from M 1 to M 3, does not accord with the situation in other mammals, where M^2 is frequently larger than M^1, and M_3 can be the most complex lower molar. Lumsden (1979) compares the growing end of the dentition to the progress zone of a limb bud, which contributes successively more distal segments to the growing limb. In the limb, as in the dentition, distal elements are not necessarily simpler than those that develop more proximally, and the causative mechanisms underlying the sequence remain problematic.

Anterior to the stem precursor Osborn (1978a, 1978b) postulates a forward growth that produces deciduous molars in sequence. He refers to an anterior progress zone (1978b), but there is no histological indication of a region of growth such as occurs at the posterior end. The early development of deciduous molars has not been investigated experimentally, but the histological picture (e.g., Gaunt 1959a) gives the impression that mesenchyme cells, at first distributed along the lamina, become concentrated at the locations of tooth development, the rapidity of the process suggesting cell movement rather than growth. The stem precursor, from which Osborn derives all the mesenchyme of the cheek teeth, ranges from dP 3 (man, cat, horse) to dP 4 (mole) and M 1 (mouse). Acceptance of the homology of all these teeth involves the writing of new dental formulas that are very difficult to reconcile with paleontology. Moreover, the stem precursor has no constant place in the sequence of cheek-tooth patterns: it ranges from a simple premolariform tooth in the cat to a fully molariform tooth in the horse and the mouse.

Though modified in specialized mammals, a basic sequence of cheek tooth patterns can be traced throughout the class (Butler 1939); yet the order in which the teeth develop is not always the same. It seems that the complex cheek teeth of mammals require a long time to develop and have to start early, when the jaws are too short to contain the full dentition. The posterior part of the dentition is therefore delayed until space becomes available. How many teeth are delayed will differ in different mammals. An analogy might be drawn with the development of somites in vertebrate embryos (Cooke 1975). In the trunk region somites develop from mesoderm formed during gastrulation, and the series is continued posteriorly by the formation of a tail bud. Even when the tail bud is very small, removal of its tip reduces the number of somites that will develop from it, suggesting that the information necessary to produce an entire tail is already present. It is conceivable that the organization of the whole dentition is determined over a short period of time at an early stage, but in the posterior region the effect of the determination becomes apparent only with subsequent growth in the progress zone. In a similar way, the replacing dentition could be determined at the same time as the deciduous dentition.

If much remains to be elucidated about the initiation phase of development, the process of morphogenesis can hardly be said to be better understood. The problem here is to explain how the mesenchyme of a papilla can control the folding of the inner dental epithelium in such a way as to produce a definitive cusp pattern, which differs from species to species and from tooth to tooth. An essential prerequisite is the integrity of the basement membrane: reagents which interfere with collagen synthesis

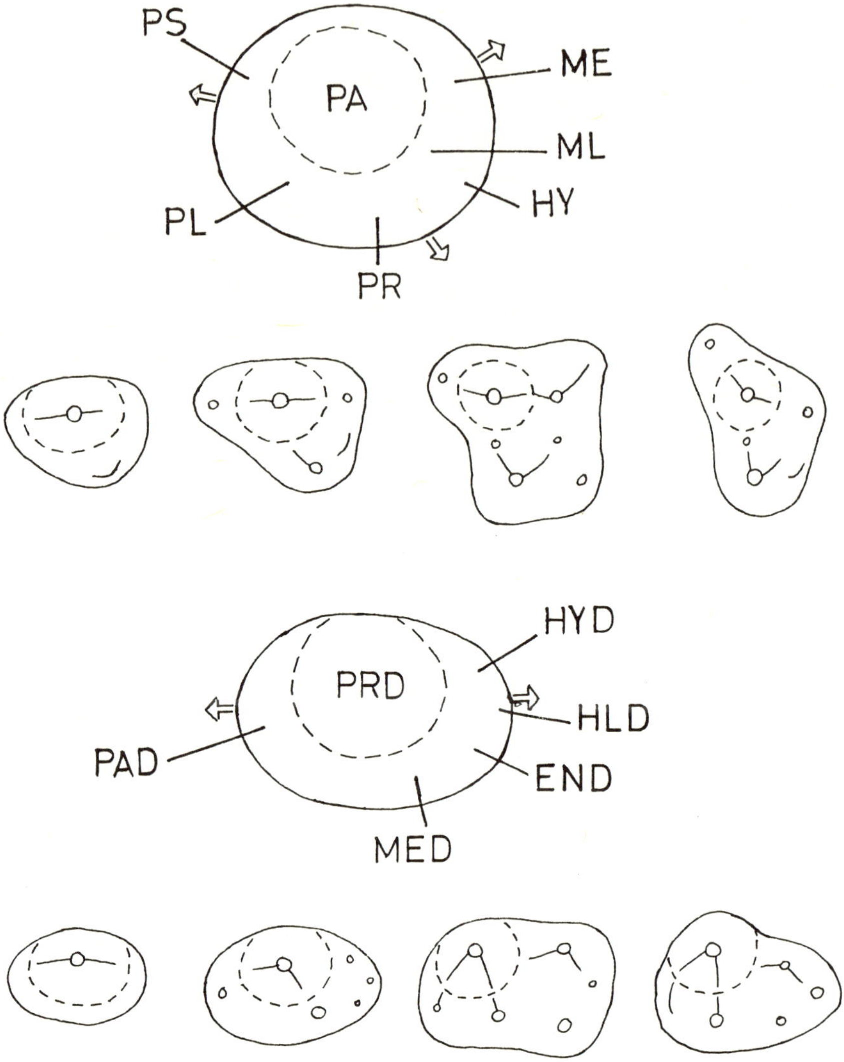

Figure 4.1. Diagrams showing situations of presumptive cusps of upper and lower teeth in the cap stage, and crown views of generalized mammalian teeth.

Key: (upper tooth) HY, hypocone; ME, metacone; ML, metaconule; PA, paracone; PL, paraconule; PR, protocone; (lower tooth) END, entoconid; HLD, hypoconulid; HYD, hypoconid; MED, metaconid; PAD, paraconid; PRD, protoconid.

Arrows indicate the directions of maximum horizontal outgrowth of the tooth germs, corresponding to the locations of future roots.

Cusps arise in a marginal zone around the paracone or protoconid. Their relative development varies with the position of the tooth in the dentition.

prevent the formation of the membrane and destroy the morphology (e.g., Galbraith and Kollar 1976). The processes involved in the formation of a cusp are still speculative. It is known that development proceeds from tip to base: epithelial cells at the tip of the cusp are the first to cease dividing and become columnar. At first there is only one cusp, the paracone or protoconid, standing on a base that is approximately circular in crown view. Other cusps arise in order in different positions in a marginal zone, while the base of the tooth grows out into lobes related to the future root pattern. (Butler 1956, 1967, 1971; Gaunt 1955, 1959b; Ramadan-Sadek 1962).

By extrapolating backwards in time it is possible to draw a diagram of the cap stage showing the locations where future cusps will develop (fig. 4.1). Such a diagram can be compared with the prospective fate map of a gastrula. As the distribution is two-dimensional, at least two factors are required to explain it. A uniform mesenchyme operating on a uniform epithelium could not produce a pattern with mesiodistal and buccolingual polarities.

All the teeth of a dentition can be traced back to a similar arrangement of prospective cusps in the cap stage, but as the teeth develop differences become apparent. In the upper canine the marginal zone remains in abeyance, producing only rudimentary cusps at the base of the paracone. In the premolars of most mammals the protocone forms from the lingual part of the marginal zone, but development of the distal cusps (metacone and hypocone) is weaker. These distal cusps reach their largest size on the molariform teeth, but they are again retarded on the posterior molars.

Mouse tooth germs cultured in isolation develop their normal cusp patterns from a very early stage, showing that cusp development is an intrinsic process. It follows that tooth germs which develop into different sorts of teeth are already different before cusps appear. One might suppose that teeth in the cap stage differ in the number of cells that are predestined to develop into a given cusp. When few cells are involved the cusp is rudimentary and it can be detected, if at all, only late in development; larger cell populations produce visible cusps earlier and result in larger cusps. Whether these cells are in the mesenchyme or in the enamel organ is an open question: perhaps an interaction between the two tissues is involved.

Investigation of tooth development comes up against what is perhaps the most fundamental problem of embryology: how does a genetic constitution produce a specific morphological pattern? At the present stage of our understanding it is easier to ask questions than to find means of answering them. The technical difficulties of experimenting on the earlier phases of tooth development are formidable. Perhaps we are not even asking the right questions.

References

Butler, P. M. 1939. Studies of the mammalian dentition. Differentiation of the post-canine dentition. *Proc. Zool. Soc. Lond.* (now *J. Zool.*) 109 B:1–36.

Butler, P. M. 1941. A comparison of the skulls and teeth of the two species of *Hemicentetes. J. Mammal.* 22:65–81.

Butler, P. M. 1956. The ontogeny of molar pattern. *Biol. Rev.* 31:30–70.

Butler, P. M. 1967. The prenatal development of the human first upper permanent molar. *Archs. Oral Biol.* 12:551–563.

Butler, P. M. 1971. Growth of human tooth germs. In A. A. Dahlberg, ed., *Dental Morphology and Evolution*, pp. 3–13. Chicago: University of Chicago Press.

Colyer, F. 1936. *Variations and Diseases of the Teeth of Animals.* London: John Bale and Danielsson.

Cooke, J. 1975. Control of somite number during morphogenesis of a vertebrate, *Xenopus laevis. Nature* 254:196–199.

Galbraith, D. B. & E. J. Kollar. 1976. In vitro utilization of exogenous procollagen by embryonic tooth germs. *J. Exper. Zool.* 197:135–140.

Gaunt, W. A. 1955. The development of the molar pattern of the mouse. *Acta Anat.* 24:249–268.

Gaunt, W. A. 1959a. The vascular supply to the dental lamina during early development. *Acta Anat.* 37:232–252.

Gaunt, W. A. 1959b. The development of the deciduous cheek teeth of the cat. *Acta Anat.* 38:187–212.

Green, M. 1977. Neogene Zapodidae (Mammalia: Rodentia) from South Dakota. *J. Paleont.* 51:995–1015.

Grüneberg, H. 1965. Genes and genotypes affecting the teeth of the mouse. *J. Embryol. Exp. Morph.* 14:137–159.

Grüneberg, H., B. M. Cattanach, A. McLaren, H. G. Wolfe, and P. Bowman. 1972. The molars of tabby chimaeras in the mouse. *Proc. Roy. Soc. Lond. B.* 182:183–192.

Kollar, E. J. and G. E. Baird. 1969. The influence of the dental papilla on the development of tooth shape in embryonic mouse tooth germs. *J. Embryol. Exp. Morph.* 21:131–148.

Kollar, E. J. and A. G. S. Lumsden. 1979. Tooth morphogenesis: the role of the innervation during induction and pattern formation. *J. Biol. Buccale* 7:49–60.

Kraus, B. S. and R. E. Jordan. 1965. *The Human Dentition Before Birth*. pp. 200–218. Philadelphia: Lea & Febiger.

Lumsden, A. G. S. 1979. Pattern formation in the molar dentition of the mouse. *J. Biol. Buccale* 7:77–103.

Miller, W. A. 1978. The dentitions of tabby and crinkled mice (an upset in mesodermal-ectodermal interaction). In P. M. Butler and K. A. Joysey eds., *Development, Function and Evolution of Teeth*, pp. 99–109. New York: Academic Press.

Osborn, J. W. 1978a. Morphogenetic gradients: fields versus clones. In P. M. Butler and K. A. Joysey, eds. *Development, Function and Evolution of Teeth*, pp. 171–202. New York: Academic Press.

Osborn, J. W. 1978b. A cladistic interpretation of morphogenesis. *J. Biol. Buccale* 6:327–337.

Ramadan-Sadek, A. A. 1962. Some aspects of the growth and differentiation of the molar tooth germ in the mouse (*Mus musculus*). Ph. D. thesis, Univ. of London.

Sofaer, J. A. 1969a. Aspects of the tabby-crinkled-downless syndrome. I. The development of tabby teeth. *J. Embryol. Exp. Morph.* 22:181–205.

Sofaer, J. A. 1969b. Aspects of the tabby-crinkled-downless syndrome. II. Observations on the reaction to changes of genetic background. *J. Embryol. Exp. Morph.* 22:207–227.

Sofaer, J. A. 1977. Tooth development in the "crooked" mouse. *J. Embryol. Exp. Morph.* 41:279–287.

5. A 3-Dimensional Computer Synthesis of Cephalograms and Dental Study Casts

BHIM SEN SAVARA, FUMIHIRO MASAKI, RICHARD JAMES SHADEL, AND SUSAN NIELSEN

Since Broadbent's first introduction of cephalometric roentgenography (1931), numerous applications for his technology have been developed in both research and diagnosis. The practice of cephalometrics as a means of relating the teeth to each other, and their bases to craniofacial bony structures, has a long and distinguished history among anthropologists, physicians, dentists, and orthodontists. As currently carried out, this type of analysis is generally performed on a single lateral cephalogram, though Broadbent's technology allows for a frontal (postero-anterior) orientation as well. Means exist whereby the lateral and frontal cephalograms may be quantitatively incorporated into a single synthetic three-dimensional view (Savara 1965). This is done by locating landmarks representing size dimensions of a specific bone or bones on both lateral and postero-anterior cephalograms. Landmarks are measured in x, y, z coordinates which are then corrected for magnification and distortion, and size dimensions between any landmarks are computed. This computerized method has been in use for a number of years and has demonstrated its reliability as a tool in tracking growth and development of craniofacial bones. See Tracy, Savara, and Brant, 1965; Savara, Tracy, and Miller 1966; Miller, Savara and Singh 1966; Tracy and Savara 1967; Singh and Savara 1966; Savara and Tracy 1967; Savara and Singh 1968; Nakamura, Savara and Thomas

This study was supported in part by the National Institute of Dental Research grant DE3343-01-03 and National Institute of Child Health and Human Development grant HD00157-01-16, National Institutes of Health.

1972; Sekiguchi, Savara and Arya 1973; Takeuchi, Savara, and Shadel 1980.

It is inherently difficult to visualize three-dimensional dental structures (the teeth) on two-dimensional lateral or frontal cephalograms. This problem can be overcome with stereo cephalograms to some degree, but the teeth overlap and thereby reduce visibility. Frequently this problem is solved by the use of study models. Such models are an accurate portrayal of dental structures, i.e., the teeth in relationship to each other, individually and in occlusion. However, they do not relate directly to the whole head and face complex, and to date have not been incorporated into the technique of three-dimensional reconstruction of head and face from lateral and frontal cephalograms.

Similarly, despite some excellent research by other investigators, it has not previously been possible to monitor precisely the interactions between the variables studied by different types of data acquisition techniques, e.g., cephalograms, study casts, photographs, etc.

This paper reports an attempt to devise a system whereby two invaluable diagnostic aids, the cephalograms and the study casts, are used to study growth of the head, face, and dentition simultaneously and quantitatively for an individual subject over a number of years.

Such a method would provide the following advantages:

1. Any measurements of the teeth from study casts would be compatible with measurements of bones from cephalograms, as the latter would have been corrected for magnification and distortion errors.
2. Movement of teeth relative to each other and the occluding members would be discernible.
3. Changes in tooth position associated with the underlying bones would be perceived quantitatively.
4. Changes due to differential growth of the mandible resulting in forward or downward rotations may become apparent in shorter intervals, e.g., within one or two years.
5. Role of eruption, i.e., rate, sequence, and timing, if related to differential growth of the mandible and the maxillae and occlusion, will be discerned more accurately.

In outline, the proposed method operates in three steps, beginning with the use of two cephalometric views which produce three-dimensional Cartesian coordinates for craniofacial landmarks, and a computer that recreates the skull in three-dimensional space. Second, points are located and marked with a delible pencil on the plaster study cast of the dentition (which matches the cephalograms for age and date) and measured, also in a three-dimensional Cartesian coordinate system. Finally, matching points in the two systems are used to transfer coordinates from the study cast model to the cephalographic space. These coordinates are then available to measure any angle or distance desired within or between tooth landmarks on the study cast and bone or soft tissue landmarks on the cephalograms.

Methods

The cephalograms and study casts used in this paper were obtained from the files of the Child Study Clinic, School of Dentistry, University of Oregon Health Sciences Center, Portland. These resources comprise a longitudinal sample which has been under continuous observation for more than twenty-five years. Both the cephalographic materials and the study casts (as well as a body of physical measurements, dental data, and additional films) were collected at intervals of 6 months until the subjects reached the age of 14 years. Subsequent observations were carried out annually as near the subjects' birthdays as possible. The actual method of data reduction from the cephalometric films is described in Part A, whereas Part B presents the method of digitizing the dental study casts or models.

A. Cephalograms

The method by which both frontal and lateral cephalograms are utilized to obtain reliable three-dimensional Cartesian coordinates and distances between features has been in use at the Child Study Clinic since 1965 (Savara 1965). Films are exposed by two X-ray sources placed so that their central rays intersect at an angle of 90°. The distance from each source to the intersection of the two central beams is 152.4 cm. This point of intersection provides the "origin" for the Cartesian coordinate system. Let the frontal central ray line be the x-axis and let the lateral central ray line be the y-axis (the z-axis being the perpendicular at their intersection); then the horizontal or xy-plane is defined by the two intersecting rays. The subject's head is oriented to Frankfort horizontal position parallel to this xy-plane (fig. 5.1), and frontal and lateral cephalograms are obtained. Figure 5.1 shows that the yz plane defined by the coordinate axes is parallel to the frontal film, and that the xy-plane lies parallel to the lateral cephalograms.

These cephalographic images make up the raw materials of a synthetic three-dimensional view of the face. Like all roentgenographic images, however, they are not accurate to size and proportion in the original anatomic structures. Because X-rays diverge as they are emitted from the source, the film image is magnified proportionally to the distance of the focal plane from the midsaggital plane, and of the midsaggital plane to the film.

The method for correction of values obtained from such films, and for three-dimensional treatment of landmarks derived from them was pre-

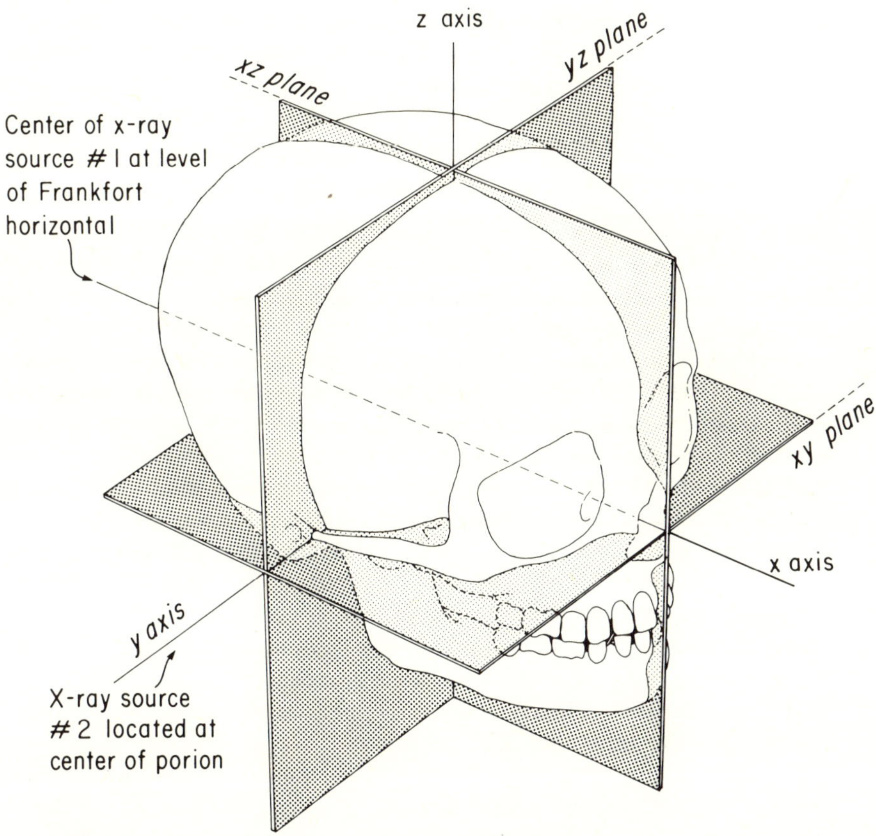

Figure 5.1. Diagram of subject's head showing intersection of x-, y-, and z-axes.

sented by Schwartz (1943). Savara (1965) elaborated a computer application of that method; it has been applied consistently since then to the analysis of growth and development of the face and head.

The procedure requires overlaying the cephalograms with a transparent acetate paper. The landmarks to be measured are located onto acetate paper and then an xy digitizer measures the xy coordinates and punches out the measurements onto IBM cards. The cards pass that information to the computer. The transformation equations are applied to these landmarks to correct for the magnification and distortion errors.

In order to "correct" the locations of landmarks on the grid of the x-, y-and z-axes, the following formula is applied:

$$x = 152.4\ x_p\ (n\text{-}y_p)/mn\text{-}xy$$
$$y = 152.4\ y_p\ (m\text{-}x_p)/mn\text{-}x_p y_p$$
$$z = 152.4\ z_p\ (n\text{-}y_p)/mn\text{-}x_p y_p$$

Where x_p, y_p, and z_p are the projected or apparent values as recorded on the X-ray film, *m* and *n* are the distances from the midsaggital and transverse vertical planes to the film plane.

B. Study Casts

Analyses of tooth size and positions, as measured in two dimensions on plaster study casts, have been performed at the Child Study Clinic since 1969 (Savara and Sanin 1969). Measurements are taken on a comparator devised by special modification of a traveling microscope. The comparator now displays a dental cast holder which can be moved horizontally and vertically. The comparator (fig. 5.2) has a horizontal lead screw axis joined to a vernier, which permits the cast holder to move along a fixed millimeter scale. Another lead screw axis perpendicular to the former moves the cast holder antero-posteriorly along the millimeter scale. Both verniers permit readings to the nearest 0.1 mm. The cast holder can rotate or move the dental cast vertically, facilitating visual orientation. Located above the cast holder is a clear plastic horizontal plate, 1.5 cms thick, marked with four fine lines. Two lines intersect each other on the upper surface and two on the lower surface of the plastic plate. The intersection point, viewed from above, eliminates parallax and is used for landmark location on the cast. Two linear action potentiometers on the comparator are connected via a flexible cable to the decimal converter, which reads the x and y coordinates and punches the data onto IBM cards.

To read the third dimension of the model, that is, tooth height, a third linear action potentiometer was installed on the comparator. It records the axis designated as z. Vertical travel was incorporated into the cast-bearing frame. Insertion of a micrometer gauge through the clear plastic plate makes precise measurement of the height of dental landmarks from a designated zero at the base of the cast holder. The micrometer scale acts as an indicator of the point of contact between the gauge and the model. The values from this axis (z) are digitized in the same manner as the x- and y-axes.

Mathematical Basis

The cast-and-cephalogram measuring device provides three sets of independently oriented three-dimensional Cartesian coordinates: the cephalograms, the mandibular cast, and the maxillary cast. These three sets can be merged if three equal points, or two common points and the sense of measurement, are known in each system. Considering that two or more three-dimensional Cartesian coordinate systems are measuring the same object, the sense of measure is as follows:

If one axis, say the x, increases in value when measuring from left to right

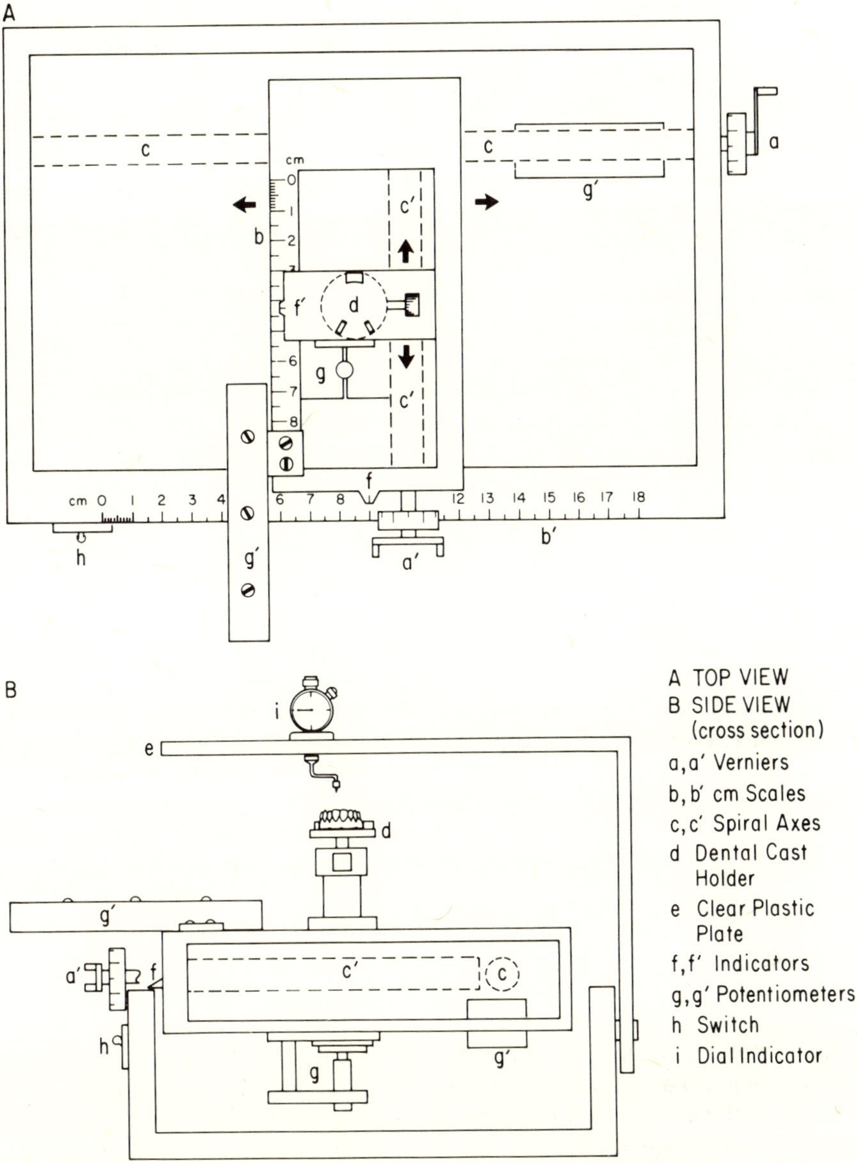

Figure 5.2. Comparator to measure tooth size and positions.

in one system, the same relationship must hold for all systems. For example, when measuring the cast in the cast-measuring device, the resulting measurements had the same sense as the cephalometric measurements (by chance). Now consider the maxillary cast, when placed in the same measuring equipment. Left and right are reversed, up and down are also technically reversed, but front and back are not.

As it is difficult to locate tooth landmarks on the cephalograms, the two point method was selected. (The sense of measure can be kept the same by repositioning the dental cast in its holder or multiplication of the x-, y-, or z-axis components by −1 as necessary.)

The coordinate line-up is accomplished by first transferring the origin of both systems to the first of the two common points (the central incisor in most cases). This is accomplished by simply subtracting these point coordinates from all points in the system. Then, a three-dimensional rotation of the cast axis based on the second line-up point (the left most posterior molar) completes the line-up.

This rotation is as follows:

$$\begin{aligned} x^1 &= \lambda_1 x + \mu_1 y + v_1 z \\ y^1 &= \lambda_2 x + \mu_2 y + v_2 z \\ z^1 &= \lambda_3 x + \mu_3 y + v_3 z \end{aligned}$$

Where x^1, y^1, and z^1 are cast coordinates after rotation to the cephalometric system; x, y and z are the original cast coordinates; and the cephalometric x^1-, y^1-, and z^1-axes have directional cosines (λ_1, μ_1, v_1), (λ_2, μ_2, v_2), (λ_3, μ_3, v_3) respectively, in the cast system of axes.

Computer Program

The translations, rotations and other mathematical calculations are performed by a Fortran computer program. The program operates as follows:

First, the lateral and anterior-postero cephalometric landmarks are read from storage (disk, tape, or cards). The coordinates are corrected for magnification and distortion (as previously described) and stored.

Next, the coordinates from the mandible and the maxillae are utilized. The two common points in both jaws are used to generate two translations and rotations from the dental cast system to the cephalogram system. Then, the individual three-dimensional cast coordinates are transformed to the cephalographic space. Practically, it was found easier to use a rotation in three two-dimensional plans in sequence (xy, xz, and yz) than the three-dimensional rotation described earlier. The line-up points which are used are usually a point on the central incisor and a point on the most posterior left molar, although different points could be used if they are easier to locate (such as a prominent restoration).

Finally, the various distances between any two points or angles between three points are printed out. The actual output is controlled at program execution time and may be as extensive or meager as desired.

Accuracy and Validity

The method was first tested on a dry skull. The skull was considered the best means for testing accuracy of the method. Postero-anterior and lateral cephalograms and plaster study casts of the skull were obtained and used. A second test was carried out, based on the cephalograms and casts from an actual subject at ages 9, 11, 13, and 15 years. The longitudinal materials illustrated actual ease or difficulty of pinpointing various landmarks at different ages, and thus established the validity of the method.

Craniofacial Dimensions

One millimeter metal balls were cemented to ten landmarks on the skull (fig. 5.3). The points chosen for this exercise were nasion, sella turcica, and four landmarks on the teeth of both the mandible and the maxillae. The dental landmarks selected (for ease of location on cephalograms) were the left central incisor tip, left central incisor neck, incisal tip of the cuspid, and a point on the left second molar. (Features on the left side were utilized, as it was nearest to the lateral film plane and presented the clearest image.)

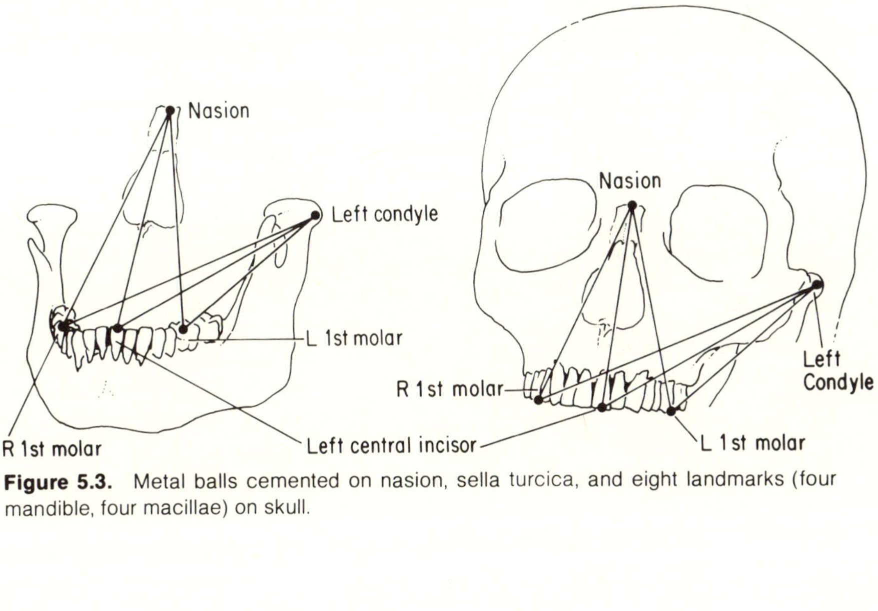

Figure 5.3. Metal balls cemented on nasion, sella turcica, and eight landmarks (four mandible, four macillae) on skull.

Postero-anterior and lateral cephalograms were made as well as study casts of both mandibular and maxillary teeth. During processing of the data, the incisor neck and the second molar were used as line-up points for the computer's integration of cephalograms and study casts. It has to be kept in mind that the point of origin for the Cartesian coordinates of the cephalograms is the point of bisection of the lateral and postero-anterior cephalograms, whereas the measurements of coordinates on the study casts are based on a point of origin which is related to the study casts but is unrelated to the cephalograms. To integrate the two coordinate systems, the landmarks known as incisor neck and the second molar (which could be seen in the cephalograms and the study casts) are aligned, and the program rotates the casts and integrates the two systems.

Five measures which were accessible to caliper verification were measured directly and also computed independently from Cartesian coordinate values processed by the computer; the results of a comparison of computer values and caliper values is shown in table 5.1. Additional points were used to verify that the rotations had been correctly performed. The distances given in table 5.1 were derived from easily determined landmarks. They represent the best results obtainable through this method and this technology.

Longitudinal Subject

The landmarks used in the longitudinal series were somewhat different from those used on the skull. In addition to the line-up, landmarks on the left central incisors and the left second molars, the nasion and the left condylion were located.

Landmarks on the study casts were: left central incisor neck, left second

Table 5.1 Comparison of Direct and Computer Measurements on the Human Skull.

Measurement	Computed Distance	Caliper Distance	Absolute Difference
Second molar to incisor cervical line, mandible	4.48	4.48	0.0
Second molar to nasion, mandible	7.90	7.94	0.04
Second molar to incisor cervical line, maxilla	3.46	3.51	0.05
Second molar to nasion, maxilla	7.65	7.64	0.01
Nasion to sella	6.28	6.29	0.01

All measurements in centimeters.

Table 5.2. Computer measurements between landmarks on cephalograms and study casts on one subject from 9 to 15 years.

Description	Measurements on the Mandible Age in Years 9	11	13	15
Nasion to left molar	7.22	7.25	7.84	8.09
Nasion to right molar	7.13	7.17	7.78	7.92
Nasion to left incisor	6.87	6.86	7.43	7.56
Left condyle to left molar	6.56	6.57	7.61	7.82
Left condyle to right molar	8.72	8.65	9.69	9.79
Left condyle to left incisor	9.49	9.67	10.50	10.63
	Measurements on the Maxilla Age in Years 9	**11**	**13**	**15**
Nasion to left molar	7.59	7.42	8.01	8.36
Nasion to right molar	7.22	7.35	7.89	8.06
Nasion to left incisor	6.85	7.30	7.58	7.86
Left condyle to left molar	6.36	6.28	7.16	7.42
Left condyle to right molar	9.66	9.72	10.54	10.63
Left condyle to left incisor	9.86	9.99	10.71	10.88

All measurements in centimeters.

molar, highest point of the mesio-buccal cusp of both left and right first molars, and a prominent point on the occlusal surface of the left central incisor. The landmarks which were not used for line-up were selected as points which could be followed longitudinally at all four ages.

Cranial-to-dental landmarks (table 5.2) are listed in the first column, with the ages at which the measurements were made listed across the top. The table shows that the measurements consistently increased with age. As these measurements cannot be verified directly, we decided to measure the distances between the line-up points on the study cast, as their accuracy could be verified. The measurements were made on the casts and are shown in table 5.3.

Conclusion

This demonstrates the feasibility of combining synthetically a three-dimensional study cast and a set of lateral and postero-anterior cephalograms to portray the relationships between the dentition and facial structures. The study was carried out using modified equipment that was intended for other purposes. Direct measurements on the actual study cast and on a skull used in preliminary experiments would indicate that the degree of accuracy between actual structures and recreated (computer

Table 5.3. Computed and direct measurements on the study casts.

	Distance from Left Molar to Left Central Incisor on the Mandible Age in Years			
Method of Measurement	9	11	13	15
Integrated cephalogram and cast distance (computed)	3.52	3.62	3.55	3.47
Comparator distance for casts	3.53	3.53	3.44	3.39
Caliper distance for casts	3.50	3.55	3.41	3.38
	Distance from Left Molar to Left Central Incisor on the Maxilla Age in Years			
	9	11	13	15
Integrated cephalogram and cast distance (computed)	3.82	3.85	3.72	3.81
Comparator distance for casts	3.87	3.84	3.81	3.73
Caliper distance for casts	3.85	3.79	3.75	3.70

All measurements in centimeters.

model) structures is good. Although the technique presented here is primitive in relation to that possible using newer hardware designed for graphic representation, the results are encouraging. It is probable that in application to a larger working sample, the measurement error in reconstructing landmarks in the dentition can be routinely kept under 0.5 millimeter.

In the past it has not been possible to visualize in detail all the structures of the face and teeth. Such techniques are controversial, for instance among orthodontists, for a number of reasons; any aid to better visualization of the total face should be considered beneficial for clinical treatment. As an example, the effect of manipulation of the teeth on surrounding structures can be better observed. Cephalographically "hidden" structures may be incorporated into an analysis and treatment plan. Specific structures, such as the curve of spee, may be measured easily on a cast model and then integrated into the general face. Hence, with modifications to make this method easily workable, a number of applications are foreseeable.

References

Broadbent, B. H. 1931. A new X-ray technique and its application to orthodontia. *Angle Orthod.* 1:45–66.

Miller, P. A., B. S. Savara, and I. J. Singh. 1966. Analysis of errors in cephalometric measurement of three-dimensional distances on the maxillae. *Angle Orthod.* 36:169–175.

Nakamura, S., B. S. Savara, and D. R. Thomas. 1972. Norms of size and annual increments of the sphenoid bone from four to sixteen years. *Angle Orthod.* 42:35–43.

Savara, B. S. 1965. A method for measuring facial bone growth in three dimensions. *Hum. Biol.* 37(3):245–255.

Savara, B. S. and C. Sanin. 1969. A new data acquisition method for measuring dentitions and tests for accuracy. *Am. J. Phys. Anthrop.* 30(2):315–318.

Savara, B. S. and I. J. Singh. 1968. Norms of size and annual increments of seven anatomical measures of maxillae in boys from 3 to 16 years of age. *Angle Orthod.* 38:104–120.

Savara, B. S. and W. E. Tracy. 1967. Norms of size and annual increments of five anatomical measures of the mandible in boys from 3 to 16 years of age. *Archs. Oral Biol.* 12:469–486.

Savara, B. S., W. E. Tracy, and P. A. Miller. 1966. Analysis of errors in cephalometric measurements of three-dimensional distances on the human mandible. *Archs. Oral Biol.* 11:209–217.

Schwartz, H. 1943. A method of measuring points in space as recorded by the Broadbent-Bolton Cephalometer technique. M.S.D. thesis, Northwestern University, Chicago.

Sekiguchi, T., B. S. Savara, and B. S. Arya. 1973. Norms of size and annual increments of six anatomical measures of the cranium in boys and girls from four to fifteen years of age. *Am. J. Phys. Anthrop.* 39:49–56.

Singh, I. J. and B. S. Savara. 1966. Norms of size and annual increments of six anatomical measures of the maxillae in girls from 3 to 16 years of age. *Angle Orthod.* 36(4):312–324.

Takeuchi, Y., B. S. Savara, and R. J. Shadel. 1980. Norms of size and biennial increments of eight anatomical measures of the temporal bone in boys and girls from 4 to 20 years of age. *Angle Orthod.* 59(2):107–113.

Tracy, W. E. and B. S. Savara. 1967. Norms of size and annual increments of five anatomical measures of the mandible in girls from 3 to 16 years of age. *Archs. Oral Biol.* 12:469–486.

Tracy, W. E., B. S. Savara, and J. W. A. Brant. 1965. Relation of height, width and depth of the mandible. *Angle Orthod.* 35(4):269–277.

6. Aberrant Maxillary Third Molars *Morphology and Developmental Relations*

R. M. S. TAYLOR

ABSTRACT

In a collection of extracted teeth lacking case histories a number were found to be obviously malformed, and some of these have been conveniently described as "compressed." (Taylor 1978).

A study of anatomical relationships during development of third molars showed that such structures as the maxillary sinus, the maxillary tuberosity and the tooth itself were themselves subject to variation in stage of development and form, at different ages, and that the developing tooth crown might sometimes be obstructed.

It was surmised that varying combinations of growth stages might lead to some degree of deformity even in cases where the mature palate was well formed, and examples were recognized in both contemporary and prehistoric skeletal material of various ethnic groups. Examples of grossly deformed teeth were found only in comtemporary material, and it was seen how these could occur in cases of palatal deformity.

Thus a reasonable explanation is offered for certain morphological variations, by relating them to aspects of maxillo-facial development which do not seem to have received attention elsewhere. It is hoped to publish elsewhere an extended report with illustrations of many teeth presenting compression characters.

In my recently published book (Taylor 1978), some morphological variations of maxillary third molars were designated as "compressed," since this adjective conveniently described their difference from the more

I am grateful to Professor J. B. Carman, Department of Anatomy, University of Auckland, for advice with the manuscript; to I. R. MacDonald, photographer in the same department, who made the slides; and for editorial assistance from H. D. Erlam, formerly librarian at the School of Medicine, University of Auckland.

usual form. Third molars, maxillary and mandibular, show variation in many other aspects such as size, number of cusps and root branches, and age of eruption or failure to develop. This has no doubt led to the hypothesis that great variability indicates a genetic trend towards elimination of this tooth from the human dentition. I reject such hypotheses as convenient evasions. Admittedly we do not know, but this paper presents evidence that many morphological variations seen in maxillary third molars may well arise through obstruction of the developing tooth.

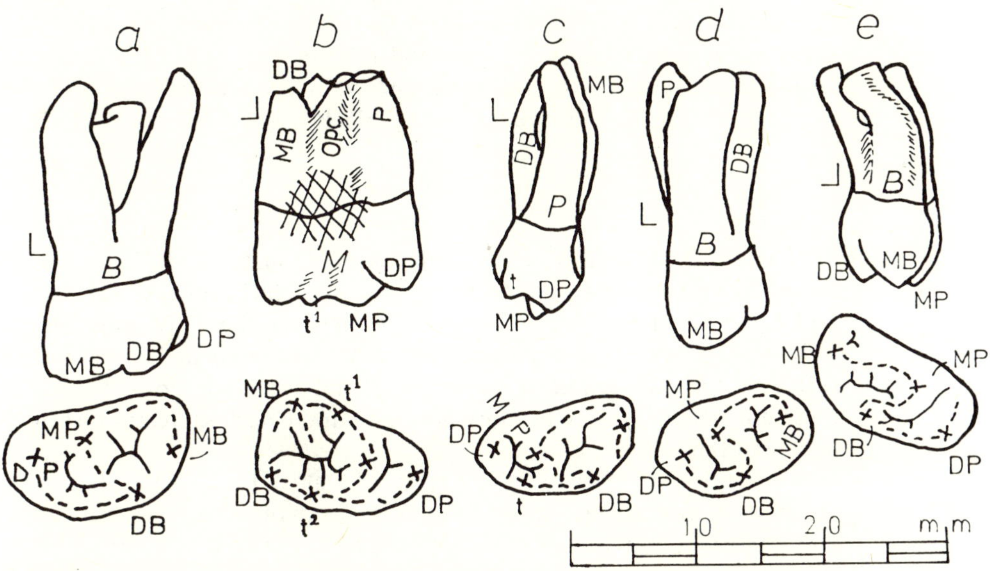

Figure 6.1. Teeth with style A1 compression characters. The main cusps are prominent and recognizable, but an unusual arrangement affects the usual quadrilateral outline of the occlusal view. The DP cusp is displaced mesially so that it is almost in line with the MB and MP cusps, and the oblique transverse axis is increased while the M-D dimension is reduced. Buccal cusps tend to be displaced anti-clockwise in right teeth and clockwise in left teeth. See figures 6.6 and 6.7

Key: Figures 6.1 to 6.4 are drawn and lettered in the manner employed in my work on variation (Taylor 1978), where the symbol Opc for additional elements in crown or root is explained. For outlines other than occlusal, the aspect is indicated in italic capitals, and cusps and root elements or short branches are identified in bold capitals. A tubercle as distinct from a cusp is indicated by t, although there may be uncertainty whether the tubercle is a reduced DP cusp or Opc. In occlusal views an x indicates the position but not the size of a cusp tip, tubercle or Opc; broken lines represent ridges, and continuous lines indicate grooves or pits. Parallel lines indicate flattening of occlusal surfaces, and cross hatchings show that a mesial surface is concave. The light shading to locate shallow grooves on crown or root, and the irregular outlines of carious cavities are easily recognized. Whether the tooth is a right or left third molar is shown by the symbols ⅃ and | respectively.

Although detailed attention cannot be given here to the lack of normal elements or to the presence of additional elements in crown or root in so far as they affect the potential for variation, it will be shown that some teeth which hitherto were presumed to be within the range of normal variation may in fact be cases of compression, differing in degree or some other respect from the more obvious examples. Thus, although compressed teeth contribute to and share in the great variety of form that has been noted, they present characters which are additional to (rather than an integral group within) the range of the presumed potential for normal variation. As far as I am aware they have been identified almost exclusively in maxillary third molars, although among second molars there are a few doubtful cases. These could arise through uncertainty in identification of an extracted tooth, such as one figured by Black (1901).

Styles of Compression

I have defined two distinct styles of compression, A and B. In each style there are degrees of deformity which may conveniently be called substyles 1 and 2. Briefly, the distinguishing features between these divisions are the following. In style A the occlusal surface has nearly normal cusps, or has ridges, but it is altered in its relative M–D and B–P dimensions. In style B these dimensions tend to be more normal, but the occlusal features show a degree of flattening.

In substyles, A1 teeth may have cusps of almost normal form, but their positions are somewhat changed (fig. 6.1). Teeth of A2 substyle may have lost identity of individual cusps, and tend to have two longitudinal ridges separated by a groove that runs somewhat obliquely MB–DP (fig. 6.2).

In substyle B1 the flattening tends to be combined with reduction in the relative size of distal cusps, and this may be most easily seen in the buccal view of the tooth (fig. 6.3).

In substyle B2 the flattening is combined with a distortion, as though the occlusal portion had been forced to move mesially and "overhang" the cervical part of the mesial surface, so that there is apparent gross deformity of the tooth (fig. 6.4).

Although some teeth clearly belong to a particular style and substyle, there are many others which are intermediate, have characters of both styles A and B, and collectively present a considerable range of degree of compression characters. This finding suggests that the cause or causes of compression characters may involve a number of factors. The varied effects in style and degree suggest further that compression characters may well result from variety in combination and influence of such factors. In other words, explanation of compression characters must be sought among the varying influences to which the developing tooth is subjected.

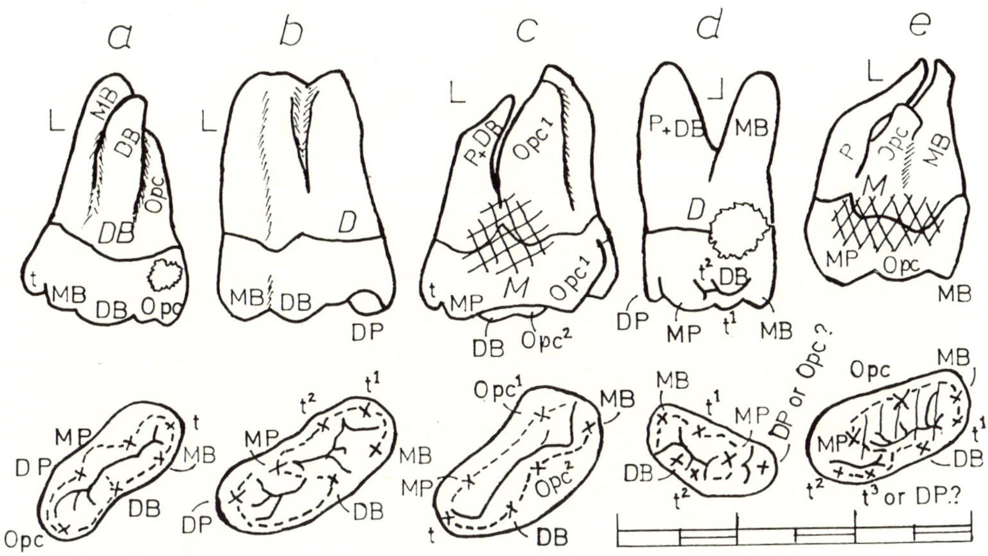

Figure 6.2. Teeth with style A2 compression characters. Some cusps may be recognized and identified, but they tend to be replaced by long, uneven ridges placed somewhat mesially and distally, and separated by an elongated groove. The change in outline tends to exaggerate that seen in style A1. Tooth *e* in this figure has also some characters of style B. Key: see figure 6.1.

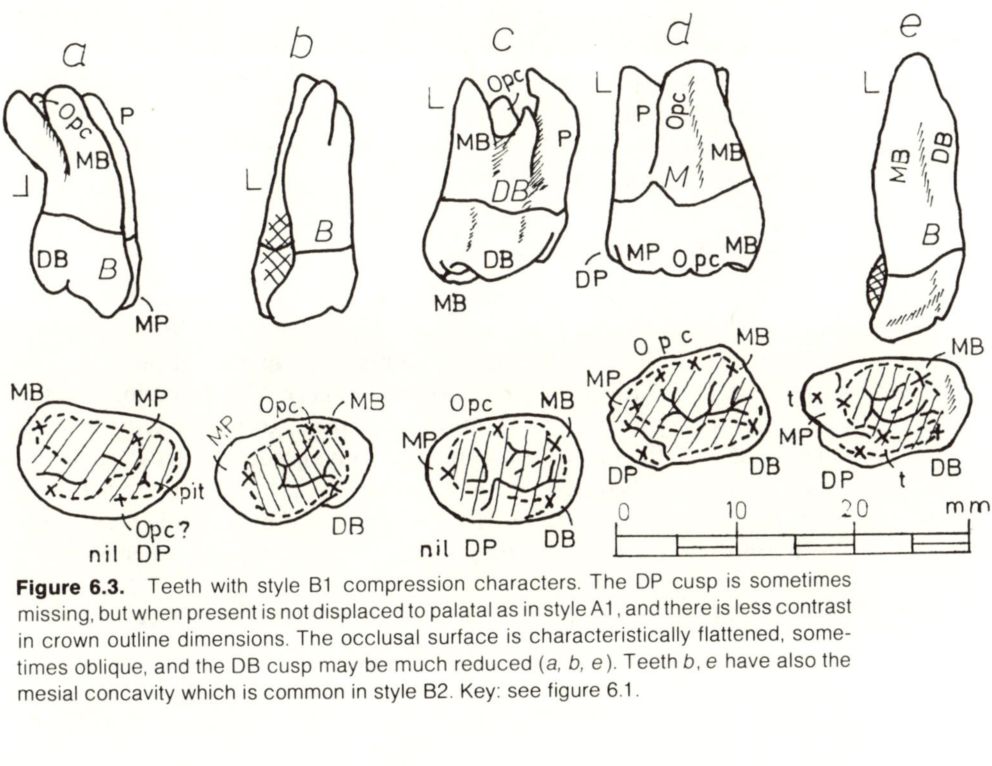

Figure 6.3. Teeth with style B1 compression characters. The DP cusp is sometimes missing, but when present is not displaced to palatal as in style A1, and there is less contrast in crown outline dimensions. The occlusal surface is characteristically flattened, sometimes oblique, and the DB cusp may be much reduced (*a, b, e*). Teeth *b, e* have also the mesial concavity which is common in style B2. Key: see figure 6.1.

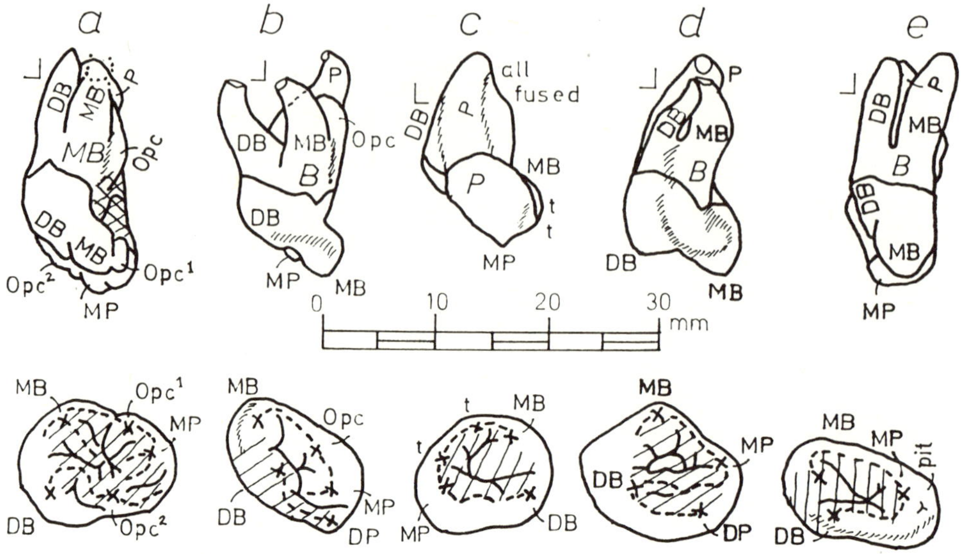

Figure 6.4. Teeth with style B2 compression characters are obviously deformed. The occlusal surface may be flattened and oblique (as in style B1), but the crown sometimes appears to be partly folded to overhang the mesial surface, forming an irregular concavity. Key: see figure 6.1.

It must be mentioned that while there are some characters which might well be considered as compression effects among variations in root morphology in third molars, these cannot be dealt with here. It is sufficient to observe that compression characters may be found in crown or root or both, and that the effect must be produced in the matrix of dental tissue while it is developing, and before calcification is complete.

Period of Crown Development

The process of crown formation and calcification extends over several years, during which time the tooth germ increases in size as its morphology becomes established. Calcification commences in cusp tips, and it is suspected that calcification in mesial cusps is ahead of calcification in distal cusps. But from the time of formation of the tooth germ and its crypt, until calcification of the crown is appreciably advanced, this growing dental tissue could be distorted. It is postulated that if confinement with obstruction to any increase in size is experienced during this period, the tooth crown will not achieve its potential morphology. In other words, it will show compression effects.

It should be emphasized that the term "compressed" conveniently describes the apparent result in the calcified crown. The process, however, is not to be seen as changed morphology resulting from pressure applied to a full-grown but still plastic body, but rather as obstruction to an increase in bulk. The latter cannot be prevented and is consequently diverted.

Studies of development of the maxillary third molar involve considerations of the period of time involved, the age of the child, and the tooth's anatomical relations and possible changes therein during that time. In a study which is concerned with the crown rather than the whole tooth, these are all very relevant. It has been shown that the elapsed period of vulnerability extends over several years. Banks (1934) found evidence of third molar crypts as early as 5 years of age, and evidence of commencing calcification of cusps about one year later. The earliest age observed for completion of cusps was 7 years (but in some it was as late as 14) with the majority between 9 and 12 years. Thus the several years involved may occur in different cases within a range of some 10 years.

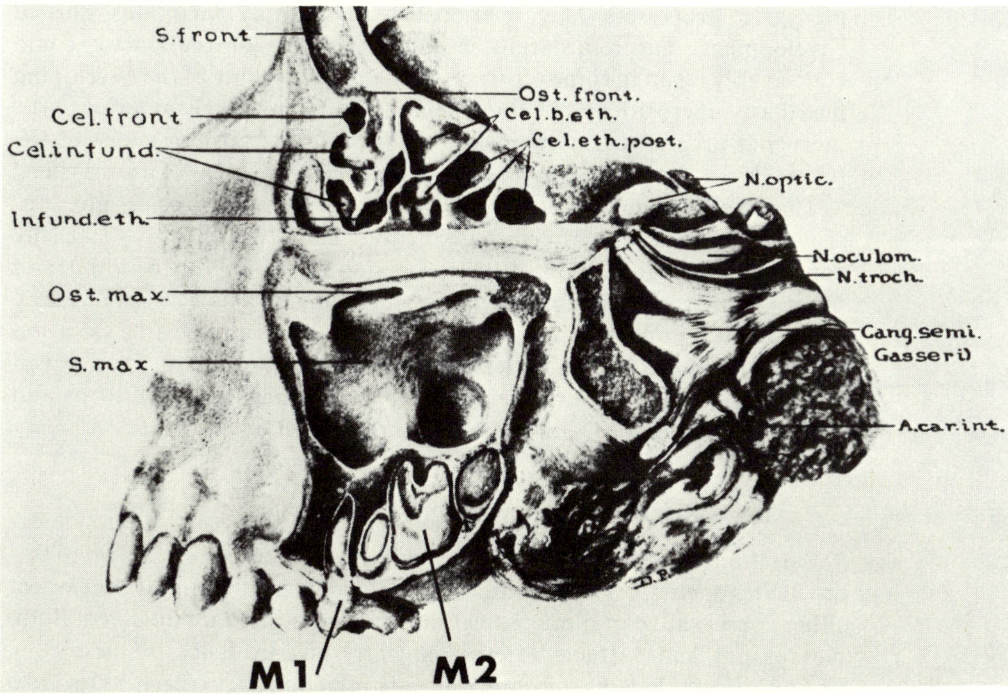

Figure 6.5. Dissection of maxillary region at age of approximately $10^{8}/12$ years. See the restricted triangular area in which the third molar is developing, and its anatomical relationships. Generally similar relationships would apply in most cases, but with variation in details and in the age at which this stage would occur. Illustration from Davis (1914).

Anatomical Relationships

As for the tooth's anatomical relationships and changes therein during this period, particular attention must be given to the maxillary sinus, the maxillary tuberosity, the pterygoid process and the second molar. The latter tooth with its root at least partly formed lies closely antero-inferior to the developing third molar crown and cannot move to "make more room" for it. The expanding maxillary sinus is superior, with its floor almost in contact with what is destined to be the distal surface of the crown (fig. 6.5). At this stage the calcifying occlusal surface of the tooth faces almost backwards and somewhat outwards, and then turns increasingly downwards during its descent in the tuberosity. In the earlier stages this "occlusal" surface is closely parallel to the posterior wall of the maxillary tuberosity, and (particularly in lateral view) to the pterygoid process nearby.

If the palates of young skulls are viewed from below, the maxillary tuberosities are seen to lie antero-laterally rather than anterior to pterygoid processes. This relationship presumably facilitates normal development, but retardation in lateral growth of the palate could conceivably result in some obstruction to the remodeling of the developing maxillary tuberosities while they remain more immediately anterior to the pterygoid processes. Enlow (1968) states that "the posterior face of the tuberosity represents one of the most rapidly growing areas in the maxilla."

That the relationship of tuberosity and pterygoid process is not consistently "juvenile" but often in some degree "adult" during growth is easily confirmed by examination of skeletal material (Taylor 1962b, 1963), and yet it is apparent that normal palatal development may be achieved by maturity. It is important therefore to note that while appreciable variation may normally occur in this respect without recognizable defects, yet in contemporary society a malrelationship may persist in a case of palatal deformity and malocclusion when it is not treated by an orthodontist.

Factors Relating to Compression Effects

Variation in form and extent of the maxillary sinus is well known to the practicing dentist, and its relation to tooth roots is often important when either conservative or more radical surgical procedures are involved. Both radiographs and extracted teeth frequently give evidence of these variations, in both skeletal and contemporary material, particularly as regards sinus morphology and its relation to molar roots and also to the post-dental alveolar process (Taylor 1962b, 1963, 1978, 1980a). The sinus may extend back to or posterior to the third molar, whose roots may appear partly to embrace the sinus or extend upwards beside it. The points to note are the facts that there is great variability and that the tooth roots conform to the sinus, not vice versa. The sinus enlarges ahead of tooth root formation so that in a sense the sinus is dominant. Thus morphology in the developing crown of the third molar may also be affected when this crown lies almost horizontally below a sinus which has already expanded above it.

In such circumstances, expansion of the developing crown upwards would be hindered by the sinus floor in close proximity to the potential distal surface of the tooth. As the tooth germ enlarged, its potentially most distal portion, the DP cusp, would be obliged to move palatally and mesially. It is hardly to be expected that such movement would take place without some adjustment in other cusps, logically adjustment clockwise, in occlusal view of a left tooth. That such changes in dimension and cusp position did occur is well seen in the maxillary fragment from a Maori skull (figs. 6.6 and 6.7).

Other cases, in Moriori skulls, were shown in earlier publications (Taylor

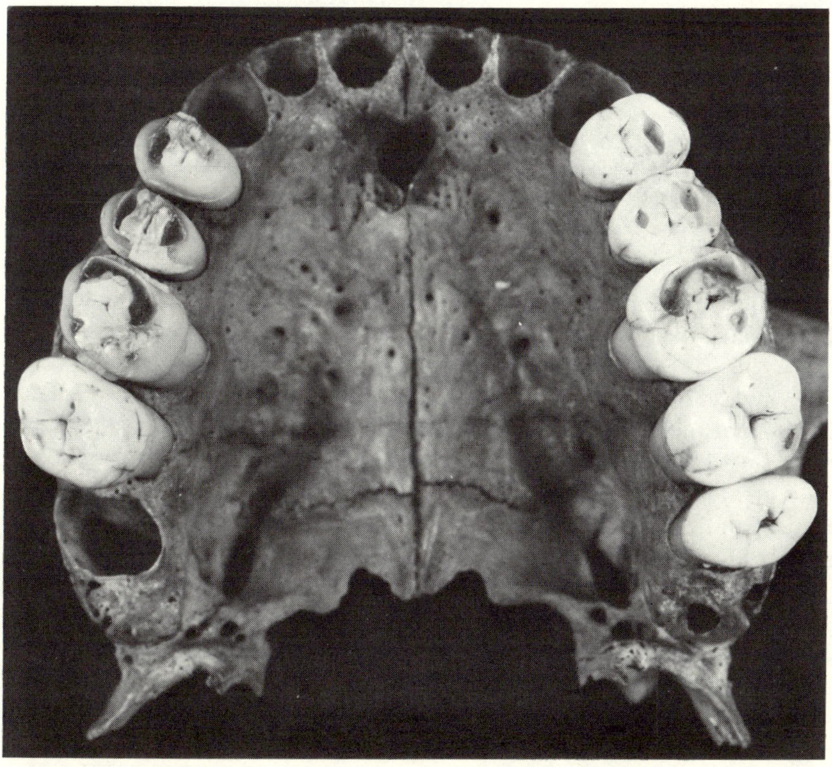

Figure 6.6. Maxillary fragment from pre-European Maori skull with compressed third molar of style A1, showing change in the usual relative dimensions.

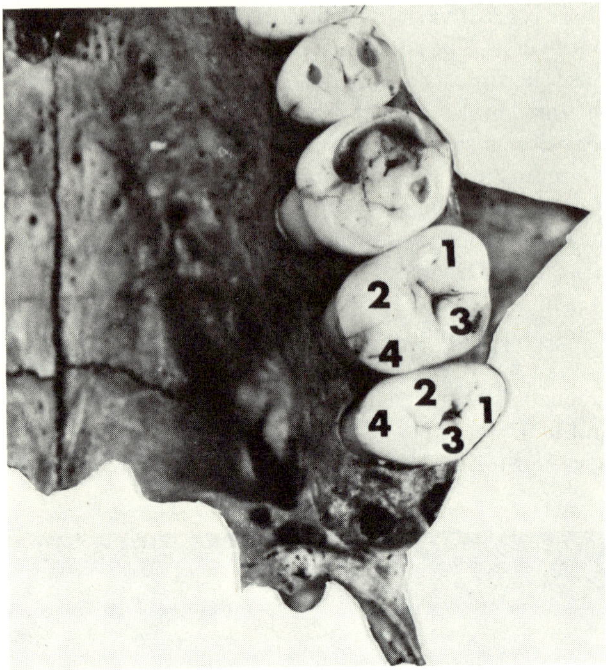

Figure 6.7. The left side molars from figure 6.6. Note the changed position of third molar cusps compared with the usual, as marked in the second molar.

1962b, 1963), and in Australian aboriginal skulls by Campbell (1925, Pl. XVII, fig. 4 and p. 29). All these, together with the extracted teeth of Caucasian stock shown in (fig. 6.1), show compression characters of style A1.

In style A2 the apparent tendency for recognizable cusps to be replaced by two parallel ridges is envisaged as resulting from conditions described for style A1, but operating either at an earlier stage of crown formation or with greater severity of obstruction. Whether such conditions may in part result from earlier or greater expansion of the sinus must at present be a matter of speculation.

Compression characters of style B (figs. 6.3, 6.4) appear to arise from somewhat different circumstances, which the writer suggests could be caused by interference with normal remodeling during growth changes in the maxillary tuberosity. Such interference, it is suggested, would be most likely to occur where the defective bone formation is associated with a narrow deformity of the palate and with dental malocclusion. It is suggested also that failure of the tuberosities to attain the normal "juvenile" lateral relationship, and their consequent close association with the pterygoid processes, would result in the kind of interference mentioned.

Since tooth development does not appear subject to interruption in the presence of palatal deformity (although antero-posterior maxillo-facial growth may clearly be affected), one can visualize an abnormal obstruction "facing" the developing occlusal surface. It is submitted that this could account for the observed flattening of occlusal surfaces in style B cases, and that this must commence at an early stage in their development.

As the tooth germ in such cases endeavors to expand, it must of necessity impinge against the posterior wall of the tuberosity, causing the potential occlusal surface to adjust downwards or mesially. If it is true that calcification in mesial cusps is in advance of that in posterior cusps, these posterior or distal cusps would tend to lose identity to a greater extent than the mesial cusps. Such a departure from normal relative size between these cusps is a readily identifiable character in style B compression.

In summary, attention is again drawn to the use of the term "compressed," because of its convenient description of the appearance of many of these teeth. It must be emphasized that the term is in no way intended to imply that the growing tooth was distorted by pressure such as might be applied to an established plastic body. On the contrary, it is suggested that the enlarging tooth germ is obstructed, or confined in some degree, so that its potential bulk is attained and calcified, but with a misshapen morphology, as figured extensively elsewhere (Taylor 1980b).

This paper has shown how normal variations in development of related structures may sometimes be combined to influence the morphology of maxillary third molars. It has also presented evidence of gross deformity of these teeth, which was presumably associated with disturbance of normal palatal development. In short, it has considered certain variables in dento-maxillary development, and some possible effects which do not seem to have received attention elsewhere.

References

Banks, H.V. 1934. Incidence of third molar development. *Angle Orthodont.* 4:223–233.

Black, G.V. 1901. *Descriptive Anatomy of the Human Teeth*. 4th ed. Philadelphia: S.S. White Dental Manufacturing Co.

Campbell, T.D. 1925. *Dentition and Palate of the Australian Aboriginal.* Adelaide: Hassell Press.

Davis, W.B. 1914. *Development and Anatomy of the Nasal Accessory Sinuses in Man*. Philadelphia: W.B. Saunders.

Enlow, D.E. 1968. *The Human Face.* New York: Harper and Row.

Taylor, R.M.S. 1962a. The human palate. *Acta Anat.* Suppl. 43.

—— 1962b. Non-metrical studies in the human palate and dentition in Moriori and Maori skulls. Parts I and II. *J. Polynes. Soc.* 71:83–100 and 167–187.

—— 1963. Cause and effect of wear in teeth. *Acta Anat.* 53:97–157.

—— 1978. *Variation in Morphology of Teeth*. Springfield, Ill.: C.C. Thomas.

—— 1980a. The maxillary sinus and molar roots. *J. Dent. Assoc. Sth. Afr.* 35:51–56.

—— 1980b. Compressed third molars—a suggestion. *Acta Anat.* 107(2):129–142.

7. The Logarithmic Spiral and Allometry as Expressions of Odontogenic Thermodynamics

MELVIN L. MOSS

It is commonly agreed that the morphological and morphometric attributes of tooth form (i.e., shape and size) provide useful data in many aspects of investigation in physical anthropology, vertebrate paleontology and taxonomy. This utility is a reflection that dental form, in normal conditions, is expressed in a highly predictable manner; that species, and often sexually in a specific dental form, display a high degree of "order." While this orderliness is both recognized and used, the more fundamental bases underlying the phenomenological data of dental morphology, which are of low explanatory value in themselves, are lacking at present. The present essay, while not providing experimental data to substantiate any hypotheses, does attempt to suggest one possible avenue or progam of future research; a program based upon the properties of biological thermodynamics (see Lehninger, 1971 for an introduction).

Order-Structure-Information and Entropy.

If we believe that all biological growth is a problem of thermodynamics, the basic question remains: how do we achieve order (structure) in a universe in which there is increasing disorder (entropy), as postulated by one expression of the second law of thermodynamics. A somewhat poetic metaphor considers living systems as analogous to islands in a sea of entropy, islands where entropy decreases as structure (information) is accumulated (Trincher 1965). More precisely, it can be shown that local

Supported, in part, by a research grant 1RO1DE05145-01, N.I.D.R.

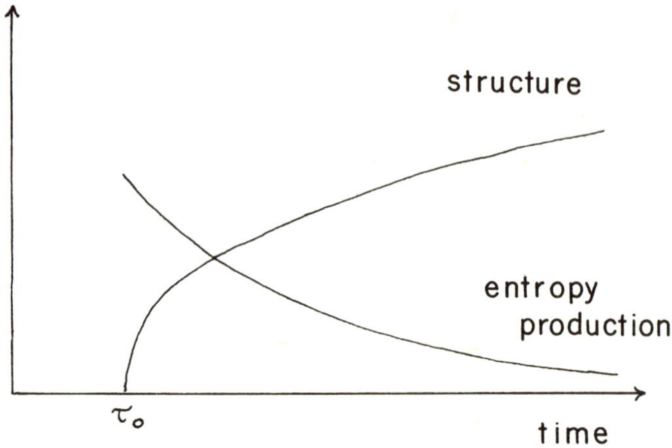

Figure 7.1. An arbitrary schematic representation of the increase in "structure," or order, in a biological system as the specific rate of entropy production decreases during time. Adapted from H.J. Hamilton 1977, fig. 1.

order in any system increases as the specific rate of its entropy production decreases (fig. 7.1) (Hamilton 1977). Parenthetically, entropy produced by any system and lost by that same system increases the entropy of the global universe within which the local system is placed, as postulated by the first law of thermodynamics.

It may be helpful to clarify the relationships between the terms "order," "structure," and "information." Order is used in the strict sense as the antithesis of disorder. More precisely, order is equivalent to the term negentropy, as opposed to entropy. Order can be perceived at many levels of observation as structure, defined tentatively as evidence of some recognizable spatial or spatio-temporal array of constituent parts; a pattern if you will. Structure, in turn, is a representation of information (see Elsasser 1958 for a complete discussion); for example, a hyperbolic curve contains the same information as the mathematical formula for hyperbolas, and this informational content is identical no matter in what projection the hyperbolic curve is graphically presented. "There is a concrete, physical, material relationship between negentropy and information," and information has a "content-bearing" nature that is an attribute of matter (Trincher 1965); and in another formulation, ". . . the very complexity of form and function of living organisms is in itself a kind of energy, according to thermodynamic principles," and "information . . . is actually a form of energy" (Lehninger 1971). In another and complementary sense, structure represents information which is utilized by the local system itself in the regulation of that same system (comprehensive reviews of the relationships between structure and information are given by Kubat and Zeman 1975 and Trincher 1965).

It is of interest in the present context to note that the orderliness of tooth form is most evidently expressed by the outer enamel surface of the crown, to a lesser extent by the form of the dentino-enamel junction and far less by the form of the pulpal chamber outline (Moss and Moss-Salentijn 1977, 1978). For this reason, in this paper we will examine some of the factors possibly related to the regulation of amelogenesis in the attempt to demonstrate biological processes that may have a thermodynamic expression.

The expression of order.

Local order in the universe is frequently shown by the appearance of a particular type of structure, a logarithmic spiral. This specific type of curve is observed at many levels of organization, both biological and nonbiological. These extend from the shape of the spiral galaxies to the shape of the "armored" amoebas, Foraminifera (Spindler and Röttger 1973).

The well-known spiral pattern of a hurricane and the classic shape of the chambered nautilus furnish other examples. The pervasiveness of logarithmic spirals in nature is well illustrated by many (Thompson 1942; Cook 1979; Huntley 1970; Ghyka 1977; Hambridge 1967), and the many and often fascinating, mathematical properties of this family of curves has attracted attention since classical times (Purce 1974).

More universally, the following axiomatic statements are possible concerning any object described by a logarithmic spiral.

1. It is a determined system (not stochastic).
2. It is a stable system (resists small perturbations).
3. It is the strongest system (in an engineering sense).
4. It is associated with the concept of motion.
5. It is the most efficient way of converting energy.
6. It is an energy conservative system.
7. It is a configuration associated with the minimal energy concept.

These seven statements, suggested to me Dr. G. Dasgupta of my university, may be resolved as structural (1–3), kinematic (4) and energetic (5–7) in nature; at present we stress the energetic implications with respect to amelogenesis, although all of them provide significant modes of connection between the structure of biological systems and the diverse modes of morphological expression of the logarithmic nature of their form and growth.

Having established that the structure of many natural objects is easily

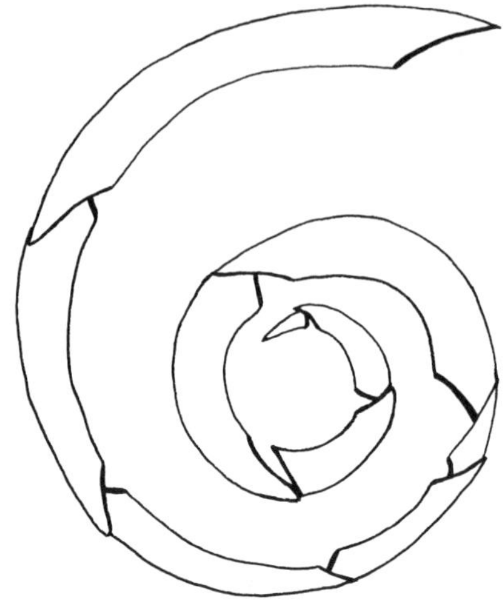

Figure 7.2. The shapes of the maxillary and mandibular incisions of the rat, at increasing postnatal ages, are successively arrayed. Note that their contours form a single logarithmic spiral. Adapted from M.L. Moss and L. Salentijn 1970, fig. 12.

perceived as consisting of a logarithmic spiral, it is equally evident that many other natural objects, and specifically biological systems, do not appear at casual inspection to possess such a geometric property. In some, instances such as the growth of rat incisor teeth (and of the human lower jaw or rather the growth in form of the inferior alveolar canal) can easily be shown to be capable of demonstrating a logarithmic spiral (figs. 7.2, 7.3) (Moss and Salentijn 1970).

In other biological systems, alternative manifestations of the mathematical information content of a logarithmic spiral are perceived; conical shape, phylotaxis, and allometry, or relative growth, to specify but three.

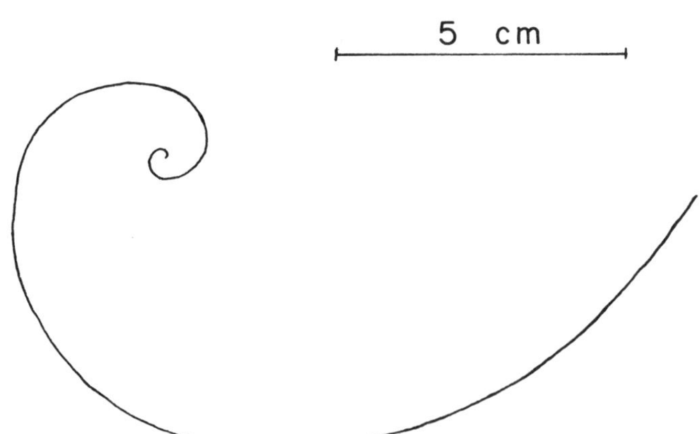

Figure 7.3. The logarithmic curve (spiral) that has been shown previously to describe the location of the foraman ovale, mandibular, and mental foramina of human mandibles at all ages. Adapted from M.L. Moss, and L. Salentijn 1970, fig. 9.

These *similarity transformations* express identical statements about the order of the same biological system. When it is remembered that a logarithmic spiral is a two-dimensional projection of a three-dimensional spiral, the relationship between spiral and cone becomes obvious (fig. 7.4). Phylotaxis is the study of the arrangement in space of the several structural components of plants, and all of these several patterns can be described either directly by a logarithmic spiral or by one or another of its similarity transformations (see Jean 1978a, for a thorough treatment and an extensive bibliography).

Allometry.

This is the particular similarity transformation we wish to stress presently. While allometry is capable of both broad and narrow definitions (see Gould 1966, 1975), it is most generally understood as a quantitative

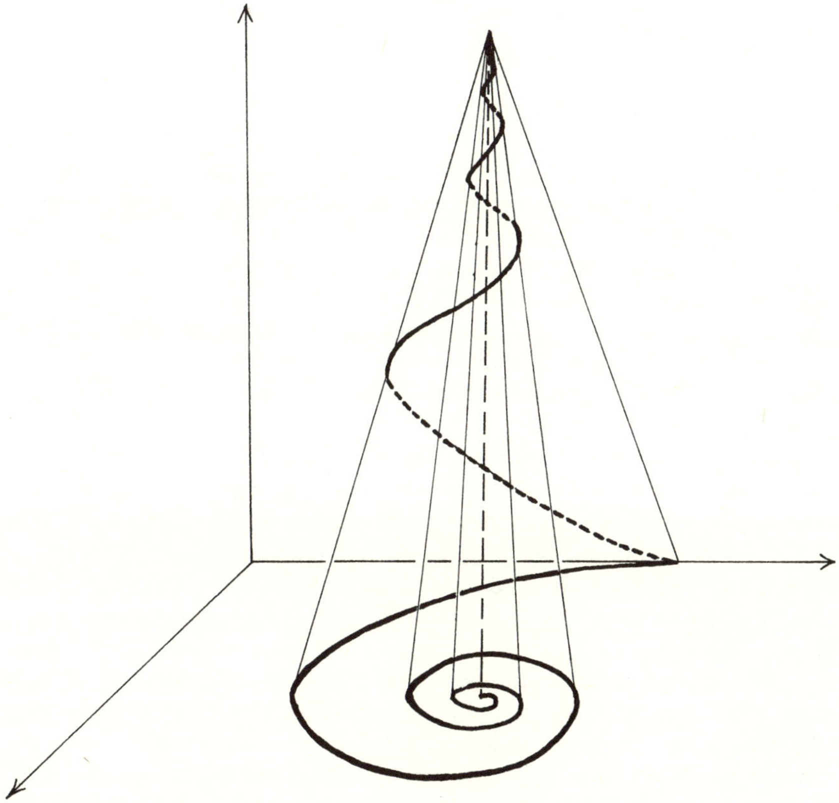

Figure 7.4 The graphic relationship between a cone and a logarithmic spiral is seen in this arbitrary figure.

expression of the very frequent observations that a constant relationship exists, between the specific growth rates of two measurable parameters, in a growing biological system, over a reasonably extended period of time. This relationship is customarily expressed in terms of the allometric or relative growth equation, popularized by Huxley (1972), where the relation between two, growing, parameters x and y is expressed by

$$y = b\,x^k, \tag{1}$$

where k is the ratio of the specific growth rates of x and y.

The relationship between allometry and the logarithmic spiral may be shown as follows. Assume a dimension l, whose specific growth rate is

$$\frac{dl}{l} = k\,dt, \text{ and assume} \tag{2}$$

$$\ln l = x, \text{ then} \tag{3}$$

$$\frac{dx}{dt} = k, \text{ and} \tag{4}$$

$$x = kt + c, \text{ hence} \tag{5}$$

$$\ln l = kt + c. \tag{6}$$

Here the l-t relationship is logarithmic, and this logarithmic description (an integral form) is synonymous with the allometric relationship (a differential form).

In the general case, it can be shown that in any growing system, wherever two parameters of that system "are growing exponentially, they are necessarily related by an allometric law" (Rosen 1967). From this, it follows that such growth is capable of both a logarithmic (integral form) and allometric (differential form) type of expression, and that these are similarity transformations.

Allometry is the subject of a vast literature, and allometric relationships have been demonstrated in all areas of biological growth (see Cook 1966 for a review of vertebrate and invertebrate examples). We have previously published on the allometry of the cranial and post-cranial skeleton, and of the brain (Moss 1954, 1955; Moss and Baer 1956; Moss et al. 1955, 1956; Moss and Salentijn 1970, 1971, Noback and Moss 1971; Salentijn and Moss, 1956).

Teeth also demonstrate allometry, in terms of growth data as well as in interspecific comparative studies (Moss and Applebaum 1957; Moss and Chase 1966; Herzberg and Schour 1940; Butler 1967a,b, 1968). From the considerations above it is possible to suggest that the conical shapes of some teeth, and of the cusps of other teeth, are similarity transformations of the allometry descriptive of their growth, and that logarithmic spirals would also be expected to be perceived in certain cases (as in Herzberg and Schorr 1940 and Moss and Salentijn, 1970); indeed Kawai (1952, fig. 2) depicts such a spiral formed by the bands of Schreger in the enamel of carnivore teeth. More recently, it has been shown that in human enamel, "cuspal prisms are arranged in a series of interlocking tapered cones" (Osborn 1968), and at a lower hierarchal level, that the axes of the enamel apatite crystallites "are distributed in a cone-like fashion about the tooth axis" (Davis et al. 1966) (see Schroeder 1976, for an excellent, concise review). Having established that dental form is capable of expression in terms of a logarithmic spiral, it is pertinent to consider the implications of the energetic implications of such a description.

Biological Thermodynamics

It is necessary to distinguish classical from biological thermodynamic systems. Thermodynamically, systems can be isolated, closed or open. In the first, neither energy nor matter can be exchanged between the system and the surroundings; in the second, energy but not matter may be exchanged, while both energy or matter can exchange in the third (Callen 1960; Pippard 1961; or any text of classical thermodynamics). Both isolated and closed systems are characterized, classically, by reversible processes which tend toward equilibrium and maximization of entropy. Open, biological systems entail many irreversible processes, where organismic order increases, and there is a tendency toward the constant elimination of structural defects which continuously and randomly arise at vital temperatures (Trincher 1965). Biological thermodynamics consider living systems incomprehensible by the methods of chemical kinetics, equilibrium statistical thermodynamics or classical thermodynamics (Glansdorff and Prigogine 1971; Nicolis and Prigogine 1977; Prigogine 1969; Prigogine and Nicolis 1971; Prigogine et al. 1972; Trincher 1965; Zotin 1972). Rather, it is necessary to consider specific non-linear kinetic laws, far from equilibrium conditions. Order is destroyed near thermodynamic equilibrium, but created far from it; order is created by thermodynamic "branch systems," within which a hierarchy of more and more complex and organized states is successively created by "dissipative structures," defined as having "continuous exchange of energy and matter with the outside world (Prigogine and Nicolis 1971; Davies 1975). The "sudden" formation of such a branch system may fruitfully be considered within the context of one of the elementary catastrophies of Thom (1975).

The possible relationship between open, biological thermodynamic systems and the logarithmic spiral, or its similarity transformations, is indicated in several ways. "Growth, and aging, represent the reliable foundation upon which the theory (of Prigogine) has been built" (Zotin 1972). Zotin also notes that this foundation rests primarily on the fact that for all animal species studied, growth is accompanied by a decrease in respiratory rate, in a relationship described by an equation mathematically similar to that of allometry. Further, the order created by open thermodynamic systems is characterized by their stability and by being associated with the minimal energy concept. The relationship between biological thermodynamics and the logarithmic spiral, and its specific botanical expression of phylotaxis, is explicitly presented by Turing (1952) and by Jean (1978b). These characteristics are graphically shown in the depiction of both the energetics of developing biological systems (fig. 7.5) (Zotin 1972) and of the "epigenetic landscape" (fig. 7.6) (Waddington 1956). Both analogously show how developmental pathways are associated with both increasing structural order and stability.

Amelogenesis.

Data cited above show that tooth crown form in general, and enamel forms specifically, are capable of description by allometry or logarithmic spirals; and that there are certain energetic implications to such mathematical properties. It is useful to examine the biological processes and structures capable of permitting the exchange of energy and matter, postulated to occur in an open thermodynamic system, related to the formation of one of the most highly ordered biological structures, dental enamel.

Hexagonal packing. Viewed in transverse section, Vertebrate ameloblasts are individually hexagonal near their apical end, and show hexagonal packing (Boyd 1960), as do odontoblasts, epidermal cells (Christophers 1971; Potten 1974; Mackenzie, 1975; Menton, 1976; Green and Thomas 1978), and, very closely approximated by, osteoblasts (Jones and Boyd 1976, 1977; Zallone 1977). Found also in plant tissues, this packing closely approximates that proposed on theoretical grounds by Kelvin (1887, 1894) (see Menton 1976; Toth 1969).

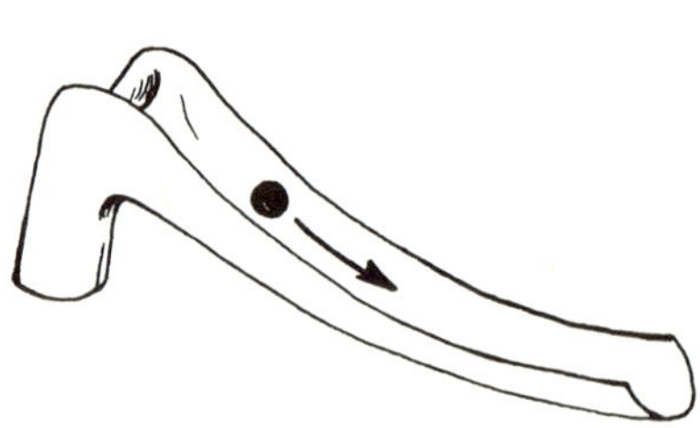

Figure 7.5. In this figure a developing biological system is depicted as a sphere moving down a slide. The curvature of this trough is analogous to the curve showing a decreasing rate of specific entropy production in figure 7.1. As the sphere descends, the "structure" or "order" of the system will increase. Adapted from A.I. Zotin 1972 (fig. 2).

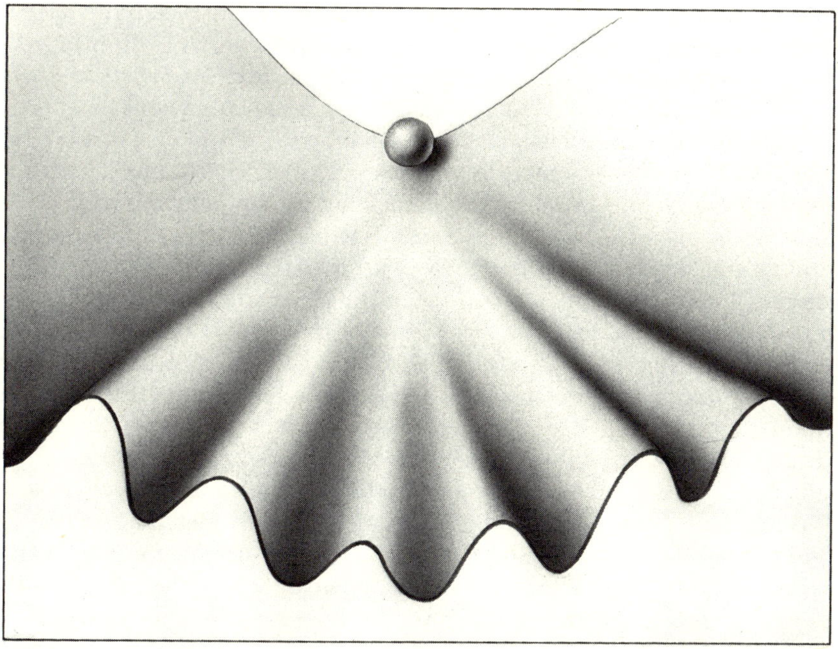

Figure 7.6. This figure attempts to show the ontogenesis of a biological system, indicated by a sphere. Here, the developmental pathways become increasingly evident as ontogenesis proceeds and the sphere descends. The walls of the several developmental pathways, or "chreods," increase in height and so serve to strengthen the tendency to a channeled development; these walls act as boundary conditions. In this figure "structure" or "order" is increased as the sphere descends and development continues. Adapted from C. Waddington 1956.

While a thermodynamic analogy of the formation of hexagonal cell outlines is provided by the Bénard cell phenomenon, produced by the uniform heating of a thin layer of fluid (Gmitro and Scriven 1966), it is more likely that the geometrical concepts of Kelvin (1887, 1894) are regulative. In this regard it should be noted that the exclusion of the *thermal* basis for hexagonal packing (say of ameloblasts) does not exclude an *energetic* basis for the same phenomenon; however, I am unaware of any studies of this topic. Surely increasing order here can be achieved only by a dissipation of entropy to the surround.

Prism Formation. Dental enamel expresses the highest degree of structural organization, or information, of all dental tissues at the macroscopic as well as at the microscopic, level. The formation and structure of enamel prisms, and the spatial array of hydroxyapatite crystallites within the enamel, give evidence of an exceptionally great degree of order compared to other biological systems.

A number of physical processes have been suggested as causally related to these phenomena, and these were renewed by Osborn (1970), where he compared the concepts of Fearnhead (1965) and Boyd (1965) with his own. The former considered crystallite orientation as due to the effects of both the concentration gradients between ameloblasts and the enamel matrix and the orientation produced by the streaming of organic molecules into that matrix, acting as powerful organizing factors. Boyd invokes the kinematics of ameloblasts, and considers that their centrifugal motion during amelogenesis as productive of a "stroking factor" that regulates enamel microstructure. Osborn interestingly invokes the Bénard cell phenomenon as the basis for an essentially rheological agreement (1970; reviewed in Osborn and Ten Cate 1976). Here a shearing force, postulated to be produced by ameloblast movement, is exerted upon the forming enamel matrix, whose specific vectorial properties are the postulated as regulative of the well known, species specific, enamel microstructure array. This hypothesis is felt by Osborn (1970) to have additional energetic bases. It is postulated that the ameloblast acts a "tube," in which the most rapid "flow" of enamel precursor substances occurs at the middle, being associated with a higher temperature than that at the periphery of the tube. Further, it is considered that an important factor in crystallite orientation lies in the relationship between the "energy" of the materials secreted and reabsorbed. However, no quantitative values have been proposed, nor any mathematical formalism yet presented. These comments do not imply either rejection or acceptance of such hypotheses, but rather an indication of the rigorous work yet to be performed. It is noted that this topic is not currently under study, as judged by recent symposia (Nylen and Termine 1979).

The relationship between precursor secreting skeletal or dental tissue cells and the structural array, or order, of their products is a topic of considerable theoretical interest. Such a cause and effect relation has been suggested for bone, but rejected for dentin (see Jones and Boyde 1977), a conclusion somewhat unsatisfactory, at least theoretically, from the viewpoint of parsimony of scientific explanation; but a fuller discussion of this point is beyond the scope of the present paper.

It may be more helpful to examine another potentially involved process, that of the dynamic structure occurring under certain conditions of freezing of a solid out of molten solution (Gmitro and Scriven 1966; Chalmers 1964). This transformation process "is a solidification with an accompanying heat release. Both heat and solute are transported. There is a tendency for the solute to be redistributed in a regular hexagonal pattern at the freezing face of the solid . . . This results in a permanent hexagonal-column concentration pattern" (Gmitro and Scriven 1996) (fig. 7.7).

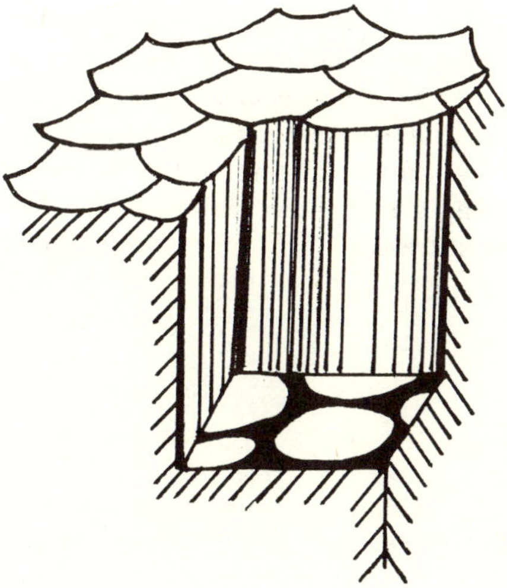

Figure 7.7. This depiction of the freezing of solid out of molten solution is presented as a possible analogy to enamel prison formation. Note the tendency for solute to be redistributed in a regular hexagonal pattern at the freezing face of the solid. In this process both heat and solute are transported. Adapted from J.I. Gmitro and L.E. Scriver 1969.

The morphological basis in the enamel organ that might possibly be involved in such energetic exchanges is provided by the vascularity of that organ during active amelogenesis.

The enamel organ is avascular during its earlier formative stages. In eutherian mammals, however, the enamel organ becomes vascularized by extension of peripheral vessels either just prior to or just at the onset of amelogenesis (Glasstone 1962; Greene 1937; Saunders and Rochert 1967). A capillary network or plexus of considerable density is noted (Gaunt 1959), and it has been observed that the shape of this plexus, adjacent to the ameloblasts, alters during amelogenesis to assume a configuration reportedly capable of carrying out an increased velocity of blood flow, elevated (local) blood pressure and aerobic transport (Iwaku and Ozawa 1979). It is suggested here that future study of the energetics of the morphological apparatus might prove useful. This "thermodynamic suggestion" should be viewed as complementary to, and not exclusive of, the earlier concepts of the role of ameloblast movement in the formation of biological order.

Summary

Dental crown size and shape, and enamel structure, represent an extremely ordered biological system. The thermodynamic bases for order, or structure, are summarized and it is shown that local order in the universe is frequently shown by a logarithmic spiral or one of its several similarity transformations, of which allometry is one example. Allometry is briefly reviewed and dental crown form is shown to be capable of such a mathematical description. The basic concepts of biological thermodynamics are partially reviewed, and amelogenesis examined from this viewpoint. In particular, the need to permit free exchange of energy and matter between the developing system and its surround, and to permit aerobic respiration is noted; following which the vascular apparatus existing within the active enamel organ is suggested as providing a morphological basis for such reactions. The need for further study is stressed.

References

Boyd, A. 1965. The structure of developing mammalian dental enamel. In M. V. Stack and R. W. Fearnhead, eds. *Tooth Enamel*, pp. 163–167. Bristol: Wright.

Boyd, A. 1966. The developmental of enamel structure in mammals. *Proc. 3d Europ. Symp. Calcif. Tissues*, pp. 276–280. New York: Springer Verlag.

Butler, P. M. 1967a. Relative growth within the human first upper permanent molar during the prenatal period. *Archs. Oral Biol.* 12:983–992.

—— 1967b. Comparison of the development of the second deciduous molar and first permanent molar in man. *Archs. Oral Biol.* 12:1245–1260.

—— 1968. Growth of the human second lower deciduous molar. *Archs. Oral Biol.* 13:671–682.

Callen, H. B. 1960. *Thermodynamics.* New York: Wiley.

Chalmers, B. 1964. *Principles of Solidification.* New York: Wiley.

Christophers, E. 1971. Die epidermale Columnarstruktur. *Z. Zellforsch.* 114:441–450.

Cook, T. A. 1979. *The Curves of Life.* New York: Dover.

Davies, P. C. W. 1975. Cosmological aspects of irreversibility. In Kubat and Zeman, eds. *Entropy and Information in Science and Philosophy* pp. 11–28.

Davis, B. L., G. Rapp, Jr. and A. S. Posner. 1966. Application of the small-circle net method to the study of bovine bone and tooth fabrics. *Archs. Oral Biol.* 11:1241–1260.

Elsasser, W. M. 1958. *The Physical Foundation of Biology.* Oxford: Pergamon Press.

Fearnhead, R. W. 1965. The insoluble organic component of human enamel. In M. V. Stack and R. W. Fearnhead, eds., *Tooth Enamel*, pp. 127–131. Bristol: Wright.

Gaunt, W. A. 1959. The vascular supply to the dental lamina during early development. *Acta Anat.* 37:232–252.

Ghyka, M. 1977. *The Geometry of Art and Life.* New York: Dover.

Glansdorff, P. and I. Prigogine. 1971. *Thermodynamic Theory of Structure, Stability, and Fluctuations.* London: Wiley-Interscience.

Glasstone, S. 1962. Vascularization of enamel organ in the rabbit and some rodents. *Odont. Tidsk.* 70:80–87.

Gmitro, J.T. and L.E. Scriven. 1966. A physiochemical basis for pattern and rhythm. In K.B. Warren, ed. *Intracellular Transport.* New York: Academic Press.

—— 1969. A physiochemical basis for pattern and rhythm. In C.H. Waddington, ed., *Towards A Theoretical Biology*, 2:184–203. Chicago: Aldine.

Gould, S.J. 1966. Allometry and size in ontogeny and phylogeny. *Biol. Rev.* 41:587–640.

Gould, S.J. 1975. Allometry in primates, with emphasis on scaling and evolution of the brain. *Contrib. Primat.* 5:244–292.

Green, H. and J. Thomas. 1978. Pattern formation in cultured human epidermal cells: development of curved ridges resembling dermatoglyphs. *Science* 200:1385–1388.

Greene, H.L.H.H. 1937. The development and morphology of the teeth of *Ornithorhynchus. Phil. Trans. Roy. Soc. London* B228:367–420.

Hambridge, J. 1967. *The Elements of Dynamic Symmetry.* New York: Dover.

Hamilton, H.J. 1977. A thermodynamic theory of the origin and hierarchal evolution of living systems. *Zygon* 12:289–335.

Herzberg, F. and I. Schour. 1940. The pattern of appositional growth in the incisor of the rat. *Anat. Rec.* 80:497–506.

Huntley, H.E. 1970. *The Divine Proportion. A Study in Mathematical Beauty.* New York: Dover.

Huxley, J.S. 1972. *Problems of Relative Growth*, 2d ed. New York: Dover.

Iwaku, F. and H. Ozawa. 1979. Blood supply of the rat periodontal space during amelogenesis as studied by the injection replica SEM method. *Arch. Histol. Jap.* 42:81–88.

Jean, R.V. 1978a. *Phytomathematique.* Montreal: Les Presses de l'Université du Quebec.

—— 1978b. Growth and entropy: phylogenism in phylotaxis. *J. Theor. Biol.* 71:639–660.

Jones, S.J. and A. Boyde. 1976. Morphological changes in osteoblasts *in vitro*. *Cell Tiss. Res.* 166:101–107.

—— 1977. The migration of osteoblasts. *Cell Tiss. Res.* 184:179–193.

Kawai, N. 1951. The bands of Schrager observed through the enamel surface. *Acta Anat. Nippon.* 26:25–28.

Kelvin, W.T. 1887. On the division of space with minimal partitional area. *Phil. Mag.* 24:503–514.

—— 1894. On the homogenous division of space. *Proc. Roy. Soc.* 55:1–16.

Kubat, L. and J. Zeman (eds.) 1975. *Entropy and Information in Science and Philosophy.* Amsterdam: Elsevier Science.

Lehninger, A.L. 1975. *Bioenergetics*. 2d ed. New York: W.A. Benjamin.

Mackenzie, I.C. 1975. Spatial distribution of mitosis in mouse epidermis. *Anat. Rec.* 181:705–710.

Menton, D.N. 1976. A minimum-surface mechanism to account for the organization of cells into columns in the mammalian epidermis. *Am. J. Anat.* 145:1–22.

Moss, M.L. 1954. Differential growth analysis of bone morphology. *Am. J. Phys. Anthrop.* 12:71–75.

—— 1955. Relative growth of the human fetal skeleton. *Ann. N.Y. Acad. Sci.* 63:528–536.

Moss, M.L. and E. Applebaum. 1957. Differential growth analysis of vertebrate teeth. *J. Dent. Res.* 36:644–651.

Moss, M.L. and M.J. Baer. 1956. Differential growth of the rat skull. *Growth* 20:107–120.

Moss, M.L. and P. Chase. 1966. Morphology of Liberian negro deciduous teeth. 1. Odontometry. *Am. J. Phys. Anthrop.* 24:215–229.

Moss, M.L., C.R. Noback and G.C. Robertson. 1955. Critical development horizons in human fetal long bones: correlated quantitative and histological horizons. *Am. J. Anat.* 97:155–175.

—— 1956. Growth of certain human fetal cranial bones. *Am. J. Anat.* 98:191–204.

Moss, M.L., L. Moss-Salentijn and H.P. Ostreicher. 1974. The logarithmic properties of active and passive mandibular growth. *Am. J. Orthodont.* 66:645–664.

Moss, M.L. and L. Moss-Salentijn. 1977. Analysis of developmental processes possibly related to human dental sexual dimorphism in permanent and deciduous canines. *Am. J. Phys. Anthrop.* 46:407–414.

—— 1978. Analysis of developmental processes possibly related to human sexual dimorphism. In P.H. Butler and K.A. Joysey, eds., *Development, Function, and Evolution of Teeth*, pp. 135–148, New York: Academic Press.

—— 1970. The logarithmic growth of the human mandible. *Acta Anat.* 77:341–360.

—— 1971. The unitary logarithmic curve descriptive of human mandibular growth. *Acta Anat.* 78:532–542.

Nicolis, G. and I. Prigogine. 1977. *Self-Organization in Non-Equilibrium Systems.* New York: Wiley.

Noback, C.R. and M.L. Moss. 1956. Differential growth of the human brain. *J. Comp. Neurol.* 105:539–552.

Nylen, M.U. and J.D. Termine, eds. 1979. Tooth enamel III. its development, structure, and Composition. *J. Dent. Res.* 58B:674–1031.

Osborn, J.W. 1968. Directions and interrelationship of prisms in cuspal and cervical enamel of human teeth. *J. Dent. Res.* 47:395–402.

—— 1970. The mechanism of prism formation in teeth: a hypothesis. *Calc. Tiss. Res.* 5:115–132.

Osborn, J.W. and A.R. Ten Cate. 1976. *Advanced Dental Histology*. 3d ed. Bristol: Wright.

Pippard, A.B. 1961. *Elements of Classical Thermodynamics for Advanced Students of Physics.* Cambridge: Cambridge University Press.

Potten, C.S. 1974. The epidermal proliferative unit: The possible role of the central basal cell. *Cell Tiss. Kinet.* 7:77–88.

Prigogine, I. 1969. Structure dissipation and life. In M. Marois, ed., *Theoretical Physics and Biology*. Amsterdam: North-Holland.

Prigogine, I. and G. Nicolis. 1971. Biological order, structure and instabilities. *Quart. Rev. Biophys.* 4:107–148.

Prigogine, I., G. Nicolis, and A. Bobloyantz. 1972. Thermodynamics of evolution. *Physics Today*, pp. 23–28, 38–44.

Purce, J. 1974. *The Mystic Spiral*. London: Thames and Hudson.

Rosen, R. 1967. *Optimality Principles in Biology*. London: Butterworths.

Salentijn, L. and M.L. Moss. 1971. Morphologic attributes of the logarithmic growth of the human face: gnomonic growth. *Acta Anat.* 78:185–199.

Saunders, R.L. and H.O. Rockert. 1967. Vascular supply of dental tissues, including lymphatics. In A.E.W. Miles, ed., *Structural and Chemical Organization of Teeth*, 1:199–245. New York: Academic Press.

Schroeder, H. E. 1976. *Orale Strukturbiologie*. Stuttgart: Georg. Thieme.

Spindler, M. and R. Rottger. 1973. Der Kammerbauvorgang der Grossforaminifere *Heterostegina depressa. Marine Biol.* 18:146–159.

Thom, R. 1975. *Structural Stability and Morphogenesis*. Reading, Mass.: W.A. Benjamin.

Thompson, D.A.W. 1942. *On Growth and Form*, 2d ed. Cambridge: Cambridge University Press.

Toth, L.F. 1964. *Regular Figures*. Oxford: Pergamon Press.

Trincher, K.S. 1965. *Biology and Information. Elements of Biological Thermodynamics*. New York: Consultants Bureau.

Turing, A.M. 1952. The chemical basis of morphogenesis. *Proc. Roy. Soc.* 237B:37–72.

Waddington, C.H. 1956. *Principles of Embryology*. New York: Macmillan.

Zallone, A.Z. 1977. Relationships between shape and size of the osteoblasts and the accretion rate of trabecular bone surfaces. *Anat. Embryol.* 152:62–72.

Zotin, A.I. 1972. *Thermodynamic Aspects of Developmental Biology*. Basel: S. Karger.

II

STRUCTURE, SHAPE, AND FUNCTION

8. Neogene Hominoid Evolution
A Tooth's Inside View

DAVID G. GANTT

Dental remains account for nearly all the data upon which we have based our current views of hominoid evolution. Therefore, correct interpretation of these remains is critical to our understanding of adaptation, phylogeny, and systematics. Data obtained from these specimens have often led to a variety of conflicting interpretations. Morphological analysis, especially when isolated teeth are evaluated, have not provided definitive answers regarding either phylogeny or classification. For a more adequate appraisal, new analytic methods and techniques must be used to provide additional data, which will assist us in our interpretations and reconstruction of the evolutionary history of the hominoids. Scanning electron microscopic analysis (SEM) and thin section analysis of calcified dental tissues, especially the enamel and dentin, are two such approaches which have provided new insights into hominoid evolution.

Numerous authors (Gantt 1977; Jolly 1970; Pilbeam et al. 1977; Simons and Pilbeam 1972; Simons 1977a and 1977b) have proposed that the initial split and subsequent radiation of the hominids resulted from an adaptive radiation of the Middle Miocene hominoids into a savanna-like environment which required a shift in their dietary habits. This new dietary specialization is viewed as an adaptation to foods which require greater

I wish to thank Bill Miller and Dan Cring for their assistance with the SEM analyses. I am also grateful to A. Walker, G. R. H. von Koenigswald, P. Andrews, and D. R. Pilbeam for the loan of hominoid fossil material, without which this study could have never been completed. I am indebted to Alan Boyde for his comments and suggestions concerning etching techniques and stereoanalysis.

I wish to acknowledge the L. S. B. Leakey Foundation, FROM Foundation, and an Alexander von Humboldt-Stiftung Research Fellowship from the German Federal Republic that allowed me to complete the study of this project in 1979, while on leave from Florida State University.

amounts of mastication and/or increased amounts of food which resulted in an increase in the thickness of enamel. It is suggested that hominids have thicker enamel than the living pongids or other Neogene hominoids. Therefore, the thickness of enamel can provide a ready means by which hominids can be distinguished from other hominoids. However, no quantitative data have been provided to support this proposal with the exception of my own research on extant primates which has clearly shown that man and *Ramapithecus* has absolutely thicker enamel than any living pongid (Gantt 1977; Molnar and Gantt 1977).

A second hypothesis concerning the structure of the enamel concerns the ultrastructural unit of enamel, the prism. It is proposed that hominids, which includes *Homo*, *Australopithecus*, and probably *Ramapithecus*, have a "keyhole" enamel prism pattern (see figs. 8.1 and 8.2), while a "hexagonal-shaped" prism pattern is found among the extant and extinct hominoids (see figs. 8.6–8.8) (Gantt 1977, 1979a, Gantt et al. 1977).

Methods and Procedures

Dental remains of Neogene hominoids (both extant and extinct forms) were obtained for this study and divided into three groups according to Simons, Andrews, and Pilbeam (1978): Pongidae, which includes the Dryopithecinae, and Hominidae which includes *Ramapithecus* (see table 8.1).

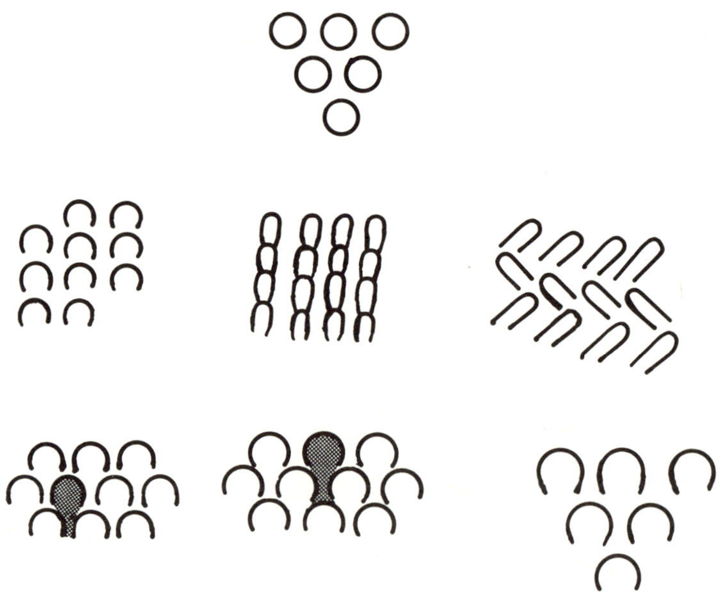

Figure 8.1. The development patterns in enamel have been documented by Boyde (1964, 1965, and 1976). Pattern 1 prisms are most common in the Insectivora, Sirenia, and some of the Cheiroptera. They are also found in some Lemuriformes and in human early cuspal enamel. In the developing enamel the Tomes' process (the secretory part of the ameloblast) lies in a pit in the floor of the developing enamel and forms the central portion of the prism pattern. In pattern 1 prisms the pits are nearly straight cylinders with flat floors. The prisms are separated from one another by a continuous interprismatic region which forms the mutual walls of the pit at the formative surface, and by the sides of adjacent Tomes' processes (Boyde 1976:341 and figs. 8.7 and 8.10a). Pattern 1 prisms are the result of the ameloblast movement which is nearly perpendicular to the developing surface, causing the crystal orientation to be the same in the centers of the interprismatic and prismatic regions.

Pattern 2 prisms are found in most ungulates, marsupials, and rodents. In pattern 2 prisms the ameloblasts are not perpendicular to the surface as they are in pattern 1 prisms. The pits enter the surface obliquely so that the floor of a pit becomes tilted and forms the missing part of the wall (Boyde 1976:341 and figs. 8.9 and 8.10b). In pattern 2 prisms the floor is the cervical side of the pit, with the ameloblast aligned in rows parallel to the longitudinal axis of the tooth. The crystallites in the floor of the pits grow parallel with the direction of movement of the ameloblast, while the majority of the crystals in the inter-row sheets grow perpendicularly to the tops of the inter-pit walls, which are in the general plane of the developing surface. In pattern 2 prisms where two ameloblasts make each prism, there are also two predominant crystal orientations—those in the prisms and those in the interprismatic sheets.

Pattern 3 prisms are most commonly found in human enamel being described as the keyhole enamel prism pattern. The pattern 3 prism pits are inclined to the general plane of the ameloblast matrix. The pits are arranged in transverse rows, though the row axis is rotated in areas where the net cell movement has a lateral as well as a cuspal component. The floor is the cervical side of the pit which merges with the interpit wall next to it instead of joining the next cervical wall as in pattern 2 prisms (Boyde 1976:341 and figs. 9 and 10c). There exists an extreme range of crystal orientations, but the orientation in the interprismatic (winged process or tail) regions of the prisms merges gradually into that of the prism body proper. The tail regions are what was the inter-pit wall and contain the crystals which grew mostly in relation to the tops of the inter-pit walls and thus normal to the general plane of the developing surface (Boyde 1976:342 fig. 15).

The crystals in the prism body (head region) centers are parallel with the prism direction, so that there must be a span of orientations between the body and the tail at least as great as the angle of inclination of the prisms with respect to the developing surface. However, the crystals in the most cuspal side of the prism may diverge cuspally toward the next prism by some 5 or 10°; and those few which grow in the limited amount of new matrix added along the long cuspal wall of the pits (i.e., in the far cervical, fan-shaped part of the tails of the prisms) diverge cervically toward that surface, so that the total fan angle of orientations from top-of-head to bottom-of-tail of pattern 3 prisms may exceed 60° (Boyde 1976:343; fig. 8.2 this text). four ameloblasts contribute to each prism.

Although Boyde has documented three basic pattern, several variations have also been identified (see fig. 8.1 and Boyde 1964). The range and extent of variation at present has not been fully documented. In addition, there have been few attempts to correlate the prism patterns of the fully mineralized enamel surface to the developmental prism patterns documented by Boyde (1964, 1965, and 1976).

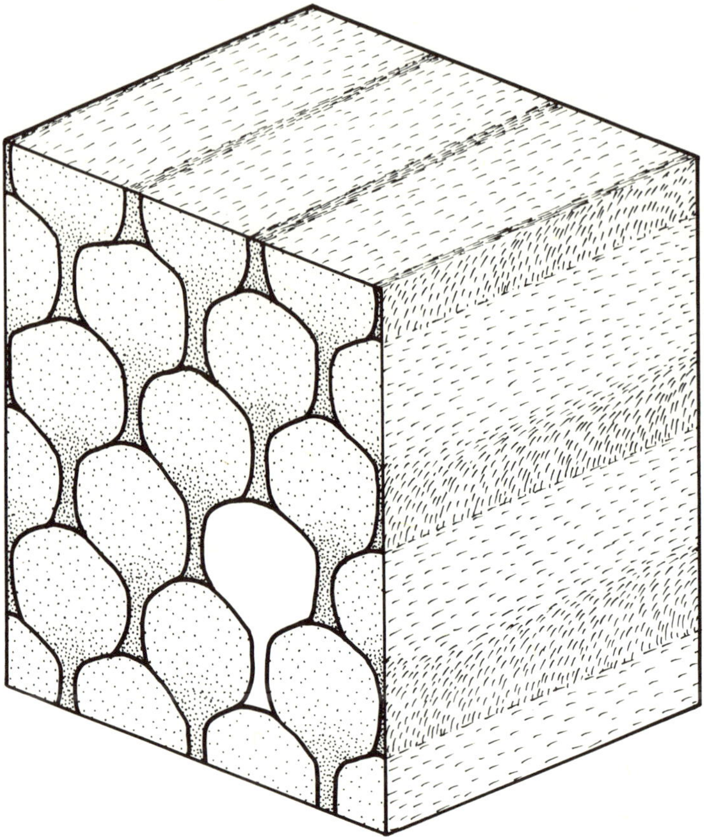

Figure 8.2. A model of the enamel prism structure of human enamel showing the outline of the "keyhole" enamel prisms (developmental enamel prism pattern 3b, fig. 8.1). The outline of "keyholes" on the front surface is a cut which is perpendicular to the direction of the prisms. The top view, a horizontal cut to the section, gives a banded appearance and is caused by cross-sectioned heads and tails which have different crystal arrangements. The side view, a vertical cut, shows the tilting of crystallites in the head to tail portions of the "keyhole" prisms (see Poole and Brooks 1961, for a more detailed explanation).

Enamel Thickness. The teeth for thin section analysis were prepared per the methods and procedures outlined by Gantt (1977) and Molnar and Gantt (1977). In addition, measurements were taken from broken fossil specimens and used instead of cutting the rare fossil hominoids specimens. These specimens were measured in the same manner as were the thin sectioned teeth (see Gantt and Gest n.d.). Data obtained from these specimens were compared to the data obtained from other primate species (Gantt 1977, 1979b).

Enamel Prism Patterns. The analysis of the ultrastructural unit of

Table 8.1. Neogene Hominoids used in this study.

Extant Hominoids	Individuals	Molars and Premolars
Pan	10	20
Gorilla	10	20
Pongo	20	30
Homo sapiens	10	20
Extinct Hominoids		
Homo erectus	5	10
Australopithecus	4	5
Ramapithecus	2	2
Sivapithecus indicus	4	5
Sivapithecus sivalensis	2	2
Gigantopithecus bilaspurensis	1	1
Gigantopithecus blacki	2	2
Proconsul africanus	2	2
Proconsul major	1	1
Proconsul nyanzae	1	1
Limnopithecus legetet	1	1

enamel, the prism, was conducted as outlined by Gantt (1979a), using the etching techniques developed by Boyde et al. (1978). A highly polished facet was placed on the mid-laterial crown and then etched with a 0.074 M solution of phosphoric acid (H_3PO_4) for 60 seconds, and then washed in water and ethanol and allowed to air dry. Each specimen was then mounted on stubs and coated with gold/palladium by evaporation. They were then examined with a Cambridge S-4 scanning electron microscope at 10 and 15 KV in the secondary emission mode.

Results

Data obtained on the thickness of enamel from the Neogene hominoids studied supports the previous published hypothesis that hominids have thicker enamel than pongids. However, there is an exception, the orangutan *Pongo*. Greenfield (1974 and 1979) has noted that among the extant pongids, *Pongo* appeared to have relatively thicker enamel than *Pan* or *Gorilla*, which is true (see table 8.2 and fig. 8.3). Based on a previous study of the thickness on enamel by Gantt and Gest (n.d.) together with the data obtained here, it is suggested that *Pongo's* excessively thick enameled teeth are inherited from a thick enameled ancestor. Analysis of the Neogene hominoids *Ramapithecus*, *Sivapithecus*, and *Gigantopithecus* indicates that the Asian hominoids recently assigned to the Ramapithecidae (*Ramapithecus*, *Sivapithecus*, and *Gigantopithecus*) are "thick" enamel forms while the African hominoids, *Proconsul*, appear to be "thin" enamel forms. (An analysis of the African hominoids is currently underway.)

Another Ramapithecid has recently been discovered and assigned to the genus *Ouranopithecus* (Pilbeam 1978) based on its morphological features and the position of "thick" enameled teeth (de Bonis, personal communication).

Enamel Prism Pattern Analysis

Enamel prism patterns and their importance to studies of mammalian phylogeny have only recently been realized, although studies of enamel

Table 8.2. Species, weights, and diets used in analysis of variables.

	Weight	Diet
Australopithecus robustus	50000	Possibly grasses, seeds.
Ramapithecus sp.	22680	Possibly grasses, seeds.
Homo sapiens	64000	Omnivorous.
Gorilla gorilla	M 163610	Leaves, bark, fruit.
	F 110685	
Pan troglodytes	M 44980	Primarily fruit, with leaves,
	F 36480	seeds, insects.
Pongo pygmaeus	M 78345	Primarily fruit, with leaves, bark.
Hylobates lar	F 5329	Primarily fruit, some leaves, insects.
Macaca nemestrina	M 9955	Fruit, seeds, insects.
	F 5430	
Macaca mulatta	M 8648	Fruit, seeds, insects.
	F 5988	
Macaca fasicualris	M 4908	Fruit, nuts.
	F 3424	
Papio anubis	F 16244	Primarily grasses, with seeds, fruit, insects, small animals.
Mandrillus sphinx	M 22735	Fruit, grasses, seeds, insects.
Theropithecus gelada	F 13620	Grasses, seeds, roots, tubers, insects.
Cercopithecus aethiops	F 4100	Primarily fruit, with grasses, insects.
Presbytis cristatus	M 6791	Buds, leaves, shoots.
Colobus polykomos	M 10000	Leaves.
	F 7422	
Alouatta villosa	M 7400	Leaves, fruit, flowers.
	F 6350	
Ateles fusciceps	F 7700	Primarily fruit, with leaves.

NOTE: Weight and diet taken from Kay (1973) for all primates except *Ramapithecus* and *A. robustus*, which were taken from the ranges given by Pilbeam (1972) of 40–80 lbs. for *Ramapithecus* and 80–140 given for *Australopithecus*. Although these fossil species were not used in the statistical analysis, they are placed on the regression line plots in figures 8.2 and 8.3, based upon a body weight of 50 lbs. for *Ramapithecus* and 110 lbs. for *A. robustus*.

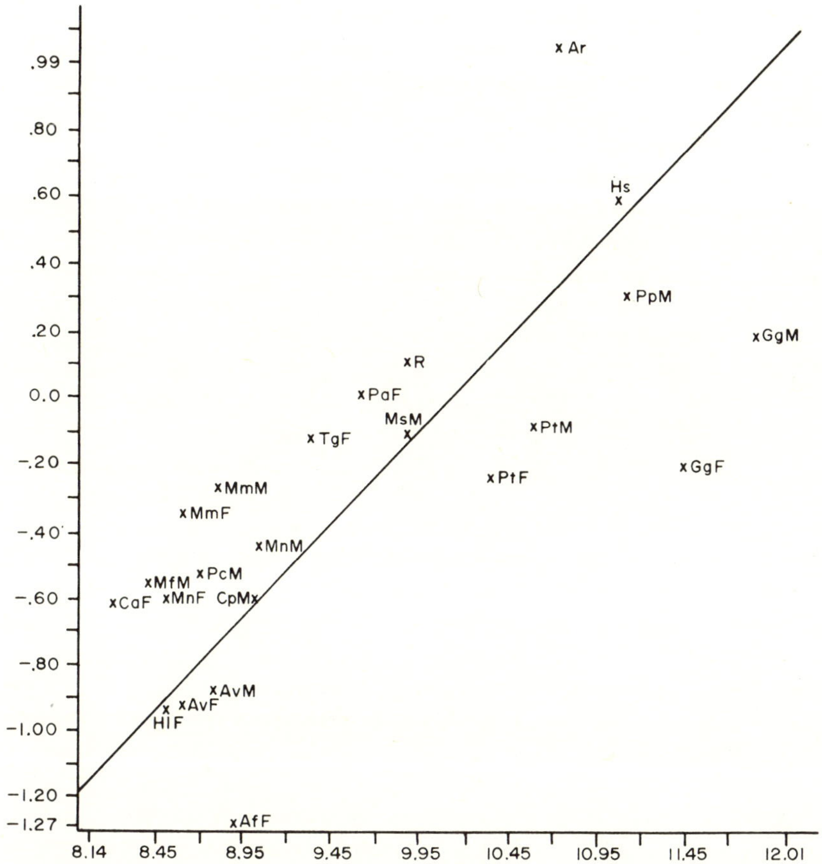

Figure 8.3. Regression of log (B) (Down) on Log (Body weight) (Across) first of genus, and sex identifies each point (see table 8.3).

structures and their implications to phylogeny and classification have long been known (Carter 1920; Korvenkontio 1934–35; Shobusawa 1952; Regan 1930; Tomes 1906; Tomes 1849 and 1850). Recent studies have documented that specific prism patterns do exist and that enamel prism patterns are an important taxonomic tool (Boyde 1964, 1965, 1971, 1978; Gantt et al., 1977; Gantt 1977, 1979a; Lavelle, et al., 1977; Koenigswald, 1977). However, a few authors have reported different enamel prism patterns for the same species of primate (Gantt et al., 1977; Lavelle et al. 1977; Vrba and Grine 1978). The differences reported by these authors resulted in their failure to recognize one or more variables.

Problems of interpretation. The first variable involves methods of analysis. The main problem in studying enamel is that its structure must be seen in three dimensions. The importance of stereoanalysis of enamel structures cannot be overemphasized. Boyde has published extensively on this subject and on the methods and techniques of applying stereoanalysis to the study of enamel prism patterns (Boyde 1973; Boyde and Ross 1975; Boyde and Howell 1977).

The second variable is concerned with the differential effects of acid etchants. Vrba and Grine (1978) used a dilute solution of hydrochloric acid (10 percent HCL) to prepare their specimens for analysis in their study of hominoid prism pattern. Acid treatment is necessary to etch the specimen to remove the prismless layer (approximately 10 to 30 micra) or the layer of mineral substitution in fossil specimens. This procedure allows one to more fully reveal the organization of the prisms and crystallite arrangement. Also, etching is of potential interest in the field of mammalian taxonomy, because enamel is the only tissue which is virtually fossilized before death.

An intensive study of the qualitative and quantitative aspects of enamel etching with acid and EDTA on human teeth has brought to light certain inadequacies and the presence of artifacts as the result of specific etchants and/or lengths of etching times (Boyde et al. 1978). The data obtained revealed that the etchants studied, including the 10 percent hydrochloric acid solution used by Vrba and Grine, will give rise to artifacts of one sort or another. Among the important artifacts of etching are the generation of either the open honeycomb pattern or its converse, in which the prism boundary crack is widened. In addition, problems of the growth of the new phase (reprecipitation of crystallites) due to acid etchants must be considered.

The results of this study led the authors to conclude that a dilute solution of phosphorate acid (0.074M = 0.5 percent vol./vol. of concentrated, 85 percent acid) was most suitable for etching polished enamel surfaces to study crystallite orientation and prism organization (Boyde et al. 1978).

The third variable is a product of the fossilization process. Fossilization subjects the enamel to a vast variety of chemical and physical agents. These agents change the surface chemical and physical composition. Therefore, treatment of the tooth surface with acid etchants will produce greatly differing prism patterns due to the remineralization products, that is, the differences in mineral content of the crystals. The fossil enamel may already be highly etched or its acid resistance altered. Acid resistance may be high, related to a high fluoride content, or lowered due to a higher carbonate content than the natural state (see fig. 8.4).

A fourth variable is the position of the prisms in the plane of view and prism decussation. Studies of the nature and development of prism decussation in mammalian enamel document that the greatest extent of decussation is in the middle portion of the tooth (Boyde 1969). It has also been shown that the position of the prism bundles vary from the cemento-

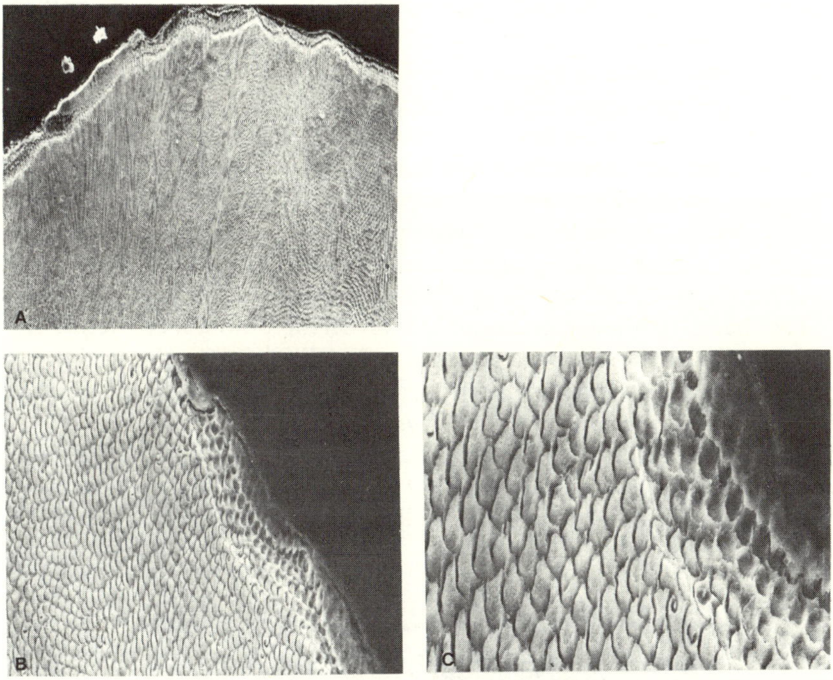

Figure 8.4. A longitudinal section of a molar from *Ramapithecus punjabicus* (AMNH 19565-B) from the Chinji Formation of the Siwalik Group in Pakistan.

The section is highly polished and etched with 0.074M H_3PO_4 for 60 seconds. *a* reveals the varying thickness of the combined nonprismatic zone and area of substitution. *b* and *c* are at higher magnifications to demonstrate the changes in prism outline and form.

enamel junction to the cusp apex (Boyde 1964, 1976). The results of these studies document the importance of prism direction and position in order to correctly interpret and describe prism pattern and organization. The area of the tooth in which prisms are best arranged to study their pattern and organization is the mid-lateral crown area.

These four variables, stereoanalysis, differential effects of acid etchants, fossilization, and prism direction, were neither discussed or evaluated by Vrba and Grine in their study of the Australopithecine prism patterns (1978). These problems and the lack of correlating the developmental patterns described by Boyde (1964, 1965, 1971, 1976; also see fig. 8.1) are presently responsible for the inconsistencies reported in the literature. The mature surface and subsurface prism patterns must be described in terms of their developmental counterparts while controlling for the previously discussed variables.

Analysis of Hominoid prism patterns. The analysis of the extant hominoids reveal two and perhaps three different prism patterns. The pattern obtained for *Homo* is the classic "keyhole" enamel prism pattern (3b) and was obtained in specimens of *Homo sapiens sapiens*, *Homo sapiens neanderthalensis*, *Homo erectus pekinensis* and *Homo erectus erectus* (see figs. 8.1 and 8.5). In *Pan* and *Gorilla* the prism pattern is as shown in figure 8.6 (pattern 3a), although *Gorilla* can be distinguished from *Pan* due to the lack of prism decussation as seen in polarized light. *Pongo* also has a pattern 3a prism pattern but differs from the pattern seen in the African apes in that the "tail" portion of the prism appears to be larger (see fig. 8.7) (the qualitative and quantitative aspects of this difference is currently being investigated by Boyde and Gantt).

The analysis of the extinct hominoids reveal three different patterns. The pattern obtained for *Australopithecus* was similar to that obtained in the genus *Homo*, the "keyhole-shaped" enamel prism. In the Ramapithecidae, pattern 3a was obtained, while *Sivapithecus indicus* revealed a prism pattern which is identical to that obtained in *Pongo* (fig. 8.8). *Giganto-pithecus* also shows a pattern similar to that obtained for *S. indicus*, with the exception that the prisms are smaller.

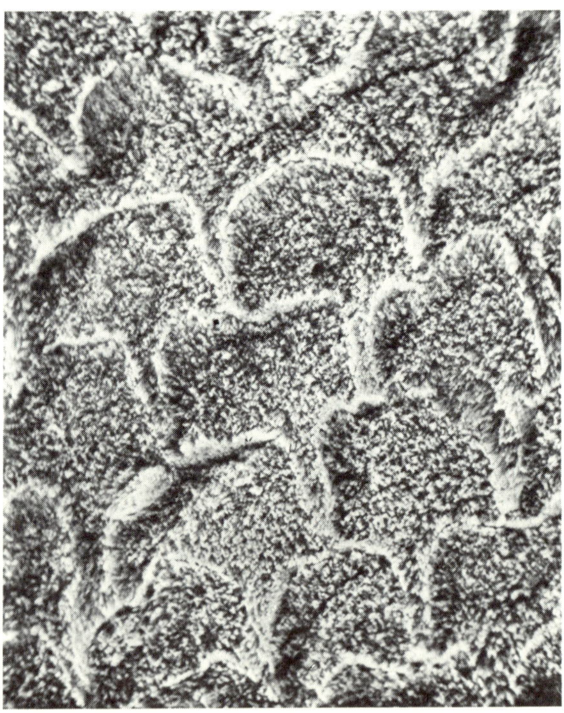

Figure 8.5. The keyhole prism pattern.

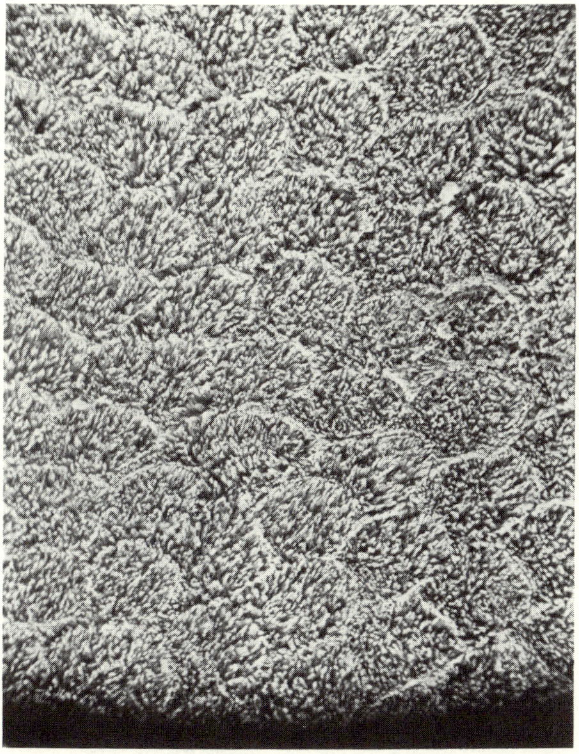

Figure 8.6. The prism pattern seen in *Pan* and Gorl *Gorilla*.

Discussion

The prism patterns obtained to date document the mature surface and subsurface enamel prism pattern. However, these patterns must be correlated to the known developmental patterns observed in various extant mammals (see Boyde 1964, 1965, 1971, 1976; see also fig. 8.1). For example the human enamel prism pattern is described as the keyhole prism pattern (see fig. 8.2) and relates to developmental pattern 3b (see figs. 8.1(3b) 8.2, 8.5). This pattern is characteristic of the genus *Homo*, encluding the fossil forms. The few specimens of *Australopithecus* studied also reveal a "keyhole" prism pattern, but the clarity and crystallite detail presented in figures 8.5 to 8.8 has not been obtained. Therefore, one can not at present determine if developmental pattern 3a or 3b is observed.

The extant apes, as well as the fossil hominoids, *Gigantopithecus*, *Sivapithecus indicus* and *Ramapithecus punjabicus*, all reveal a "hexag-

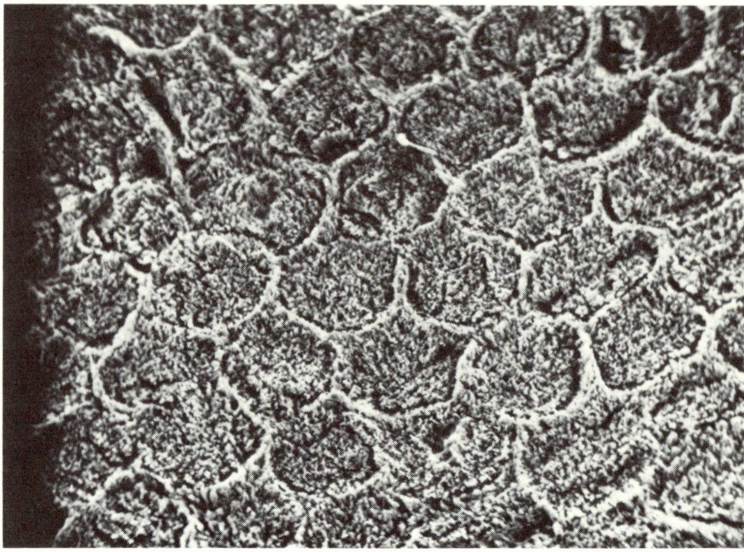

Figure 8.7. The prism pattern seen in *Pongo*.

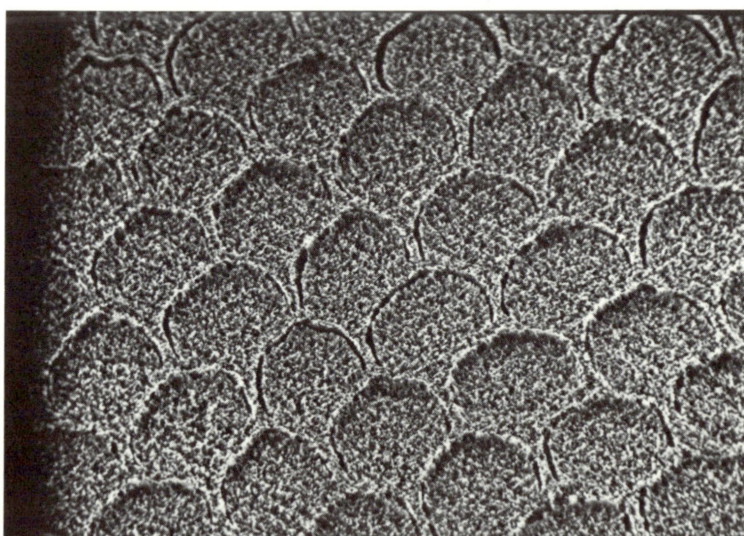

Figure 8.8. The prism pattern seen in *Sivapithecus indicus*.

onal-shaped" prism pattern with tail, being similar to the developmental pattern 3a (see figs. 8.1(3a), 8.6, 8.7, 8.8).

The two variants, 3a and 3b, appear to be significantly different in that these variants may be explained by and would have to be accompanied by slight modifications of the shape of the developing enamel surface, and would thus show a different type of transverse fracture (Boyde, personal communication). The degree of dovetailing of the common inter-pit wall derived component (the prism tail, see fig. 8.2) between the prism heads would mean that the type 3b prisms would part in a transverse fracture plane by rupture of the tail crystals under tension, in addition to separation along the prism boundary planes (Boyde 1976b). The tails of type 3a prisms would tear out in a more intact condition requiring less energy to exploit the separation of the prism boundary discontinuities (see fig. 8.9).

Conclusions

Using the new methods and techniques developed by Boyde et al. (1978) and Gantt (1979a), a much clearer understanding of the variations, problems of interpretations, and differences in prism pattern morphologies has been obtained. Previous publications (Gantt et al. 1977; Gantt 1977; Lavelle et al. 1977; Vrba and Grine 1978) must be questioned concerning the reported prism patterns obtained on hominoid material. Analysis of the

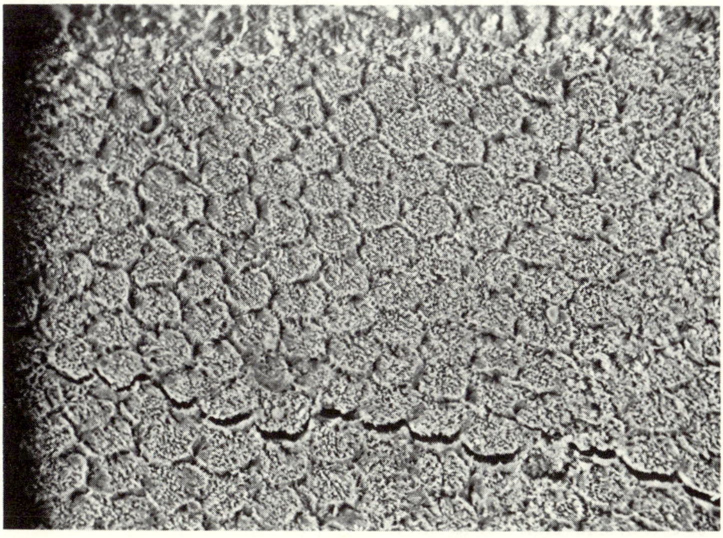

Figure 8.9. The etched surface of a molar from *Gorilla* revealing the pattern of fracture.

type specimen of *Ramapithecus* (courtesy of Pilbeam) did not reveal a "keyhole" prism pattern as was previously reported, but a pattern similar to the other ramapithecids which resembles type 3a. This evidence does not exclude *Ramapithecus* from being a hominid. Rather, it suggests that no Middle or Late Miocene hominoid, presently known, has the typical hominid enamel prism pattern—the "keyhole" prism pattern (see figs. 8.2, 8.4). Also, it is not presently clear if the Australopithecines have the "keyhole" prism pattern either.

Perhaps the most important findings in this study concern the phyletic position of the orangutan *Pongo*. Analysis of extant and fossil orang teeth document a prism pattern which is identical to that observed in *Sivapithecus indicus* (see figs. 8.7, 8.8). In addition, the recently discovered face of *S. meteai* has provided new evidence as to the affinities of *Pongo*. *S. meteai* has a deep and widely flaring zygomatic process, a marked alveolar prognathism that is combined with a short upper face, and a narrow interorbital distance, as well as large upper I 1's relative to the I 2's and large squared molars, all characters which are shared only with the orangutan among the hominoids (Andrews and Tekkaya 1980).

Therefore, based on the molecular data which document that *Pongo* is significantly different from either *Pan* or *Gorilla*, and assumming that the molecular clock is close to being correct, then the earlier suggested divergence of *Pongo* is supported by the finding of this investigation (see Goodman and Tashian 1976; Sarich 1973). Thus, it is possible to reconstruct a phylogeny which places *Pongo* in the Ramapithecidae with close affinities to *Sivapithecus meteai* and *Sivapithecus indicus*, while *Pan* and *Gorilla* would then be placed into the Hominidae. This rearrangement would for the first time bring the molecular and morphological evidence concerning the origin of the hominids into agreement. However, before this proposal can be fully accepted, an extensive survey of the order Primates, both extant and extinct, is necessary to provide for a more complete understanding of the evolution of the calcified dental tissue.

References

Andrews, P. and I. Tekkaya. 1980. A revision of the Turkish Miocene hominoid *Sivapithecus meteai*. *Palaeontology*, 23:85–95.

Boyde, A. 1964. The Structure and Development of Mammalian Enamel. Ph.D. dissertation, University of London.

—— 1965. The structure of developing mammalian dental enamel. In R. W. Fearnhead and M. V. Stack, eds., *Tooth Enamel—I*, pp. 163–167. Bristol: Wright.

—— 1971. Comparative histology of mammalian teeth. In A. Dahlberg, ed., *Dental Morphology and Evolution*, pp. 81–93. Chicago: University of Chicago Press.

—— 1973. Quantitative photogrammetic analysis and qualitative stereoscopic analysis of SEM images. *J. Microscopy* 98:452–471.

—— 1976a. Amelogenesis and the structure of enamel. In B. Cohen and I. R. H. Kramer, eds., *Scientific Foundations of Dentistry*, pp. 335–352. London: William Heinemann Medical Books.

—— 1976b. Enamel structure and cavity margins. *Operative Dentistry* 1:13–28.

—— 1978. Development of the structure of the enamel of the incisor teeth in the three classical subordinal groups of the Rodentia. In P. M. Butler and K. A. Joysey, eds., *Development, Function, and Evolution of Teeth*, pp. 43–58. New York: Academic Press.

Boyde, A. and P. G. T. Howell. 1977. Taking, presenting, and treating stereo data from the SEM. *SEM/1977* 1:571–579.

Boyde, A., S. J. Jones, and P. S. Reynolds. 1978. Quantitative and qualitative studies of etching with acid and EDTA. *SEM/1978* 2:991–1002.

Boyde, A. and H. F. Ross. 1975. Photogrammetry and the SEM. *Photogramm: Record* 8:408–457.

Carter, J. T. 1922. The structure of the enamel in the primates and some other mammals. *Roy. Zool. Soc. London*. pp. 599–608.

Gantt, D. G. 1977. Enamel of primate teeth: Its Thickness and Structure with Reference to Functional and Phyletic Implications. Ph.D. dissertation, Washington University, St. Louis, University Microfilms. Ann Arbor.

—— 1979a. A method of interpreting enamel prism patterns. *SEM/1979* 2:975–981.

—— 1979b. Taxonomic implications of primate dental tissues. *J. Biol Buccale* 7: 149–156.

—— 1979c. Comparative enamel histology of primate teeth. *J. Dent. Res., Special Issue B: Proc. of the 3rd Int. Symp. Tooth Enamel* 58:1002–1003.

Gantt, D. G. and T. Gest. n.d. Hominid evolution and enamel thickness. Manuscript.

Gantt, D. G., D. R. Pilbeam, and G. Steward. 1977. Hominoid enamel prism patterns. *Science* 198:1155–1157.

Greenfield, L. 1974. Taxonomic reassessment of two *Ramapithecus* specimens. *Folia Primat*. 22:97–115.

—— 1979. On the adaptive pattern of *Ramapithecus. Am. J. Phys. Anthrop.* 50:527–548.

Goodman, M., R. E. Tashian, and J. H. Tashian. 1976. *Molecular Anthropology*. New York: Plenum Press.

Jolly, C. J. 1970. The seed-eaters: A new model of hominid differentiation based on a baboon analogy. *Man* 5:5–26.

Kay, R. F. 1973. Mastication, Molar Tooth Structure, and Diet in Primates. Ph.D. thesis, Yale Univ., New Haven. Univ. Microfilms, Ann Arbor.

von Koenigswald, W. 1977. *Micmomys* cf. *reidi* aus der villafranchischen spaltenfullung schambach bei treuchflingen. *Mitt Bayer. Staatsslg. Palaont. Hist. Geol.* 17:197–212.

Korvenkontio, V. A. 1934–35. Mikroskopische untersuchungen an nagerincisiven unter hinweis auf die schmelzstruktur der backenzahne. Histologisch-phyletische studie. *Ann. Zool. Soc., Zool-Bot. Fenn. Vanamo (Helsinki)* 2:1–274.

Lavelle, C. L. B., R. P. Shellis, and D. F. G. Poole. 1977. *Evolutionary Changes to the Primate Skull and Dentition*. Springfield, Mass.: C. C. Thomas.

Pilbeam, D. R. 1978. Rearranging our family tree. *Human Nature* 1:38–45.

Pilbeam, D. R., G. E. Meyer, C. Badgley, M. D. Rose, M. H. L. Pickford, A. K. Behrensmeyer, and S. M. Ibrahim Shah. 1977. New hominoid primates from the Siwaliks of Pakistan and their bearing on hominid evolution. *Nature* 270:689–695.

Poole, D. F. G. and A. W. Brooks. 1961. The arrangement of crystallites in enamel prisms. *Arch. Oral Biology* 5:14–26.

Regan, T. C. 1930. The evolution of the primates. *Ann. Mag. Natural Hist.* (now *J. Nat. Hist.*) (*London*) 10:383–393.

Sarich, V. M. 1973. Just how old is the hominid line? *Yearbook of Phys. Anthrop.* 17:98–111.

Shobusawa, M. 1952. Vergleichende untersuchungen uber die form schmelzprismen der saugetiere. *Okaj. Folia Anat. Jap.* 24:371–392.

Simons, E. L. 1977. The nature of the transition in the dental mechanism from pongids to hominids. *J. Hum. Evol.* 5:511–528.

—— 1977. *Ramapithecus. Sci. Am.* 236:28–35.

Simons, E. L., P. Andrews, and D. R. Pilbeam. 1978. Cenozoic apes. In V. J. Maglio and H. B. S. Cooke, eds., *Evolution of African Mammals*, pp. 120–146. Cambridge: Harvard University Press.

Simons, E. L. and D. R. Pilbeam. 1972. Hominoid paleoprimatology. In *The Functional and Evolutionary Biology of Primates*, pp. 36–62. Chicago: Aldine.

Tomes, C. S. 1906. On the minute structure of the teeth of creodonts, with special reference to their suggested resemblance to maruspials. *Proc. Zool. Soc., Lond.* (now *J. Zool.*), pp. 45–58.

Tomes, J. 1849. On the structure of the dental tissues of marsupial animals, and more especially of the enamel. *Phil. Trans.* 139:403–412.

—— 1850. On the structure of the dental tissues of the order Rodentia. *Phil. Trans.* 140:529–567.

Vrba, E. S. and F. E. Grine. 1978. Australopithecine enamel prism patterns. *Science* 202:890–892.

9. Enamel Structure in the Molars of Arvicolidae (Rodentia, Mammalia), a Key to Functional Morphology and Phylogeny

WIGHART VON KOENIGSWALD

Enamel is the most resistant material which the mammalian body can produce. It has often been proved that the morphology of the enamel surface shows a large amount of information about the adaptation of the tooth to the diet as well as about the phylogeny of the bearer. Enamel is not amorphous in structure like glass, but consists of apatite crystallites which occur either as "enamel prisms" oriented parallel to each other or as "interprismatic substance." Prisms and interprismatic substance are the main elements of the enamel. Organic material is very limited and can be neglected here.

A general hypothesis on the differentiation of enamel under the aspects of functional morphology may be outlined. On the one hand prisms continuously become more closely packed together. This might be the reason for the change of their cross section, which is modified from circular to hexagonal and finally to the keyhole pattern. On the other hand the different elements are superimposed at angles to each other. Thus the interprismatic substance often crosses the prisms at nearly right angles. Furthermore the prisms themselves need not be parallel to each other. They may cross each other as well. But the pathway of the forming ameloblasts should not be disordered. The first step of development from the initial stage of parallel-oriented prisms seems to be the multiserial Hunter-

This study was supported by the Deutsche Forschungsgemeinschaft (Ko 627/1 + 2). My cordial thanks are due to Freiherrn G. von Falkenhausen for a generous grant, which enabled me to participate in the Turku conference and to present this paper.

Schreger bands. Layers of 4 to 10 prisms in thickness are formed upon each other. In most teeth these layers are approximately parallel to the occlusal surface of the tooth. In a section parallel to the occlusal surface, the prisms originating at the dentin enamel border turn in one layer to the right, while those of the underlying and overlying layers turn to the left. So prisms of different Hunter-Schreger bands cross each other at nearly right angles.

One functional advantage of these multiserial Hunter-Schreger bands is the reduction of damage caused by hairline cracks. When all prisms are parallel, cracks may penetrate deep into the enamel and may even be continuously expanded during mastication. But when layers of prisms with different orientation are superimposed, cracks are stopped more easily. These multiserial Hunter-Schreger bands are found in a great number of mammalian dentitions (Kawai 1955). They are clearly observable in hypsodont teeth. In cases of thick enamel, the bands are less pronounced than in teeth where the enamel is very thin. But thin enamel is required in order to obtain a continuously sharp cutting edge.

Of special interest are evergrowing incisors developed in various mammalian groups. In these teeth, as in *Vombatus* (Marsupialia), *Daubentonia* (Primates), *Myotragus* (Artiodactyla), *Lepus* (Lagomorpha) and *Cavia* (Rodentia) the multiserial Hunter-Schreger bands are very distinct. It seems to be a general capability of mammals that whenever thin enamel is required to produce a sharp cutting edge, they are able to develop multiserial Hunter-Schreger bands.

Rodents which are characterized by their ever-growing incisors are able to increase this enamel differentiation. They thin the Hunter-Schreger bands from the multiserial stage to the pauci-serial and finally to the uniserial stage. The thickness of the Hunter-Schreger bands is reduced from 4 to 10 prisms to only a single prism. The final stage, the uniserial Hunter-Schreger bands, is called lamellar enamel (Wahlert 1968, Koenigswald 1977, 1980). The structure shows prisms of one layer crossing those of the next layer at approximately right angles; the interprismatic substance is strengthening the third dimension. This orthogonal system is not only most resistant against hairline cracks but also is able to resist stress and tension from any direction. However, within a tooth several requirements have to be met. Besides the resistance against stress and cracks, which seems to be optimized in the lamellar enamel, cutting edges should be as sharp as possible. It is obvious that the lamellar enamel never forms the cutting edge of the incisor itself. An outer layer of radial enamel can always be found. In this enamel type all prisms are parallel to each other and rise to the outer surface of the tooth. It might be assumed that the radial enamel is more suitable for forming the self-sharpening cutting edge. Its resistance is high enough when it is reinforced by a thick layer of lamellar enamel.

Similar problems from the functional point of view occur in hypsodont molar teeth, when enamel forms a thin, self-sharpening cutting edge, as in most grass-eating herbivores. The requirements of the enamel become apparent by comparing a mammoth with a lemming. Both feed on grass, and theoretically the same forces per unit area of their molars are required to cut the grass. An optimum force is required to cut the grass. A force below this does not achieve the required effect; a force above is wasted energy (Rensberger 1973). The enamel in a mammoth molar is about 1.5 mm thick, but that in a lemming molar is less than 0.1 mm. Obvious differences exist in the internal structure of the enamel. Hunter-Schreger bands have a thickness in mammoth molars of about 70 prisms but in that of a lemming only one. Like other arvicolid rodents, lemmings were able to introduce the lamellar enamel, first developed for the incisors, into the molars as well. But enamel structure of arvicolid molars reflects many more functional problems, some of which will be discussed in this paper.

Enamel Types and Morphology in Arvicolid Molars

The rodent family Arvicolidae, the voles and lemmings, appears to have the highest enamel differentiation of most mammals. To obtain a survey of the various possibilities of differentiation in this group, molars of about 50 genera and subgenera from both the living and fossil fauna have been studied. A detailed documentation of the material and the results are given elsewhere (Koenigswald 1980); only the basic outline will be demonstrated here.

To investigate the enamel structure, the molars are embedded into artificial resin and ground down—parallel to the chewing surface—until the entire length of the enamel is exposed. After polishing, the section must be etched in 2-normal HCl for about 2–3 seconds to make the structure visible. After sputter coating, the section can be observed in the SEM at a magnification of 500–1500 ×. The first striking impression is that the structure of the enamel in arvicolid molars is not homogeneous. Instead, several different enamel types are involved (fig. 9.2). Enamel types are defined by different prism orientation. Three basic types can be distinguished (fig. 9.1).

radial enamel (fig. 9.1a). All prisms have parallel orientation. They rise to the outer surface of the enamel and penetrate the chewing surface at a very steep angle. The interprismatic substance crosses the prisms radially at approximately right angles. This enamel type is regarded as the least derived type.

tangential enamel (fig. 9.1b). The relationship between prisms and interprismatic substance is the same as in the radial enamel, but the whole

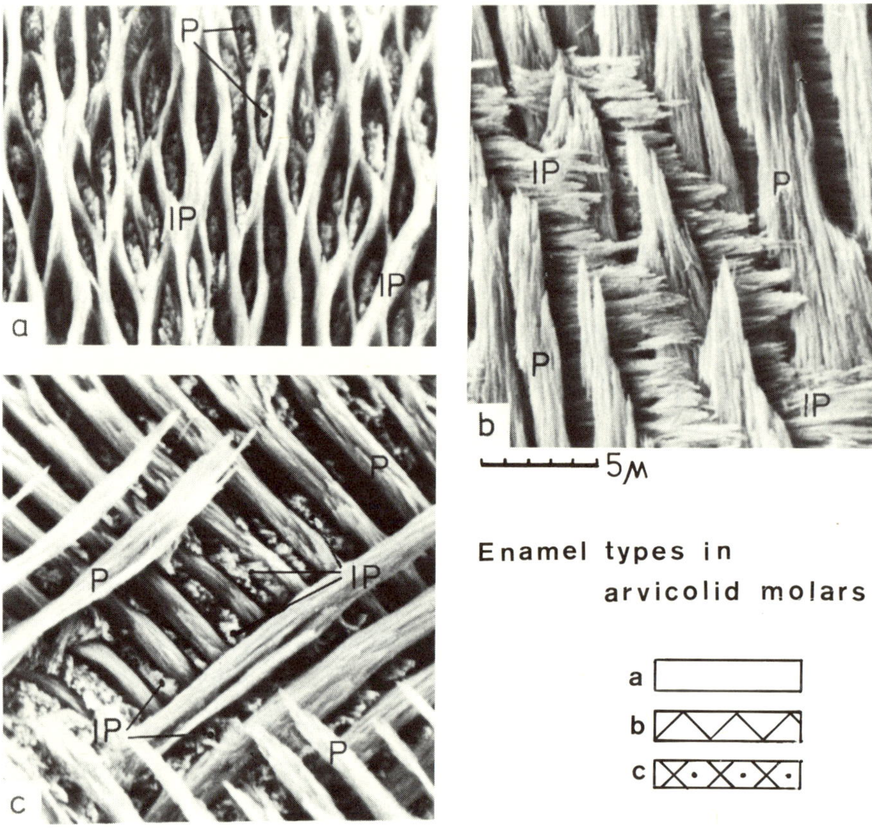

Figure 9.1. The three basic enamel types in the molars of Arvicolidae and their symbols: *a*. radial enamel (while); *b*. tangential enamel (zigzag); *c*. lamellar enamel (crosses and dots).

Abbreviations: P = prisms, IP = interprismatic substance.

structure is twisted about 90° so that the prisms as well as the interprismatic substance are oriented parallel to the chewing surface. Tangential enamel was developed from radial enamel. Intermediate stages, with prisms inclined only at about 45°, are called primitive tangential enamel.

lamellar enamel (fig. 9.1c). The prisms are organized in layers of only one prism thick. Prism of different layers cross each other at about right angles. Interprismatic substance strengthens the third dimension. In the typical lamellar enamel the layers (lamellas) are more or less parallel to the chewing surface. In less developed stages these layers may rise toward the outer surface so that they do not occur as obviously in a section parallel to the chewing surface. This form is called discrete lamellar enamel. A

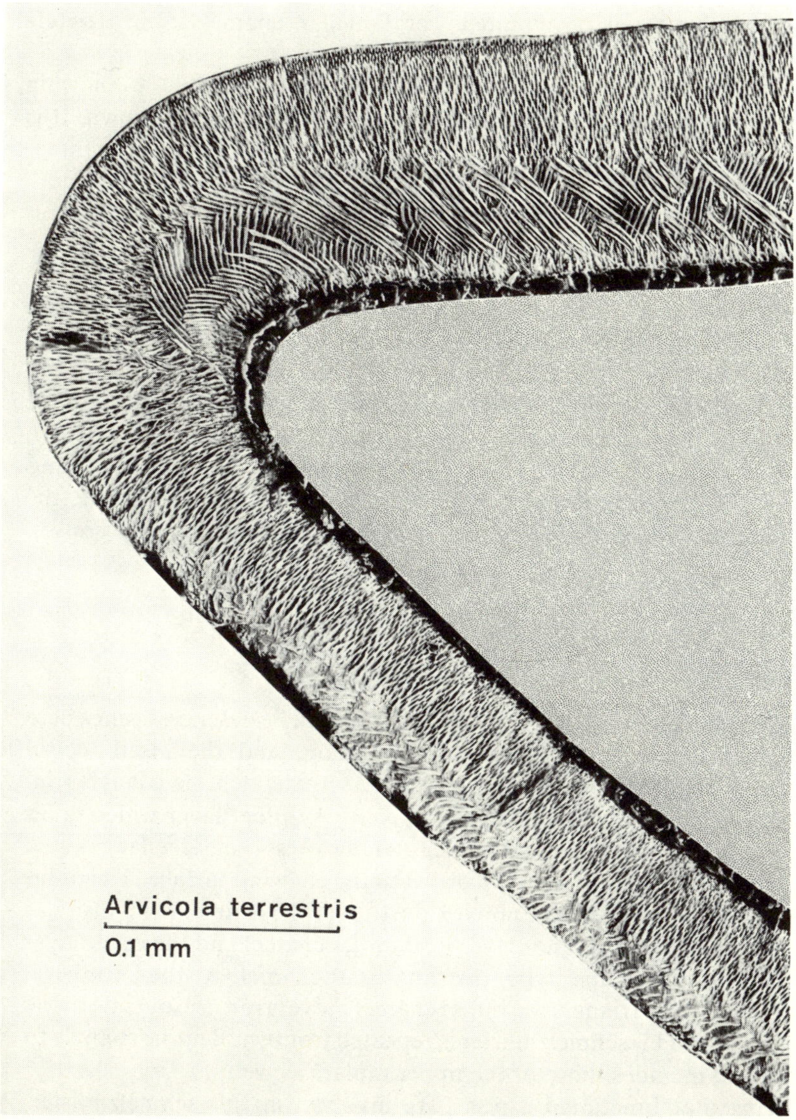

Figure 9.2. The schmelzmuster in a dentin triangle of a molar in *Arvicola terrestris* (Holocene). The section parallel to the chewing surface shows three distinct enamel types, which become visible after a short etching with HCl. The arrangment of the enamel types stays constant throughout the whole height of the molar.

hypothesis about the development of lamellar enamel from Hunter-Schreger bands is given above.

The arrangement of the different enamel types within the chewing surface remains constant throughout the whole height of the crown. It is characteristic at the genus or subfamily level. Therefore the three-dimensional arrangement of the enamel types defines the "schmelzmuster" of the relevant group (Koenigswald 1980).

The morphology of the arvicolid molars is characterized by the composition of dentine triangles. In higher derived species one can always distinguish between a concave and a convex side of the dentine triangle which is covered by the enamel band. In lower molars the concave enamel band is exposed on the mesial side of the dentine triangle and the convex enamel band on the distal side. In the upper molars, due to a reversal in the morphology, the functional importance of which will be discussed below, the concave enamel band is exposed distally and the convex enamel band mesially. This morphological distinction is of importance, since the schmelzmuster often differentiates between the concave and convex portions of the enamel band.

The Schmelzmuster in Arvicolidae

As an example, *Arvicola terrestris* (L. 1758) will be described, since here the enamel shows a very high differentiation, and the fossil record represents a sequence of evolutionary stages. In any dentine triangle (fig. 9.2) the concave enamel band has two layers. An inner layer which takes somewhat more than 50 percent of the total thickness is of lamellar enamel. The lamellas are approximately parallel to the chewing surface. The outer layer is of radial enamel. The convex portion of the enamel band also has two layers, but here the inner layer is of radial enamel and the outer layer shows a different enamel type, the tangential enamel. At the peripheral apex of the dentine triangle the radial enamel moves from the outside to the inside (fig. 9.2). This schmelzmuster is repeated from one dentine triangle to the next and includes those of the upper molars as well.

It seems that functional aspects are involved in this schmelzmuster. According to the propalinal movement of the jaw during mastication, chewing forces meet every dentin triangle in the lower jaw from the mesial side, which therefore is called the leading edge (Greaves 1973). This edge is always concave (fig. 9.3). The trailing edge on the other side of the dentin triangle shows a convex bending of the enamel bands. In the upper jaw the stress pattern is reversed. The leading edge is exposed on the distal side of the dentin triangle but, due to the reversal in morphology mentioned

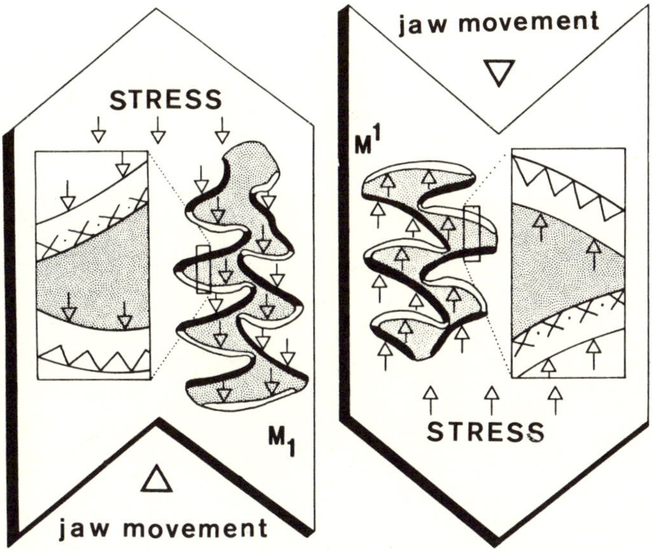

Figure 9.3. Stress pattern during mastication and schmelzmuster in upper and lower molars in *Arvicola*. Because of a reversed morphology, the leading edges of upper and lower molars are concave and the trailing edges are convex. The schmelzmuster repeats this pattern (Koenigswalk 1980, fig. 77).

above, this side is concave. The trailing edge, pointing mesially, is convex. Although there is a clear relationship between the convex or concave bending of the enamel band and the stress pattern with the schmelzmuster, the factor which controls this relationship must be discussed. The schmelzmuster on the one hand can be something like a "fabricational noise" while a concave or a convex enamel band is formed. On the other hand, the schmelzmuster may be governed by the stress pattern distinguishing between leading and trailing edges.

The solution can be found in a young specimen of the first lower molar, where the enamel band still surrounds the anterior loop in the mesial part (fig. 9.4). This portion of the enamel band is convex but exposed on the leading edge, so the combination is different from the one in the dentin triangles. The enamel parts found in this mesial portion are both lamellar and radial enamel, as in any enamel band exposed on the leading edge. Therefore the stress pattern is assumed to be the factor which controls the schmelzmuster. However, enamel can not be remodeled under external stresses such as bone. The schmelzmuster is mostly built long before any forces can act on the tooth. Therefore, the schmelzmuster has to be genetically determined, but selected according to the chewing forces.

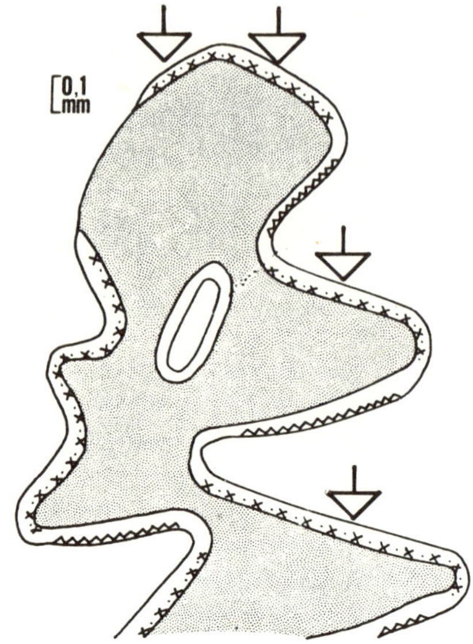

Figure 9.4. In the anterior loop of a young specimen of M_1 of *Mimomys polonicus* (Villafranchium) the mesial portion of the enamel band is convex, but it is exposed on the leading edge. The schmelzmuster follows only the stress pattern during mastication (Koenigswald 1980, fig. 71).

Phylogenetic Development of the Schmelzmuster

Since the schmelzmuster as shown in *Arvicola* is genetically determined, it should be possible to trace its evolution back into the fossil record. *Arvicola* is derived from *Mimomys*, a vole with rooted molars which has its origin somewhere in the widespread Cricetidae. As an example of the general conditions in cricetids the schmelzmuster in the genus *Cricetus* may be described. In the crown of the molars only the least derived enamel type, radial enamel, can be found. But very near to the base of the crown a basal ring of lamellar enamel surrounds the whole molar. In further evolution there are indications that the lamellar may have expanded into the height of the hypsodont crown in early Arvicolidae.

Hypsodont cricetids like *Ruscinomys* do not show any lamellar enamel within the chewing surface, but in some areas tangential enamel is introduced into the crown. The earliest member of *Mimomys* which could be investigated for enamel structures is *Mimomys (Hintonia) occitanus* (fig. 9.5) from the upper Pliocene from France. The molars already have the characteristic shape of *Mimomys* molars, but the schmelzmuster is not yet fully developed. The leading edge shows only radial enamel, but in the trailing edge some tangential enamel is found in the outer half of the enamel band. This tangential enamel does not yet reach the outer border of the enamel band. Slightly younger species like *M. cappettai* and *M. stehlini* show a higher developed schmelzmuster. In the trailing edge the tangential enamel fills the outer part of the enamel band, but in the leading edge

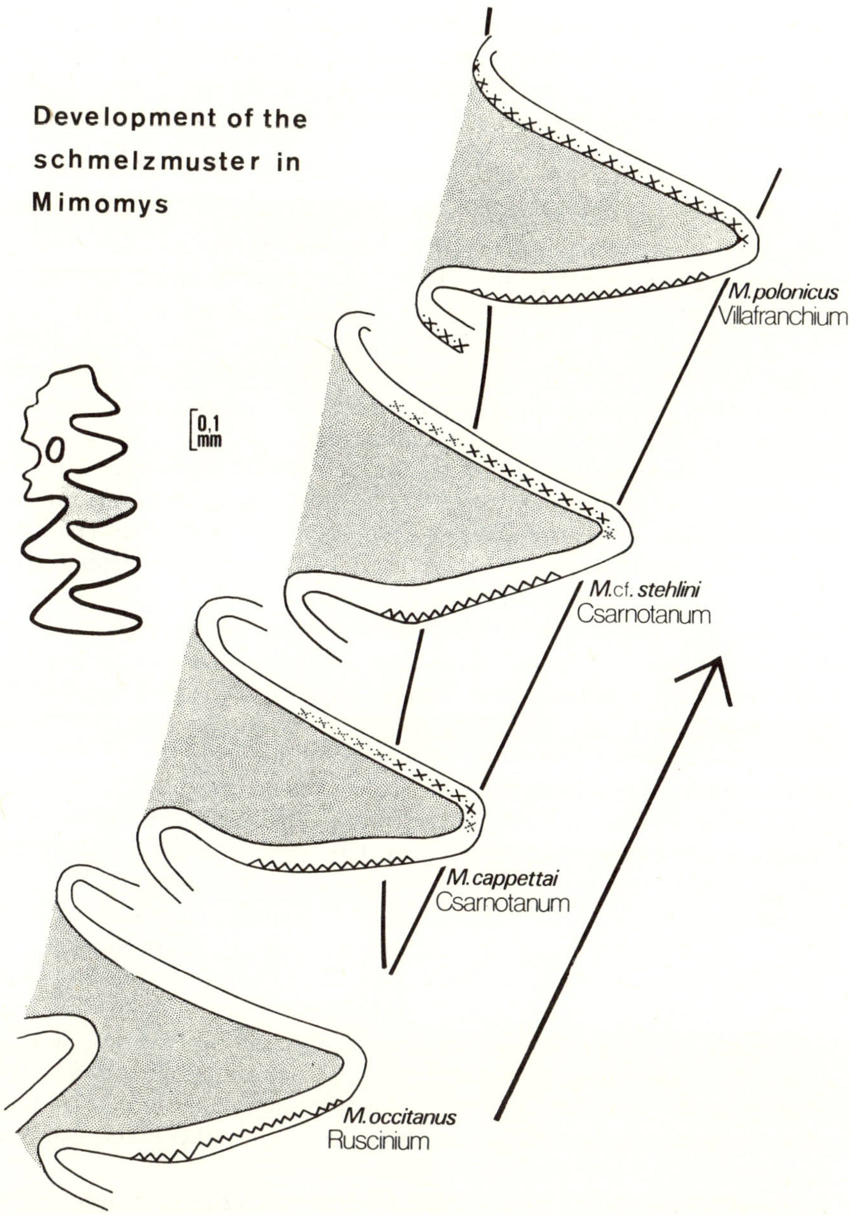

Figure 9.5. The development of the asymmetrical schmelzmuster in a phylogenetic sequence of the genus *Mimomys* (Pliocene-early Pleistocene). The lamellar enamel is introduced in the lateral part of the leading edge, expands throughout the whole length, and increases in thickness.

lamellar enamel is first introduced in the lateral part, and from there gradually penetrates throughout the whole length of the leading edge.

In this early stage the lamellar enamel is situated between an inner radial enamel and an outer, thicker radial enamel. From middle Villafranchium *M. polonicus* could be studied. Here the schmelzmuster is fully developed. The lamellar enamel not only penetrates the whole length of the enamel band, it suppresses the inner radial enamel as well.

Because of the initial differentiation between the leading edge and the trailing edge, this schmelzmuster is called an asymmetrical schmelzmuster. This asymmetry involves further biomechanical problems, which are related to the difference in schmelzmuster between *Mimomys* and the extant *Arvicola*. In *Mimomys* the enamel band in the trailing edge is markedly thicker than the leading edge. In *Arvicola* the enamel band in the trailing edge is thinner than in the leading edge.

In *Mimomys*, both the leading edge as well as the trailing edge work as cutting edges. They are exactly the same height, and attrition is the same on both sides, even when the internal structure of the enamel differs. The physical properties may differ between lamellar enamel in the leading edge and tangential enamel in the trailing edge, since the tangential enamel lacks strengthening elements in the third dimension. So it is assumed that the diminished quality of the enamel is compensated by an increase in quantity, until the balance between the leading edge and the trailing edge is established.

This interpretation is valid for *Mimomys* but not yet fit for *Arvicola*. In *Arvicola* the trailing edge is thinner and usually worn more deeply than the leading edge. In fact the trailing edge is not really worn but broken down, since it cannot resist the chewing forces. This means that the length of the cutting edge is reduced. However, according to Rensberger (1973) the efficiency of a tooth is related to the length of the cutting edge. But certainly *Arvicola* does not have less efficient teeth than *Mimomys*, its predecessor. The main difference between *Mimomys* and *Arvicola* is the fact that *Arvicola* developed rootless molars. As demonstrated by Koenigswald and Golenishev (1979), rootless molars erupt faster than rooted ones. So it is assumed that the loss of the trailing edge may be compensated by a faster eruption of the tooth, which can be worn down more rapidly. The trailing edge seems to be the less efficient one because of some other biomechanical reasons.

This reduction of the trailing edge is a fact which can be proved in the fossil record (fig. 9.6), since the predecessor of *Arvicola terrestris, A. cantiana* from the Middle Pleistocene, still shows a thick trailing edge like *Mimomys*. The reduction of the trailing edge is a general trend in derived arvicolids with rootless molars. *Microtus*, which might have developed

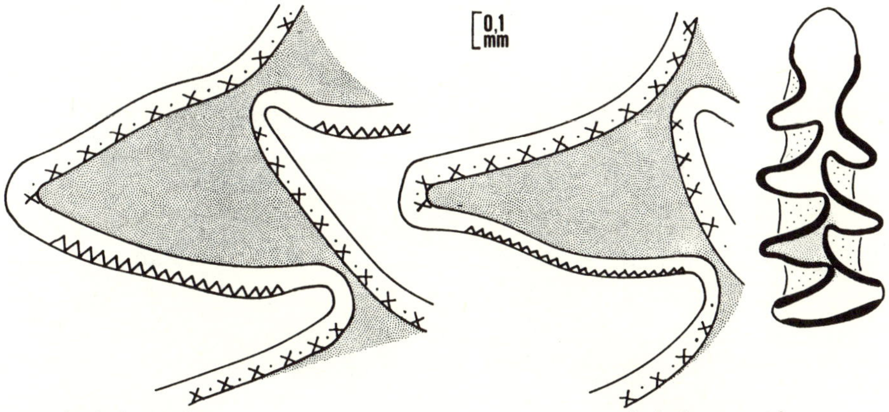

Figure 9.6. In *Arvicola* the trailing edge is gradually reduced between *A. cantiana* (middle Pleistocene) and *A. terrestris* (Holocene) (Koenigswald 1980, figs. 72 and 73).

from *Mimomys*, also shows the same reduction of the trailing edge between the early subgenus *Allophaiomys* and the extant species. All species of *Microtus* keep the basic schmelzmuster of *Mimomys*, but differ in the degree of the reduction in the trailing edge. Most derived species possess only radial enamel in the trailing edge, while the tangential enamel is reduced.

Schmelzmusters in Other Arvicolid Rodents

Even when the schmelzmuster of the *Mimomys-Arvicola* branch here described is found in the majority of the extant arvicolids, this branch is only one of several which have developed their own type of schmelzmuster. The full variety cannot be demonstrated here, so only the tribes Ondatrini and Dicrostonychini are selected since they are examples of a schmelzmuster with very great differences. *Ondatra*, the North American muskrat, originates from *Pliopotamys*. The schmelzmuster in this genus displays only radial enamel.

Further, the earliest species of *Ondatra, O. idahoensis*, presents only radial enamel. In *O. annectens* lamellar enamel was found first, at the apex of the dentine triangle. From here the lamellar enamel expands between an inner and an outer radial enamel into both sides of the dentin triangle, so the leading and trailing edges do not show any differentiation. In *Ondatra zibethicus*, the extant species, the lamellar enamel is found nearly throughout the whole length of the enamel band, but it is very often discrete in the deepest parts of the reentrant folds. The schmelz muster of *Ondatra*

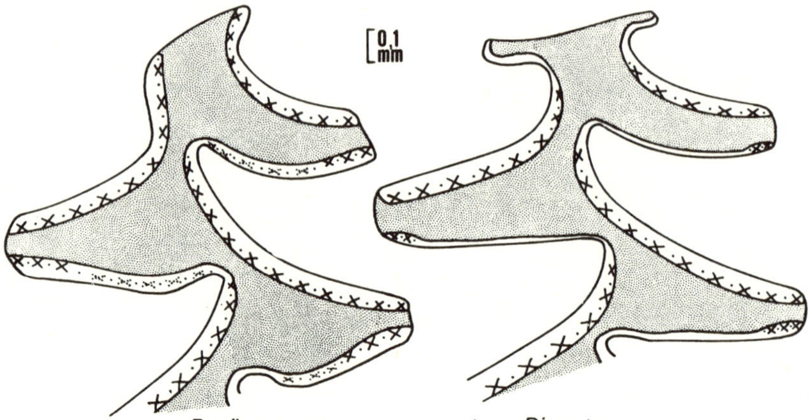

Figure 9.7. *Dicrostonyx* derived from a symmetrical schmelzmuster with lamellar enamel in the leading edge as well as in the trailing edge. Between *Predicrostonyx hopkinsi* (middle Pleistocene) and *Dicrostonyx torquatus* (recent) the reduction of the trailing edge proceeded in an obvious way (Koenigswald 1980, figs. 75 and 76).

is not very differentiated compared with that in *Mimomys*. Since the leading edge and the trailing edge are built alike, the schmelzmuster is called symmetrical. There are several fossil genera which show a primitive stage of a symmetrical schmelzmuster presenting lamellar enamel only at the apex of the dentin triangles, such as *Ungaromys* and *Nebraskomys*. The same rather primitive schmelzmuster is found in the extant fauna of Eurasia in *Ellobius* and *Prometheomys*.

This might give the misleading idea that selective pressure for developed schmelzmusters is only minor. But *Ellobius* and *Prometheomys* are relicts of an ancient radiation which survive only because they occupy a very specialized ecological niche in their subterranean habitat.

Symmetrical schmelzmusters need not stay in a primitive stage. *Dicrostonyx*, the collared lemming, shows a highly developed stage of a symmetrical schmelzmuster, which again shows a reduction in the trailing edge (fig. 9.7). The leading edge of the dentin triangles is built—as in most arvicolids—from an inner layer of lamellar enamel and an outer layer of radial enamel. The trailing edge is reduced to a very thin band of radial enamel. It is much too weak to work as a cutting edge. In the apex of the dentin triangle the enamel band is interrupted. Near this interruption the enamel band of the trailing edge shows a thickening where the radial enamel is underlaid by lamellar enamel. This schmelzmuster just described does not indicate that it was derived from a symmetrical schmelzmuster. However, in the trailing edge of the posterior loop in lower molars, lamellar enamel continues throughout the whole length. This part of the enamel band normally is more conservative and may display further information about the unreduced status. This is corroborated by the fossil *Predicrostonyx* where the reduction of the trailing edge in the dentine triangles is less. Lamellar enamel can here be traced through the entire length of the enamel band. So symmetrical schmelzmuster may become secondarily asymmetrical also when the trailing edge is reduced.

Conclusions

In some selected examples it has been shown that the three-dimensional arrangement of the different enamel types within the molars of the Arvicolidae are characteristic for groups between the subfamily and the genus level. From these schmelzmusters a wide range of information regarding functional morphology as well as phylogeny may be obtained. A survey on most genera of the Arvicolidae is given elsewhere (Koenigswald 1980).

The schmelzmuster reflects the stress pattern during mastication, but despite being highly significant in a functional aspect, it can be traced back in phylogeny. Several lines of parallel evolution are to be expected. Details on the different groups of Arvicolidae are studied by Koenigswald (1980). But there are a number of general tendencies for the family Arvicolidae. This family is presumably the only rodent family which is able to build the enamel of the main cutting edges of their molars, (always the leading edges), in a way similar to the way they build their incisors. This is with the lamellar enamel on the inside and the radial enamel on the outside. Not all taxa have achieved the full development as *Ondatra*. The trailing edge is built in different ways, and symmetrical and asymmetrical schmelzmusters are distinguishable depending upon whether or not the trailing edge is identical to the leading edge. The trailing edges are reduced in several branches and lose their function as a cutting edge, as shown in *Arvicola* and *Dicrostonyx*.

The Arvicolidae are one of the youngest families within mammalian evolution. But they have had a rapid evolution with several radiations, such that they are at the present time one of the dominant groups in the holarctic fauna. This rapid evolution may be related to the capability to produce as efficient a structure within their molars as the rodents have in their incisors.

References

Graeves, W.S. 1973. The interference of jaw motion from tooth wear facets. *J. Paleont.* 47:1000–1001.

Kawai, N. 1955. Comparative anatomy of bands of Schreger. *Okaj. Folia Anat. Jap.* 27:115–131.

Koenigswald, W. V. 1977. *Mimomys* cf. *reidi* aus der villafranchischen Spaltenfüllung Schambach bei Treuchtlingen. *Mitt. Bayer. Staatsslg. Paläont. Hist. Geol.* 17:197–212.

——1980. Schmelzstruktur und Morphologie in den Molaren der Arvicolidae (Rodentia). *Abh. Senckenberg. Naturforsch. Ges.* 539:1–29.

Koenigswald, W. V. and F. N. Golenishev. 1979. A method for determining growth rates in continuously growing molars. *J. Mammal* 60:397–400.

Rensberger, J. M. 1973. An occlusion model for mastication and dental wear in herbivorous mammals. *J. Paleont.* 47:515–528.

Wahlert, J. H. 1968. Variability of rodent incisor enamel as viewed in thin section, and the microstructure of the enamel in fossil and recent rodent groups. *Breviora, MCZ*, vol. 103.

10. Morphological Approach to Heterodonty and Homology

JEFFREY H. SCHWARTZ

Although from time to time there have been alternative suggestions, the most commonly upheld criteria for the identification of teeth have focused on the position of teeth in the upper jaw, and the occlusal relationships of lower with upper teeth. Upper incisors are those teeth anchored in the premaxilla; the canine is the first tooth in the maxilla (i.e., the first tooth behind the premaxillary-maxillary suture), behind which is some number of premolar, then molar teeth. Lower incisors are those teeth in front of the lower canine, which is defined as the tooth that occludes in front of the upper canine; behind the canine is some number of premolar, then molar teeth. Molars are distinguished from premolars in that they are morphologically more complex and are unreplaced primary teeth. Occasionally, the first tooth of the premolar series is an unreplaced primary tooth (see Butler 1978; Kindahl 1967; Ziegler 1971).

The rationale behind the use of bony landmarks and occlusal relationships appears at first glance to be justified. From what is known of various Cretaceous and early Tertiary mammals, one can readily distinguish four sets of teeth representing three tooth classes (e.g., Butler 1978, and references therein). The first tooth in the maxilla is set apart by its "caniniformity" (= being the most robust, longest, simplest, conical tooth); in front of this tooth (= in the premaxilla) is a smaller set of teeth; behind this tooth is smaller "premolariform" set of teeth, behind which is a set of

I thank W. Bock, P. Butler, L. Krishtalka, and P. Lucas for suggestions, criticism and discussion of different points of view. Most especially, I extend my gratitude to A.G.S. Lumsden for his unbounded patience in discussions relating to dental development and homology. Travel to the symposium was supported in part by the Carnegie Museum of Natural History for which I thank Craig C. Black, Director. The manuscript was typed by Genevieve Loalbo.

"molariform" teeth. Although the corresponding lower jaw is but one bone, four similar sets of teeth are distinguishable; the most "caniniform" tooth occludes in front of the upper "caniniform" tooth. Many extant mammals also possess these distinct tooth classes.

What has always been a problem is the co-existence with the above-described mammal dental type either of mammals which have fewer recognized tooth classes (e.g., rodents) or those which have enough teeth to fill the jaw, but teeth that cannot be separated morphologically into four distinct tooth types (e.g., shrews). There are many mammals, fossil and extant, in which the most caniniform tooth is not the first tooth in the maxilla but is instead the first tooth in the jaw; behind this tooth is some number of premolariform and then molariform teeth. In yet other mammals (e.g., macropodid marsupials) this latter dental complement is seen in the lower jaw while an incisor tooth class can morphologically be distinguished in the upper jaw. Still other mammals (e.g., some fossil and extant hedgehogs, "flying lemurs," artiodactyls) do not display a recognizably caniniform tooth but otherwise possess "incisiform," "premolariform," and "molariform' teeth.

Clearly, the majority of mammals are heterodont; (the derivation of homodonty in cetaceans and other mammals is an exception). Beyond this obvious statement, however, what are the elements of heterodonty that mammals share? Except for cases such as among rodents, wherein only two "types" of teeth differentiate, do mammals maintain the development of four tooth types (i.e., three tooth classes)? Except when the first premolar of some taxa is a retained primary tooth, are all antemolar teeth secondary teeth?

Are Premaxillary Teeth Always Incisors?

Although many mammals, fossil and extant, possessed or possess four distinguishable tooth types with incisors, canines, premolars, and molars occupying "expected" positions in the jaw (see above), does it necessarily follow that all mammals possess incisors, canines, premolars, and molars, and that these teeth have always been and will always be in the same positions? Because tooth identification hinges on the identification of the upper canine, the question can be restated: although the most caniniform tooth of many mammals is the first tooth in the maxilla and in front of this tooth—in the premaxilla—lies a smaller set of teeth (i.e., the incisors), is the first tooth in the maxilla, regardless of its morphology, always a canine, and are all premaxillary teeth, regardless of their morphologies, always incisors? If the answer to this is "yes," there must be something in bone development that is related to the differentiation to tooth classes.

Observations on *Homo sapiens*

If upper canine development is inextricably tied to the region of the maxilla, and that of incisors to the premaxilla, then, as is commonly believed, canines can never lie anterior to the premaxillary-maxillary suture. It must also be the case that incisor development is tied to the premaxilla. Neither of these assumptions, however, obtains in *Homo sapiens*.

In *H. sapiens*, the oral epithelium proliferates over the area of the maxillary and median nasal prominences (Andersen and Matthiessen 1967; Nery, Kraus, and Croup 1970; Ooë 1957); it may, however, be interrupted by the nasal pit. Tooth development is initially indicated by separate, localized thickenings of the epithelium, with the germs of the presumptive deciduous central incisor and anterior deciduous molar appearing prior to those of the lateral incisor and canine (Ooë 1957). Initially, the presumptive central incisor germs occur quite posteriorly over the underlying fronto-nasal process and, through time, migrate forward and approach each other at the midline. It might also occur that the central incisor epithelium originates over the "seam" of the median nasal and maxillary promiences or even over the maxillary prominence (cf. Nery, Kraus, and Croup 1970; Tonge 1976). The epithelial thickenings which correspond to the lateral incisor most frequently originate over the underlying maxillary prominence, although this has also been noted over the "seam" of the two prominences (Ooë 1957). In time, the developing lateral incisors shift forward and come to lie lingual to and are markedly overlapped by the central incisors (Ooë 1956). The presumptive canine germs originate over the maxillary prominence and end up buccally, next to the central incisors, and overlapping the lateral incisors (Ooë 1956, 1957). These positions are attained by the 23 cm CR stage. While tooth initiation occurs prior to 15 mm CR length (Ooë 1957), ossification of facial bone, beginning with the vomer, is not seen until 23 mm (Andersen and Matthiessen 1967).

In perinatal individuals the growing crowns of the deciduous incisors and canines are still in the same relative positions as in the 23 cm CR length fetus. As can be seen in figure 10.1, there are two triangular crypts, side by side and oriented buccally, which overlap a third crypt that is directed lingually. The two buccal crypts house the central incisor and canine crowns, while the middle, lingual crypt is for the lateral incisor. With the palatal bony laminations removed, it is obvious that these three crypts are contained within the premaxilla. The first tooth in the maxilla is the first deciduous molar.

In some ways it is fortuitous that it is *H. sapiens* which provides a clear-cut example that tooth formation and the eventual position of teeth-

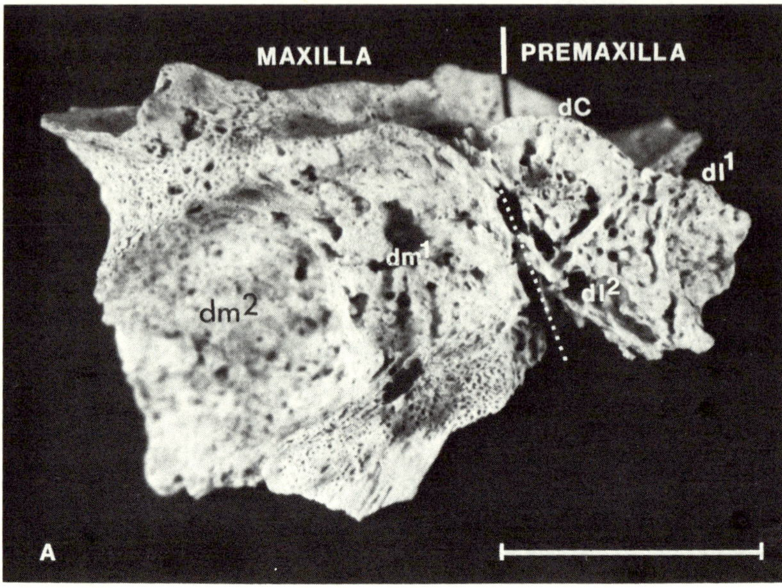

Figure 10.1a. Right osseous upper jaw of a perinatal *Homo sapiens* showing three triangular crypts (for the deciduous incisors and deciduous canine) in the premaxilla and two crypts (for the deciduous molars) in the maxilla.

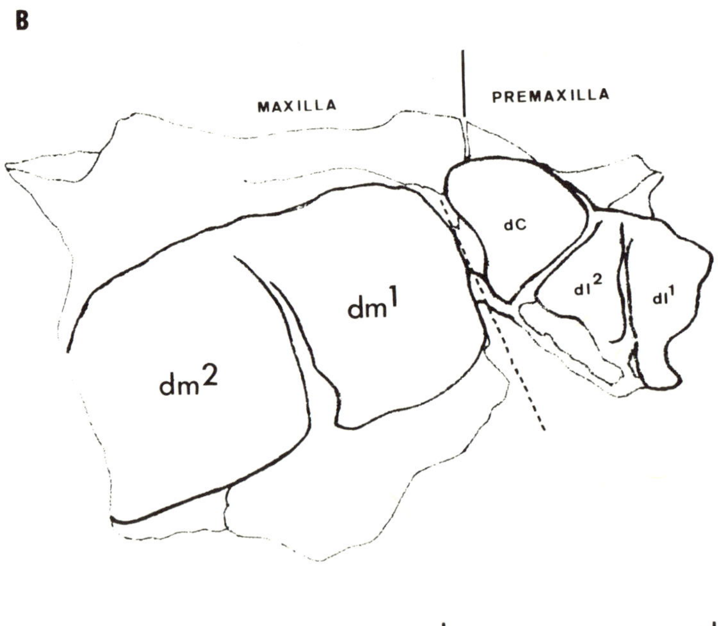

Figure 10.1b. Drawing of specimen in 10.1a.

to-bone are independent. For one thing, the teeth involved correspond morphologically to their identification. The incisors are recognizably incisiform; the canine is the most caniniform tooth in the jaw; the deciduous molars are replaced by unquestionably premolariform teeth. The canine and lateral incisor, if not at times the central incisor as well, originate over the maxillary process and move into the region that becomes the premaxilla, thereby illustrating the independence of tooth development from bony "landmark" development and falsifying the hypotheses that premaxillary teeth are always incisors and that the canine is always the first tooth in the maxilla.

It should also be pointed out that, since evidence of premaxillary-maxillary distinctiveness is obliterated quite early in *H. sapiens*, the teeth of this taxon have been identified on the basis of their *presumed* positions in the upper jaw, in other words, morphologically. Although the canine eventually comes to reside in the premaxilla, and thus erupts where it should not (and, by popular dogma, where it *cannot*), I see no reason to re-identify the teeth of *H. sapiens*. The very fact that the lateral incisor of *H. sapiens* typically originates over the maxillary prominence is reason enough to question the presumption that incisors are premaxillary teeth. Indeed, the observation that in *H. sapiens* (lateral) incisors typically originate over the maxillary process and others (central incisors) typically originate over the median nasal prominence demands that a morphological criterion for tooth identification be employed. How else can the fact that morphologically similar teeth with supposedly different origins be reconciled?

Other Taxa

Although there are quite a few studies on tooth development in various mammals, many present these data in isolation without reference to other developing structures. Furthermore, the earliest "stage" of tooth development presented is often relatively advanced, with many tooth germs already at the cap or bell stage (a function of the available material). One study by Kindahl (1957:467, fig. 3), on the tree shrew does, however, illustrate a transverse section through the upper jaw of a 13 mm embryo in which, if the apex of the primitive choana is taken as the "boundary" between the maxillary and median nasal prominences, the germ of the "second incisor" lies over the maxillary prominence. Since the "second incisor" and its successor erupt in the premaxilla, it is apparent that this tooth migrates forward in the jaw. Morphologically, the "second incisor" of *Tupaia* is not "incisiform."

And, finally, Lumsden (pers. comm.), while researching other aspects of dental development in the mouse, has found over the maxillary prominence epithelial thickenings which apparently correspond to the presumptive "incisors." This tooth, which becomes entirely encompassed by the premaxilla (Gaunt 1964), is not (strictly speaking) "incisiform."

Implications of "Migrating" Anterior Tooth Germs

Although I would predict that the phenomenon is quite common among mammals, it nonetheless is significant that in three phylogenetically distinct taxa, upper anterior teeth originate much farther back in the presumptive jaws than are the positions, and bone, into which they eventually erupt. Migration of tooth germs into anterior positions also occurs in the lower jaw in *H. sapiens* (Ooë 1956, 1957) and the mouse (Lumsden, pers. comm.); such information is not retrievable from Kindahl's study on *Tupaia*, but one might expect it to mirror the upper jaw.

The above, and especially the observations on *H. sapiens*, certainly illustrate the independence of tooth development from bone development, and would seem to demand the decoupling of tooth identification from tooth position in the jaw. As seen in *H. sapiens* and the mouse, upper and lower teeth develop similarly, regardless of the fact that two bones eventually comprise the upper jaw. This, of course, should not be surprising. Different types of teeth develop in the lower jaw without there being sutural demarcation between tooth classes. In fact, if there were special "properties," which "coded" for premaxillary teeth always becoming incisors, this still would not explain the development of different types of teeth in the maxilla. Let us return to the original discussion of heterodonty.

There are many mammals, fossil and extant, with teeth that we would on morphological grounds identify as incisors, canines, premolars and molars, and these teeth are in their expected positions in the upper and lower jaws. We could thus reasonably ascribe this condition to the ancestral mammal. There are also, however, many mammals which have teeth in the same positions; but the morphologies of these teeth do not correspond to the primitive mammalian tooth morphology/tooth position condition. Thus, for example, there are mammals with incisiform teeth in the premaxilla and others with perhaps the same number of teeth in the premaxilla; but one of the teeth in the latter group of mammals may be caniniform and the others premolariform. Are the teeth homologous because they are in the premaxilla or is it merely that mammalian premaxillae are homologous and teeth—*different teeth*—come to reside in the premaxillae of different mammals? I suggest it is the latter.

If an animal has its most caniniform tooth at the front of the jaw, this tooth is the canine, the incisors having been lost. In mammals with a caniniform tooth at the front of the jaw, this tooth is often followed by some number of premolariform teeth and then by the molars (e.g., soricids, various plesiadapiform-tarsiiform primates). I would identify these premolariform teeth as premolars, even though some may have come to lie in the premaxilla; one of these is the first tooth in the maxilla, and the remainder are in their "expected" positions. There are other mammals which possess some number of similarly incisiform teeth followed by some premolariform teeth and then the molars but lack a caniniform tooth (e.g., some fossil and extant erinaceids and dermopterans). I would identify these teeth as incisors, premolars, and molars, and conclude that the canine had been inhibited (i.e., "lost"). In cases where there are more teeth in one jaw than in the other (e.g., *Tarsius*, various artiodactyls), or where the antepremolariform teeth in one jaw are markedly different, morphologically, from the antepremolariform teeth in the other jaw (e.g., *Tarsius*, macropodid marsupials), I would identify the teeth of each jaw independently. Thus, artiodactyls which have an edentulous premaxilla and lack teeth immediately behind the premaxillary/maxillary suture, have only premolars and molars in the upper jaw and incisors, premolars, and molars in the lower jaw. Macropodid marsupials possess incisors, premolars, and molars in the upper jaw, but a canine, premolars and molars in the lower jaw.

Why Retain Three Tooth Classes?

Except for mammals which do not possess enough teeth for the identification of incisors, canines, premolars, and molars (e.g., rodents), mammals are typically thought to possess teeth representative of the three tooth classes. As reviewed above, these teeth are commonly believed to occur in the same positions in the jaws of all mammals. To explain, therefore, why incisors may be caniniform of premolariform or why canines may be incisiform or premolariform, Butler (1939, 1963) has suggested that there are incisor, canine, and molar "fields" which can expand or contract and thus can influence the shapes of developing teeth. By accepting that tooth position is constant among mammals and that drastic alteration of tooth morphology is the common expression of evolutionary change, one must conclude that "in many modern mammals shape and size are no longer reliable bases for dental homology" (Butler 1978:218, citing Owen 1868).

In contradistinction to the stabilist theory of tooth position, Osborn (1978) has argued against the "maxillary suture system" of tooth identification. Rather, Osborn would identify "tooth classes, by the boundaries between gradients of shape, and then individual teeth" (1978:190). A tooth class is generated by a clone which determines the shape and number of teeth in that tooth class. But, while Osborn's Clone Model differs from

Butler's Field Model in how tooth shape, tooth number and tooth position are controlled, the two models are similar in that they both embrace the belief that virtually all mammals develop incisor, canine, and molar tooth classes. Thus, while Osborn remonstrates adherents of the Field Model for relying on tooth position and ignoring tooth shape, his adherence to the stability of three tooth classes leads, for example, to the identification of the most caniniform tooth in the upper jaw of the talpid *Desmana* (the Russian desman) and the shrew *Sorex* as an incisor, because it is the first tooth in the jaw. Certainly, there is a morphological "break" between the caniniform anterior tooth of *Desmana* and *Sorex* and the smaller teeth behind it. But, is it a "break" between "incisor" and thus "canine" tooth classes, or is it between canine and molar tooth classes?

Both the Field and Clone models begin (at least in part) with the recognition of three tooth classes in various Cretaceous mammals. Each then presumably assumes that because these "early," "primitive" mammals possessed the genetic constitution for this particular type of heterodonty, *all* mammals must have retained the elements of incisor, canine and molar tooth classes. But, is the "heterodonty" of a shrew (with a caniniform anterior tooth followed by some number of premolariform teeth and then molars) homologous with the "heterodonty" of a canid (with incisiform, caniniform, premolariform and molariform teeth in their "expected" positions)? Clearly, if evolution involves change, then a viable conclusion is that we cannot equate "heterodonty in mammals" with "all heterodont mammals possess incisors, canines and molars." Some mammals retained the primitive mammalian disposition of three tooth classes, in particular the positions relative to the premaxilla and maxilla, although the number of teeth in a class may be reduced. Other mammals "lost" a tooth class or tooth classes and developed the remaining teeth along the jaws. The data already citied on anterior tooth development in *H. sapiens*, on the tree shrew and on the mouse clearly demonstrate that teeth of differing morphologies can migrate to the front of the jaw.

Determination of Tooth Shape and Position

Both the Field and Clone models and the suggestions proposed here attempt to explain heterodonty. We are all aware of the fact that tooth development involves the interaction of neural crest-derived ectomesenchyme and the oral epithelium (e.g., Kollar and Baird 1969, 1970; Milaire 1959; Miller 1971; Pourtois 1961). Furthermore, as Miller (1978) and Tonge (1976) point out, the timing of the interaction between ectomesenchyme and epithelium is equally important for the initiation of tooth development. On the other hand, how is tooth shape determined? Butler (1978) suggests that the tooth shape determining properties of ectomesenchymal cells are dependent on the influence of the "fields" into which these cells migrate, while Osborn's (1978) model proposes that these cells have differentiated in shape producing potential by the time they enter the jaws.

As reviewed above for *H. sapiens*, the canine and typically the lateral incisor originate over the maxillary prominence; if this does not also sometimes characterize the central incisor, this tooth begins development posteriorly over the median nasal prominence. These three teeth end up in the premaxillary region. For the Field Model to account for this phenomenon as well as the development of incisiform, caniniform and molariform teeth in *H. sapiens*, the incisor and canine "fields" must also be farther back in the presumptive jaws and either migrate forward with the tooth germs or dissipate after the ectomesenchymal cells have been organized. If the incisor "field" does not retreat anteriorly or "shut down," the forwardly migrating canine tooth germ would come under its influence and *H. sapiens* would presumably develop three incisiform teeth. If on the other hand, the ectomesenchymal cells had differentiated into "incisor-producing," "canine-producing" and "molar-producing" primordia (*sensu* Osborn 1978) prior to invading the presumptive jaws, the tooth "type" would be unaffected by changing positions during development. That tooth development in *H. sapiens* (and also at least in *Tupaia* and the mouse) begins relatively far back in the presumptive jaws ties in nicely with Kollar and Lumsden's (1979) and Lumsden's work on innervation (this volume, pers. comm.). Precisely, maxillary and mandibular nerve fibers contact or penetrate the oral epithelium and thus form pathways, along which neural crest cells can migrate.

If we review Ooë's (1957) work in light of Kollar and Lumsden's and Osborn's Clone Model, we see that the central incisor, canine and anterior deciduous molars are initiated first. These sites would presumably have been "located" by nerve fiber endings and by ectomesenchymal cells deposited therein via these nerve fibers. At later stages, the presumptive lateral incisor and then the presumptive posterior deciduous molar swellings appear. This could reflect the "cloning" of the incisor and molar tooth classes.

The model of tooth identification I have suggested proposes that not all mammals have retained the primordia for three tooth classes. Restated in terms of development, the ectomesenchymal cells determining a tooth class either do not differentiate or do not make contact with or successfully interact with the oral epithelium. Above, for example, I suggested that the anterior caniniform tooth of soricids (shrews) is a canine, behind which are premolars, and then molars. Soricids would thus have "lost" the incisor tooth class and develop only the canine and molar tooth classes.

The best available information on tooth development in soricids comes from Kindahl's (1959a) work. Although the earliest stage of tooth development Kindahl presents for *Sorex* shows five variously developed buds, the first and the last are the most advanced. The first will become the "caniniform anterior incisor" and the last will become the last "premolar" (see below). Between these two most developed buds are smaller buds corresponding to what will become three premolariform teeth: the so-called second incisor, canine, and P3; at this earliest stage of Kindahl's sample "the P3 is not distinctly separated from the next one" (1959a: 205) (i.e., the last "premolar"). One can thus reasonably extrapolate backward and suggest that there were originally two swellings, one corresponding to the "caniniform incisor" and the other to the stem progenitor (*sensu* Osborn 1978) of the molar tooth class. In comparison with initial tooth development in *H. sapiens* (in which there is an epithelial swelling corresponding to an incisiform tooth, a caniniform tooth and a molar tooth, i.e., *three* initial "primordia"), in *Sorex* there are only *two* (corresponding to a caniniform tooth and a molar tooth). If we take the three initial epithelial swellings in *H. sapiens* as representing incisor, canine, and molar tooth classes (and, indeed, *H. sapiens* possesses recognizably incisor, canine and premolar/molar teeth), then we can only conclude that *Sorex* develops but two "centers" of tooth development because it only has two tooth classes represented in its dentition: one of the original (= primitive mammalian) three has been "lost." Since the two most advanced tooth buds noted in the development of the teeth of *Sorex* correspond morphologically to a tooth representative of a canine tooth class and a tooth representative of a molar tooth class, I conclude that the anterior-most tooth *is* a canine behind which are "premolars" and molars (= the molar tooth class), and that the incisor class has been "lost."

The "Adult" Dentition: Always "Permanent" Teeth?

I raise this question because it is of prime importance to the study of dental homology if teeth are primary or secondary. In clearcut cases where, postnatally, one set of teeth is replaced by another in a one-to-one relationship, the terms primary and secondary are synonymous with "deciduous" and "permanent." As we all know, the molars of an adult animal are primary teeth, but are referred to as "permanent" teeth because they are not shed. From paleontological evidence we know that mammal-like reptiles replaced their molars (Osborn 1974; Osborn and Crompton 1973) and that suppression or inhibition of these successors yielded the single set of "permanent" molars of therian mammals. Ooë (1979) has recently described abortive "successors" of M1 and M2 in *H. sapiens*. It may just seem like an issue in semantics, but it might be more productive to refer to therian "permanent" molars as retained deciduous teeth.

The distinction of "primary and secondary" from "deciduous and premanent" teeth takes on especial importance when antemolar teeth are involved. It is commonly accepted that in the adult animal, all antemolar teeth are permanent teeth, except in some cases when the "P1" is a retained deciduous tooth (Butler 1978; Ziegler 1971); because this tooth is retained in the adult and is unreplaced it is, by definition, a "permanent" tooth. But, if the other antemolar teeth are secondary teeth and are also referred to as "permanent" teeth because they are not shed, the details relevant to understanding homology are obscured. If, in one animal, the "P1" is a retained deciduous tooth (i.e., is a primary tooth) and, in another animal, the "P1" is a secondary tooth, comparisons between these taxa because in each case the "P1" is a "permanent" tooth are meaningless. Yet all too often, the comparison is made and statements of homology and phylogeny are generated on the basis of the fate of a tooth rather than on its developmental context.

A good example comes from Kindahl's (1959a) study of *Sorex*. Quite often, *Sorex* has been cited as a mammal whose deciduous dentition is represented by vestigial teeth and whose functioning dentition is the permanent dentition. If the "deciduous" dentition of *Sorex* corresponds to the "primary" dentition and the "permanent" to the "secondary" dentition, then among other things, this certainly has profound implications for understanding tooth loss. One would otherwise think that, because a successor tooth develops from the dental lamina or the external dental epithelium of its predecessor primary tooth (e.g., Tonge 1976), tooth loss would proceed by inhibiting the successor tooth, as is the case in Ooë's (1979) study of M1 and M2 in *H. sapiens*. While it is true that the vestigial teeth of *Sorex* are shed *in utero* (and, by definition, are thus "deciduous" teeth), and the set of teeth which erupt and function are not replaced (and by definition are thus "permanent" teeth), developmentally, the "permanent" teeth are primary teeth and the "deciduous" teeth are aborted successors.

As seen in figure 10.2, the earliest embryo Kindahl (1959a) studied had five variously developed tooth buds. At a later stage, two of these buds have moved apart (the "I2" and "C") and a new bud ("I3") appears between them; M1 has also developed. This, with subsequent development of M2 and M3, is the primary dentition. Stemming from the side of "I1," "I2," "C," "P3" and "P4" are the swellings that will become the vestigial teeth. These represent the aborted secondary dentition, and indeed Kindahl's transverse section through the developing "I1" and "Id1" corresponds perfectly to the way in which a successor tooth (the so-called Id1) develops from its predecessor tooth (the so-called I1) (fig. 10.3).

A similar, but in the end, more complicated development characterizes the "premolars" of marsupials (Archer 1974; Berkovitz 1966, 1968). The

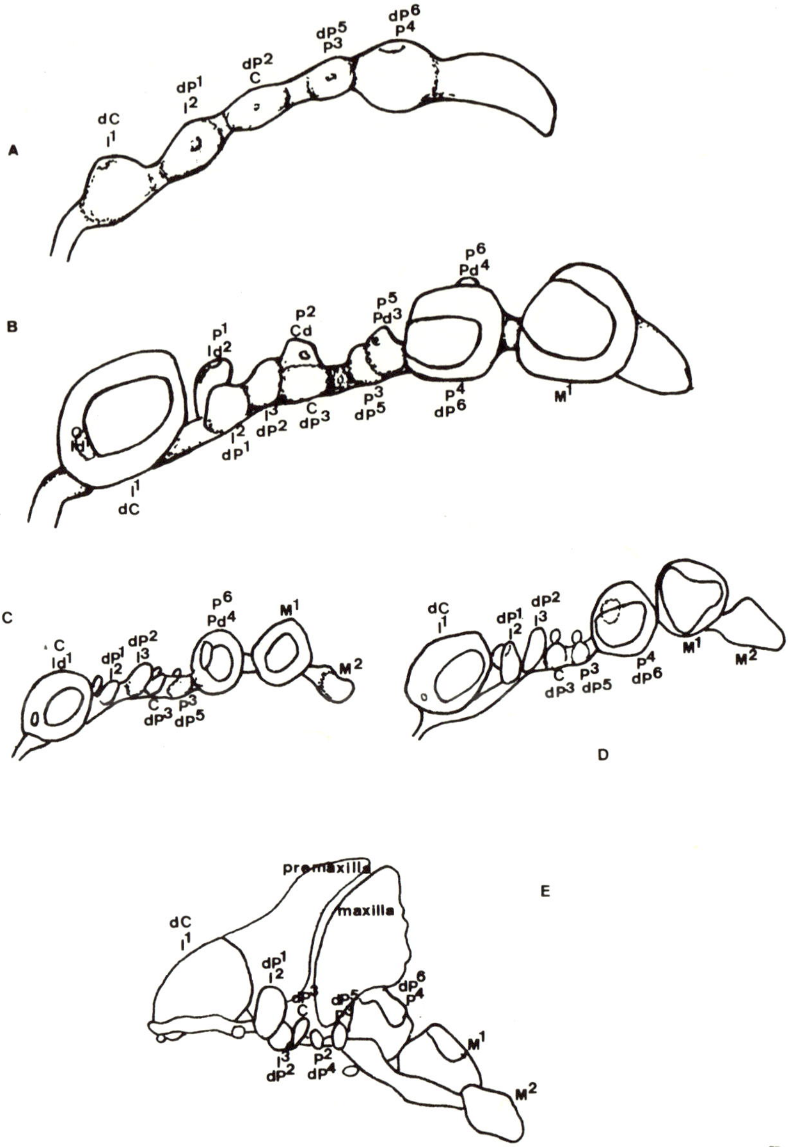

Figure 10.2. Increasingly advanced stages of dental development in *Sorex* (Kindahl 1959a). Tooth bud identification nearest bud is from Kindahl; reidentification is above Kindahl's. In *c* and *d*, only the primary dentition is labeled. Note that between dC (Kindahl's I^1) and M^1 there develop six tooth buds labeled, respectively, dp^1–dp^6. This developmental configuration in the upper jaw is not uncommon amongst insectivores.

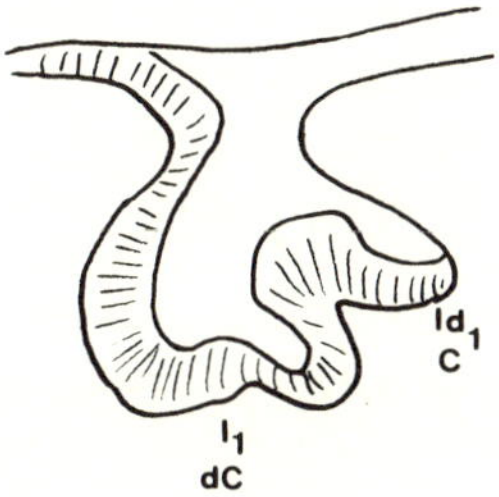

Figure 10.3a. Cross-section through primary tooth bud showing lateral development of aborted successor tooth in *Sorex* (redrawn from Kindahl 1959a). Kindahl's identification closer to figure outline; reinterpretation below.

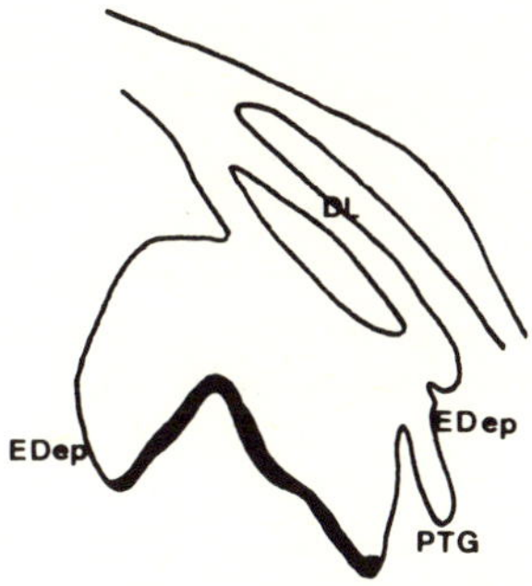

Figure 10.3b. Cross-section through human primary lower incisor bud showing development of its successor from it (redrawn and reversed from Tonge 1976). Comparison with tooth bud configuration in *a.* is compatible with present author's interpretations. PTG = permanent tooth germ; EDep = external dental epithelium; DL = dental lamina.

"P1" and "P3" are "permanent" teeth in that they are not shed but develop as primary teeth along with the "dM4"; this latter tooth is developmentally a primary tooth which is later shed. During further growth the "P3" and "dM4" buds move apart and another bud, the presumptive "P4," develops. It is this tooth, the so-called P4, which replaces "dM4" and comes to be the posterior of three marsupial "premolars." In effect, this is the same phenomenon seen in the development of the premolariform teeth of *Sorex*: tooth buds move apart and others develop in the available interstices. In the case of marsupials where there is replacement of "dM4," the "successor" tooth is also a primary tooth. In fact, it replaces the tooth developmentally posterior to it.*

While soricids (other shrews studied by Kindahl [1959a] are similar to *Sorex*) and marsupials have a "permanent" dentition composed of primary teeth, other taxa have a mixed dentition. For example, in the hedgehog, *Erinaceus europeus*, the "I3," "C," and "P2" in the upper and lower jaws and the lower "P3" develop as primary teeth (cf. Kindahl 1959b); these teeth have been identified as "permanent" teeth because they are not shed. Similar to the situation in soricids, *Erinaceus* develops from these teeth

*After this paper was completed, I received Archer's (1978) article on the homology of marsupial cheek teeth in which he identifies the premolars in the adult as P1–P3 and the molars as M2–M5. He identifies the only cheek tooth shed as M1 and refers to the premolars as permanent teeth because they are not. Clearly, all cheek teeth are primary teeth.

successor anlagen which are shed or resorbed *in utero*. (Kindahl's [1959b] fig. 4 clearly shows the so-called Cd developing from the lamina of "C;" the "C" is the primary tooth and the "Cd" its [eventually aborted] successor.) The remaining "incisors" and "premolars" are represented by primary teeth which erupt and are subsequently replaced by functioning secondary (successor) teeth. Thus, although all antemolar teeth are "permanent" in the sense that they are not shed, the antemolar dentition of *Erinaceus* consists of primary and secondary teeth.

While it is possible to give further examples, suffice it to say that the above-mentioned taxa illustrate clearly the fallacy in equating "permanent" with "secondary" teeth. Since comparative studies and ensuing theories of phylogenetic relationships among mammals are based on presumed homologies, it is imperative that at least the same developmental series be compared. Granted, this may be difficult to ascertain for fossil taxa—fossilized juveniles are a rare find—but thoughtful comparisons with extant taxa may prove more fruitful than ignoring the possibility that an animal may have had a mixed or totally primary antemolar dentition.

Can a Canine Be "Incisiform"?

Throughout, I have implied that a tooth should be identified as that tooth whose idealized morphology it most closely resembles. For example, a canine is "ideally" a simple, robust, trenchant tooth, as in a canid. The most caniniform tooth in a microsyopid, an Eocene primate, conforms to this picture, although it is at the front of the jaw. But, for example, the upper anterior tooth of *Sorex*, which I have identified as a canine, has a distinct lingual cusp. This tooth is still the most robust and trenchant tooth in the jaw, and it is also a primary (retained deciduous) tooth. In *H. sapiens*, the deciduous (= primary) upper canine typically has a pronounced lingual cusp and may even have marked lingual marginal swellings which appear as accessory cusps. This latter configuration is similar to the morphology of the anterior upper tooth of various plesiadapiform primates which, on the basis of other data, I have previously identified as a canine (Schwartz 1978a, 1980).

But, what of such things as "incisiform canines" or "caniniform premolars?" To address the former, we can compare the lower anterior teeth of an artiodactyl (e.g., a bovid, cervid or caprine) with those of a lemur or loris (fig. 10.4a,b). The artiodactyl lower anterior dentition is composed of eight teeth of the same morphology which we can describe as incisiform; there is no morphological "break" between the right or left first and second, second and third, or third and fourth. On the other hand, the lemur/loris lower anterior dentition (referred to as a "toothcomb") is

Figure 10.4a. Stylized half lower anterior dentition of a bovid showing identical morphology of all teeth, the lateralmost of which is commonly identified as an "incisiform canine."

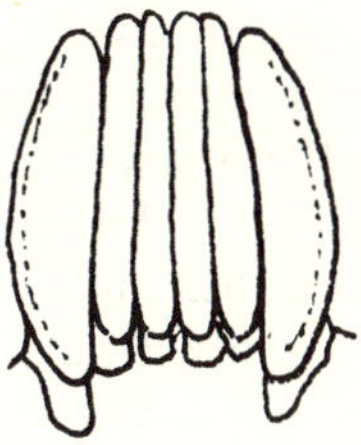

Figure 10.4b. Typical six-toothed lemur/loris toothcomb. Note morphological distinctiveness of lateral tooth.

composed of six teeth, the central four of which are morphologically similar to each other and noticeably morphologically different from the subtending lateral teeth; there is a morphological "break" between the central unit of teeth and the lateral tooth. The lateral tooth is more robust, deeper, flares laterally and has a raised lateral ridge (Schwartz 1974a, 1978b), and in these attributes is most comparable to the less procumbent caniniform tooth of other primates (Schwartz 1980). With regard to tooth growth and eruption, the lower anterior teeth of artiodactyls proceed sequentially antero-posteriorly (personal observations). In contrast, the central teeth of the lemur/loris toothcomb are markedly ahead of the lateral set in growth and eruption (Schwartz 1974a, 1978b); there is thus another "break" distinguishing the lateral from the central teeth of the toothcomb. I conclude, therefore, that the lateral tooth of the lemur/loris toothcomb is not serially homologous with the central teeth; that it belongs to a different tooth class; that, because it compares favorably with the most caniniform tooth of other primates, it is a canine.

In contradistinction, the lower anterior teeth of artiodactyls are "serial" in morphology, relative rates of growth, and eruption. I conclude, therefore, that artiodactyls with four "incisiform" teeth possess four incisors and have "lost" the canine. The paleontological "data" need not contradict this. Fossil artiodactyls ("artiodactyl" primarily by virtue of sharing derived features of the ankle) (e.g., Romer 1966) with three incisor teeth and a demonstrable canine tooth, rather than representing the "primitive" condition from which four "incisored" taxa evolved, have retained the canine but are derived in their own right in having lost an incisor.

Canines may be large or small, orthally implanted or procumbent, retained deciduous (= primary) or permanent teeth, but if they are present, they should be morphologically identifiable.

"Caniniform" Premolars?

Toothcombed primates (broadly, lemurs and lorises) are commonly described as having a caniniform premolar (e.g., Butler 1978; Gregory 1922; Hill 1953; LeGros Clark 1962). Such a description is typically given to the lower anteriormost premolar, which is the first tooth behind the toothcomb. The forked lemur, *Phaner*, has an elongate tooth immediately behind the uppermost caniniform tooth, whereas in the majority of lemurs and lorises, the tooth behind the upper canine is much smaller and its premolariformity more readily observable (e.g., Tattersall and Schwartz 1974). What, however, is the nature of the oft-described "caniniform" premolar?

As discussed above, the lower anterior dentition of lemurs and lorises is to some extent procumbent; the long axes of these teeth thus approach the axis of the mandibular corpus, effectively distinguishing themselves from the orthally implanted cheek teeth. One observes a marked morphological "break," augmented further in some cases by a small diastema, which isolates and focuses attention upon the anterior premolar in particular. This is true even if the anterior premolar does not rise much higher than the premolars behind it (see illustrations in Tattersall and Schwartz 1974). In those few taxa (e.g., the slow loris, the potto) with a "caniniform" anterior premolar, features of "premolariformity" (e.g., broad base culminating in a swollen posterior heel; an elongate, teardrop-shaped basin at least posteriorly which originates from the apex of the crown) are as discernible as they are in taxa with a shorter anterior premolariform cheek tooth. The "caniniformity" derives, therefore, not from a transformation or obliteration of premolar characters, but from an increase in crown height (fig. 10.4c). The same obtains to the "caniniform" anterior upper premolar of *Phaner*. Furthermore, it is commonly observed amongst toothcombed

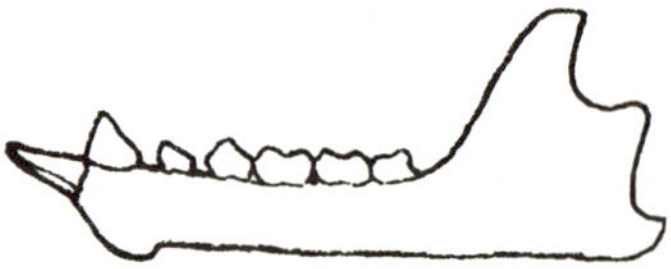

Figure 10.4c. Lateral view of typical lemur/loris lower jar. With the anterior teeth procumbent, the height of the anterior premolar is emphasized.

lemurs and lorises that the three premolariform teeth, regardless of the size of the anterior tooth, grow and erupt in the same sequence, i.e., anterior ⟶ posterior ⟶ middle (Schwartz 1974b, 1975).

Degrees of Variability

The foregoing suggestions and interpretations underlie a model of heterodonty based on flexibility in the system. Rather than seeing virtually all mammals as retaining and developing incisor, canine, and molar tooth classes, it is suggested that this is the basic ground plan, with some mammals developing all three tooth classes while others "lose" (= inhibit, do not achieve the competence to develop), for example, the incisor or canine class. It is suggested that this obtains to taxa with completely dentulous jaws as well as to seemingly more obvious taxa such as rodents and other similarly diastematic mammals. Further, in cases where the incisor class or the incisor and canine classes have been "lost," the teeth of the remaining classes can develop and erupt anteriorly in the jaws. One can characterize these aspects of heterodonty as "horizontal flexibility."

"Vertical" components of heterodonty include: the series to which the tooth belongs (primary or secondary); and the size of the tooth. With regard to the first component, the antemolar dentition of an adult mammal may be composed entirely of primary (retained deciduous) teeth (e.g. soricids); may be a composite of primary and secondary teeth (e.g. *Erinaceus*), or may be completely represented by secondary teeth (e.g., *H. sapiens*). Since we know that there tend to be differences in morphological complexity and/or size between a primary tooth and its successor, the persistence of one or the other in the "adult" dentition can contribute to the "variability" one sees in dentitions as a whole as well as to the "variability" within a tooth class. Concerning the second, superimposed upon the development of a tooth, whether it is a primary or secondary tooth, is the size that a tooth attains in different taxa. Thus there are differences among taxa in height, gracility, length, width, etc. of incisors, canines, premolars and molars.

While we may not know at this time how ectomesenchyme becomes differentiated to generate incisor, canine and molar classes, it is sufficient to accept that it does. Given the examples cited above of forward migration of developing tooth germs and the morphological approach to tooth identification proposed here, I, like Osborn (1978), suggest that ectomesenchyme enters the presumptive jaws already differentiated into incisor, canine and/or molar stem cell masses. Rather than there being one cell mass per tooth, there is one cell mass per tooth class which divides into the number of teeth that will comprise that tooth class. This can be appreciated, for example, not only in the posteriorly directed "cloning" of the (permanent) molars (Lumsden 1979) but in the spreading apart of "primary premolar" buds and the development in the interstices of other buds of that "set" (see above). Because of the latter developmental phenomenon, which is not directional, I would modify Osborn's (1978) model of cloning. The "variability" of shape and size (other than differences due to whether or not a tooth is primary or secondary) of incisors, canines and (premolars)/molars among mammals could result from the timing of ectomsenchymal migration and the competence of the cell mass (cf. Miller 1978; Osborn 1978). If we take dental sexual dimorphism as an example of variability in shape and size, differential mitotic rates and duration of amelogenesis (Moss 1978), as well as the effect of hormones (Noble and Pope 1929), emerge as additional influencing factors. Furthermore, as Osborn (1971) has suggested, elements of the developing teeth may affect the development of others.

These same latter influences would affect tooth shape and size if ectomesenchyme entered the presumptive jaws undifferentiated. According to the Field Model, ectomesenchyme is located at certain tooth forming positions, and the shape which a tooth will have is determined by the field acting upon the developing tooth germ (Butler 1978). Thus, some mammals have a caniniform tooth at the same position where other taxa have a premolariform tooth. An artiodactyl with four lower incisiform anterior teeth had its canine tooth position incorporated into the incisor field. But, even if tooth positions were programmed, we are talking about positions at which teeth develop, and not specific teeth developing at specific positions. An undifferentiated mass of ectomesenchyme, located four "positions" from the front of the jaw and "organized" to produce a tooth virtually identical in incisiformity to the three teeth anterior to it, is producing an incisor.

Conclusions

While the preceding has argued for a morphological identification of teeth, more specific details of homology (e.g., this tooth is P2 in X taxa) have not been broached. This is due partly to the enormity of the task but also to problems at this time in determining how many teeth in the incisor, canine and molar classes are primitive for mammals. For example, recent discovery of fossil Eutheria with five premolar teeth (Clemens 1973; Lillegraven 1969; McKenna 1975) and the collateral re-interpretation of the dentitions of various primates, insectivores and dermopterans (Schwartz 1978a, 1980; Schwartz and Krishtalka 1976, 1977) contradict the long held view that the primitive condition is four premolars. My preliminary review of extant insectivores leads me to believe that some may even have six "premolars" (e.g., fig. 10.2).

If we could reconstruct the dental complement of the ancestral mammal, could we identify the homologies in other mammals? It certainly would be helpful if, as is commonly believed, mammals lost teeth nearest the canine, i.e., incisors posteroanteriorly and premolars anteroposteriorly.

However, what with "primary premolars" developing between the separated buds of others, it would seem that a simple mechanism for tooth loss in this region would be the inhibition of "interstitial" tooth development. We also know that some individuals of the late Cretaceous *Gypsonictops* reduced premolar number from four to five by loss of the third of this set (Clemens 1973: Lilligraven 1969). And, from study of tooth replacement and sequences of growth and eruption correlated with morphology, it has been suggested that various primates have lost the first and third teeth of a primitively five premolared set (Schwartz 1978a, 1980). There is also the possibility, as has recently been re-discussed by McKenna (1975), Osborn (1978) and Schwartz (1978a), that premolar/molar loss could occur by inhibition of the ultimate permanent premolar and last molar with retention of the ultimate deciduous molar as the first of the molar set. While Butler (1978) has commented that there is no paleontological evidence to support this last possibility, it should be pointed out that there is little paleontological evidence of any tooth loss—only the observation that some taxa have fewer teeth than others. As it is doubtful that all mammals will retain the same teeth, or the same teeth in the same positions in the jaw, I believe it is equally untenable to expect that teeth will be lost in the same orderly fashion among mammals.

One of the attractions of the Field Model and the traditional approach to tooth identification is the rigidity of the rules, which may not allow one to identify isolated teeth with certainty, but which impose some order on what otherwise appears disorderly. Given an upper and lower jaw, one can always identify the teeth contained therein. Faced with the suggestion that upper and lower caniniform teeth at the front of the jaw are canines, it is understandable how, from a traditional stabilist point of view, one could counter this by saying that the lower canine never occludes behind the upper canine (Cartmill and Kay 1978), or that lower canines never meet side-by-side (Gingerich 1974:277). Or, given the suggestion that a mammal lacks a canine because it lacks a caniniform tooth, the response could be that we know canines become premolariform because some taxa have a premolariform tooth in the same position in which other taxa have a caniniform tooth (Butler 1978). Given the traditional rules, these are reasonable responses. But, I doubt that these are the rules by which teeth play.

References

Andersen, H. and M. Matthiessen. 1967. Histochemistry of the early development of the human central face and nasal cavity with special reference to the movements and fusion of the palatine processes. *Acta Anat.* 68:483–508.

Archer, M. 1974. The development of the cheek-teeth in *Antechinus flavipes* (Marsupialia, Dasyuridae). *J. R. Soc. W. Austral.* 57:54–63.

——1978. The nature of the molar-premolar boundary in marsupials and a reinterpretation of the homology of marsupial cheekteeth. *Mem. Qd. Mus.* 18:157–164.

Berkovitz, B.K.B. 1966. The homology of the premolar teeth in *Setonix brachyurus* (Macropodidae: Marsupialia). *Archs. Oral Biol.* 11:1371–1384.

——1968. Some stages in the early development of the post-incisor dentition of *Trichosurus vulpecula* (Phalangeroidea: Marsupialia). *J. Zool. Res.* 154: 403–414.

Butler, P.M. 1939. Studies of the mammalian dentition. Differentiation of the postcanine dentition. *Proc. Zool. Soc. Lond.* 109:1–36.

——1963. Tooth morphology and primate evolution. In D.R. Brothwell, ed., *Dental Anthropology*, pp. 1–13. Oxford: Pergamon Press.

——1978. The ontogeny of mammalian heterodonty. *J. Biol. Buccale* 6:217–227.

Cartmill, M. and R. Kay. 1978. Craniodental morphology, tarsier affinities, and primate suborders. In D.J. Chivers and K.A. Joysey, eds., *Recent Advances in Primatology*, vol. 3., *Evolution*, pp. 205–213. London: Academic Press.

Clemens, W.A. 1973. Fossil mammals of the Type Lance Formation. Part III. Eutheria and summary. *Univ. Calif. Publ. Geol. Sci.* 94:1–102.

Gaunt, W.A. 1964. The development of the teeth and jaws of the albino mouse. *Acta Anat.* 57:115–151.

Gingerich, P.D. 1974. Cranial anatomy and evolution of early Tertiary Plesiadapidae (Mammalia, Primates). Ph.D. dissertation, Yale University.

Gregory, W.K. 1922. *The Origin and Evolution of the Human Dentition.* Baltimore: Williams and Wilkins.

Hill, W.C.O. 1953. *Primates, Comparative Anatomy, and Taxonomy,* vol. 1, *Strepsirhini*. Edinburgh: Edinburgh University Press.

Kindahl, M.E. 1957. On the development of the tooth in *Tupaia javanica*. *Ark. Zool.* 10:463–479.

——1959a. Some aspects of the tooth development in Soricidae. *Acta. Odont. Scand.* 17:203–237.

——1959b. The tooth development in *Erinaceus europaeus*. *Acta Odont. Scand.* 17:467–489.

——1967. Some comparative aspects of the reduction of the premolars in the Insectivora. *J. Dent. Res.* 46 (suppl.):805–808.

Kollar, E.J. and G.E. Baird. 1969. The influence of the dental papilla on the development of tooth shape in embryonic tooth germs. *J. Embryol. exp. Morph.* 21:131–148.

——1970. Tissue interaction in embryonic mouse tooth germs. II. The inductive role of the dental papilla. *J. Embryol. exp. Morph.* 24:173–186.

Kollar, E.G. and A.G.S. Lumsden. 1979. Innervation and tooth initiation in the mouse embryo. *J. Biol. Buccale* 7:49–60.

LeGros Clark, W.E. 1962. *The Antecedents of Man*, 2d ed. Edinburgh: Edinburgh University Press.

Lillegraven, J.A. 1969. Latest Cretaceous mammals of upper part of Edmonton Formation of Alberta, Canada, and review of marsupial-placental dichotomy of mammalian evolution. *Univ. Kansas Paleontol. Contr.*, Art. 50, 12:1–122.

Lumsden, A.G.S. 1979. Pattern formation in the molar dentition of the mouse. *J. Biol. Buccale* 7:77–103.

McKenna, M.C. 1975. Toward a phylogenetic classification of the Mammalia. In W.P. Luckett and F.S. Szalay, eds., *Phylogeny of the Primates*, pp. 21–46. New York: Plenum Press.

Milaire, J. 1959. Prédifferenciation cytochimique de diverses ébauches céphaliques chez l'embryon de Souris. *Arch. Biol.*, Liège 70:587–730.

Miller, W.A. 1971. Early dental development in mice. In A.A. Dahlberg, ed., *Dental Morphology and Evolution*, pp. 31–44. Chicago: University of Chicago Press.

——1978. The dentitions of tabby and crinkled mice (an upset in mesodermal: ectodermal interaction). In P.M. Butler and K.A. Joysey, eds., *Development, Function, and Evolution of Teeth*, pp. 99–109, New York: Academic Press.

Moss, M.L. 1978. Analysis of developmental processes possibly related to human dental sexual dimorphism. In P.M. Butler and K.A. Joysey, eds., *Development, Function, and Evolution of Teeth*, pp. 135–148. New York: Academic Press.

Nery, E.B., B.S. Kraus, and M. Croup. 1970. Timing and topography of early human tooth development. *Archs. Oral Biol.* 15:1315–1326.

Noble, G.K. and S.H. Pope. 1929. The modification of the cloaca and teeth of the adult salamander, *Desmognathus*, by testicular transplants and by castration. *Brit. J. exp. Biol.* 6:339–411.

Ooë, T. 1956. On the development of position of the tooth germs in the human deciduous front teeth. *Okaj. Folia Anat. Jap.* 28:317–340.

——1957. On the early development of the human dental lamina. *Okaj. Folia Anat. Jap.* 30:197–211.

——1979. Development of the human first and second permanent molar, with special reference to the distal portion of the dental lamina. *Anat. Embryol.* 155:221–240.

Osborn, J.W. 1971. The ontogeny of tooth succession in *Lacerta vivipara* Jacquin (1787). *Proc. Roy. Soc. Lond.* B179:261–289.

——1978. Morphogenetic gradients: fields versus clones. In P.M. Butler and K.A. Joysey, eds., *Development, Function, and Evolution of Teeth*, pp. 171–201. New York: Academic Press.

Osborn, J.W., and A.W. Crompton. 1973. The evolution of mammalian from reptilian dentitions. *Breviora* 399:1–18.

Pourtois, M. 1961. Contribution à l'étude des bourgeons dentairs chez la Souris. I. Périodes d'induction et de morphodifferenciation. *Arch. Biol.* 72:17–95.

Romer, A.S. 1966. *Vertebrate Paleontology*, 3rd. ed. Chicago: University of Chicago Press.

Schwartz, J. H. 1974a. Observations on the dentition of the Indriidae. *Am. J. Phys. Anthrop.* 41:107–114.

——1974b. Dental development and eruption in the prosimians and its bearing on their evolution. Ph.D. dissertation, Columbia University.

——1975. Development and eruption of the premolar region in prosimians and its bearing on their evolution. In I. Tattersall and R. W. Sussman, eds., *Lemur Biology*, pp. 41–63. New York: Plenum Press.

——1978a. Dental development, homologies, and primate phylogeny. *Evol. Theory* 4:1–32.

——1978b. Homologies of the toothcomb. *Am. J. Phys. Anthrop.* 49:23–29.

——1980. A discussion of dental homology with reference to primates. *Am. J. Phys. Anthrop.* 52:463–480.

Schwartz, J. H. and L. Krishtalka. 1976. The lower antemolar dentition in *Litolestes ignotus*, a late Paleocene erinaceid (Mammalia, Insectivora). *Ann. Carnegie Mus. Nat. Hist.* 46:1–6.

——1977. Reinterpretation of Picrodontidae (Primates: Plesiadapiformes): dental homologies and relationships. *Ann. Carnegie Mus. Nat. Hist.* 46:55–70.

Tattersall, I. and J. H. Schwartz. 1974. Craniodental morphology and the systematics of the Malagasy lemurs (Primates, Prosimii). *Anthrop. Pap. Amer. Mus. Nat. Hist.* 52:139–192.

Tonge, C. H. 1976. Morphogenesis and development of teeth. In B. Cohen and I. R. H. Kramer, eds., *Scientific Foundations of Dentistry*, pp. 325–334. London: William Heinemann Medical Books.

Ziegler, A. C. 1971. A theory of the evolution of therian dental formulas and replacement patterns. *Quart. Rev. Biol.* 46:226–249.

11. Quantitative Characterization of Teeth from Serial Sections using a Computer-Based Three-Dimensional Recording Technique

J. G. Clement and E. Dykes

ABSTRACT

A computerized three-dimensional recording technique is described which can give volumes of any structure after it has been serially sectioned. Preliminary results of a study in dental asymmetry are presented where the volumes of enamel, dentin, and pulp are compared for two pairs of contralateral human premolars.

Sullivan (1972) described a method for the study of jaw growth in dogs, using a computer-based three-dimensional recording technique. In his study, dogs which had been injected with tetracycline at known time intervals were sacrificed, and hemi-mandibles were embedded in a polyester resin block. Two parallel grooves and one diagonal groove were milled in one surface of each block so that when cut each section could be accurately related to the other sections by superimposition of the two parallel fiducial markers. The position of any particular section within the unsectioned block (z-axis parameter) was calculated from the position of the diagonal fiducial marker in relation to the two parallel fiducial markers. Each section was projected at 10 × magnification onto an x, y coordinate plotting table, and the positions of the three fiducial markers and outlines of bone and teeth were transcribed to punched paper tape. From the x, y coordinates of the traced outlines, Sullivan was able to compute the volumes of tissue laid down between the tetracycline markers.

The method described by Sullivan (1972) has been used and the programming extended by Kimura, Dykes and Fearnhead (1977) to compute the surface areas of the enamel and dentin-enamel junction of several human teeth. From their surface area and volume measurements, they concluded from the small number of teeth studied that although there was a large variation in size between teeth of the same type, the ratio of the enamel surface area to the dentin-enamel surface area was constant for a particular tooth type. A tooth of a different type had a different surface area ratio. As a further extension to the technique, Kimura, Dykes, and Fearnhead were able to program the computer to draw back on microfilm stereo pairs of the original tooth crowns and dentin-enamel surfaces, using the x, y, z coordinates of the traced outlines.

It was pointed out by Dykes and Clement (1979) that one of the problems encountered by Sullivan (1972) was that in addition to tracing the projected image of the section he had to photograph each section in ultra violet light in order to reveal the bone marking fluorescent dye. This photographic slide then had to be projected onto the x, y coordinate plotting table so that the fluorescent outlines could be traced. In the opinion of Dykes and Clement, any inaccuracies introduced in the photographing and subsequent projecting to produce an image sufficiently large for accurate tracing to be achieved could be eliminated by the construction of an x, y coordinate plotting microscope. They described the construction and use of such an instrument where tracings were achieved by movement of a microscope micrometer stage, thus eliminating the need to project the section or photographic slide. Furthermore, it is a simple procedure to substitute ultraviolet illumination for a white light source so that fluorescent lines can be traced directly.

This paper presents a few results on the comparison of volumes of enamel, dentin, and pulp of contralateral pairs of human premolars in an attempt to quantify dental asymmetry.

Material and Method

Contralateral pairs of developing human upper first premolars and lower first premolars, extracted for orthodontic reasons, were dehydrated over a period of one week in an ascending acetone/water series. The polyester resin used was mixed with a catalyst paste (1 g of paste to 25 ml resin)* and the teeth infiltrated using an ascending resin/acetone series over a further period of seven days. In the final of the two 100 percent resin

*Resin CL 20L (without pre-accelerator) supplied by Trylon Ltd., Thrift Street, Wollaston, Wellingborough, Northants. Catalyst paste 8 supplied by Automobile Plastics Co., 7 Henry Road, New Barnet, Herts.

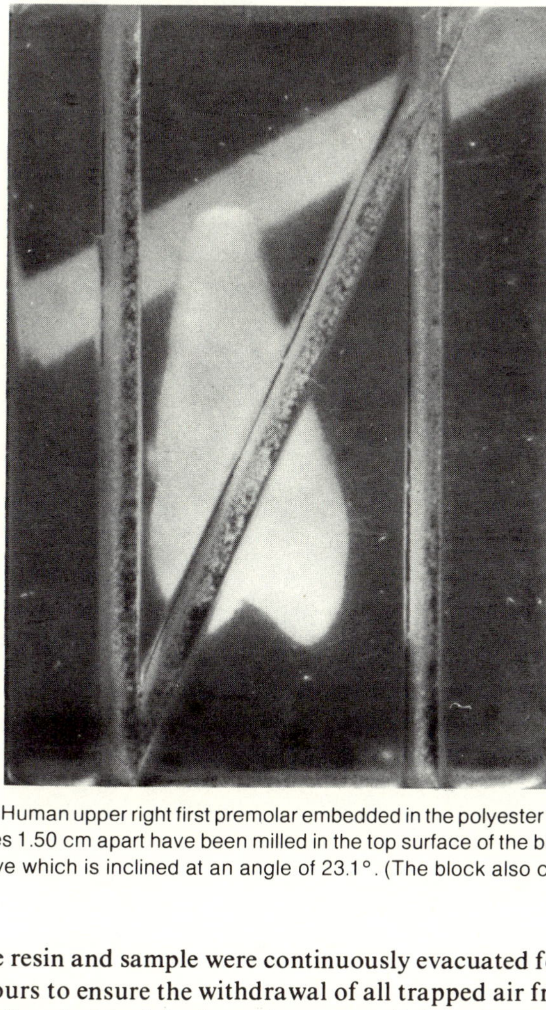

Figure 11.1. Human upper right first premolar embedded in the polyester resin block. Two parallel grooves 1.50 cm apart have been milled in the top surface of the block as well as a diagonal groove which is inclined at an angle of 23.1°. (The block also contains a paper label.)

changes, the resin and sample were continuously evacuated for a period of about six hours to ensure the withdrawal of all trapped air from the tooth and resin. After the block had cured, two parallel grooves and a diagonal groove were milled in one surface of the block using a 30° angle cutter (fig. 11.1). The whole block was then sectioned using a water-cooled rotating diamond disc. Thirty-two transverse sections (each about 200 μm thick) from each of the four premolar teeth used in this study were obtained (fig. 11.2).

Each section was traced using an x, y coordinate plotting microscope (Dykes and Clement 1979) which has a point to point resolution of 2 μm. In addition to tracing the contours of the enamel, dentin, and pulp, the positions of the three fiducial markers for each section were also recorded. At the magnification employed for tracing (40×) the apicies of the fiducial

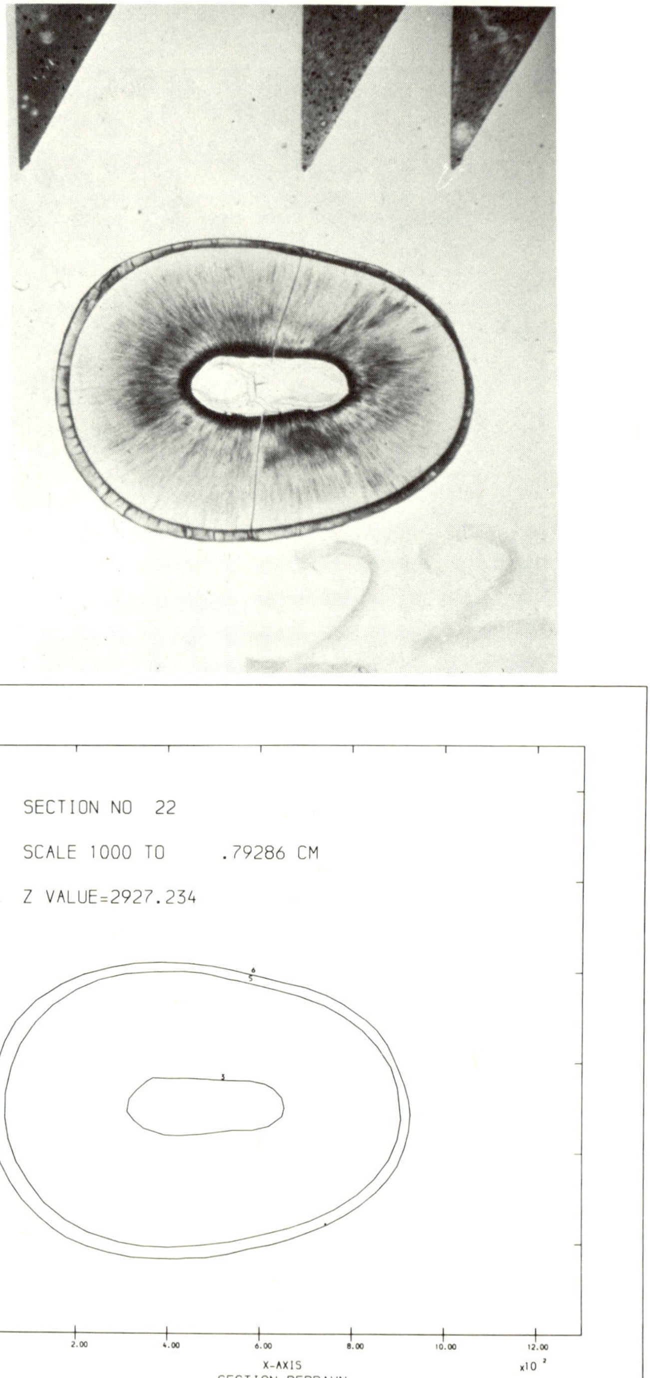

Figure 11.2. Transverse section (no. 22) from a human lower left first premolar together with a computer-generated drawing of the contours actually traced on this section using the x, y coordinate tracing microscope. The outline of the pulp (line 3) is defined by 26 x, y coordinates while those denoting the enamel-dentine junction (line 5) and enamel surface (line 6) consist of 45 and 55 x, y coordinate points respectively. The calculated postion of this section in the original block is denoted by the z value.

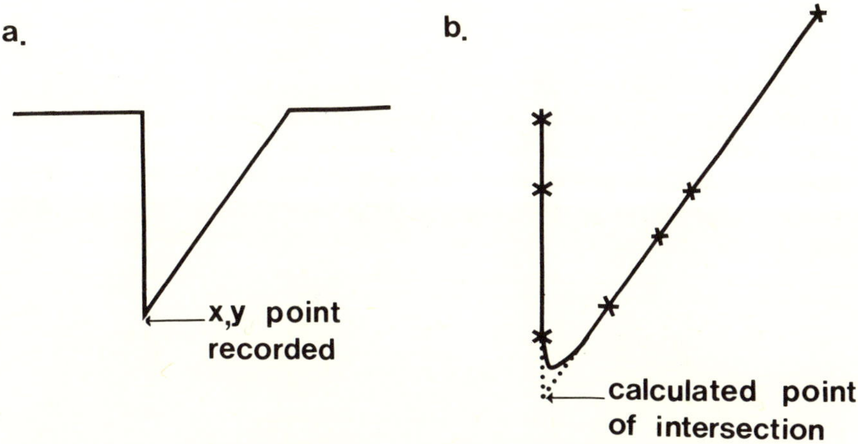

Figure 11.3. *a.* Schematic diagram of a low power view of one fiducial marker groove showing the position of the marker, whose x, y coordinate is recorded.

b. Schematic diagram of a magnified view of part of the marker shown in *a*. The true x, y position of the marker is achieved by plotting a number of x, y coordinates (minimum 2) which define the edges of the marker (crosses). The best straight lines through the recorded points are calculated using a least squares technique, and the calculated point of intersection of these two lines denotes the true x, y coordinate of any particular marker.

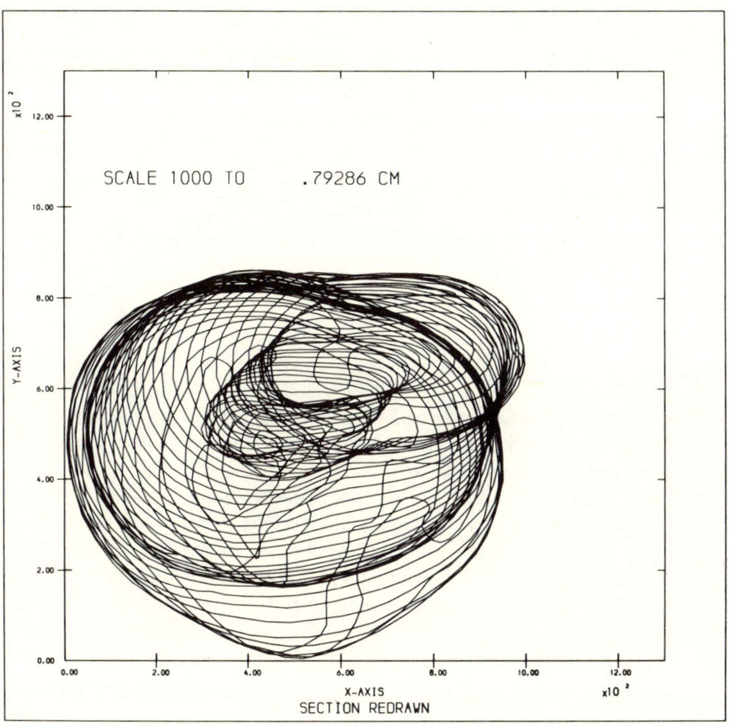

Figure 11.4. Computer-generated drawing of all the traced outlines from the 32 sections of the lower left first premolar after each section has been scaled, and rotated so that all the sections refer to a common origin.

markers appeared rounded, thus making selection of the x, y position of the marker uncertain (fig. 11.3). To overcome this uncertainty in the position of each fiducial marker, each side of a marker was traced as a straight line. In the computer processing of the data the equations of the best straight lines joining the x, y coordinates defining the sides of the marker were calculated, using a least squares fitting procedure and the true x, y position of the marker taken as the point where the two calculated lines intersect (fig. 11.3).

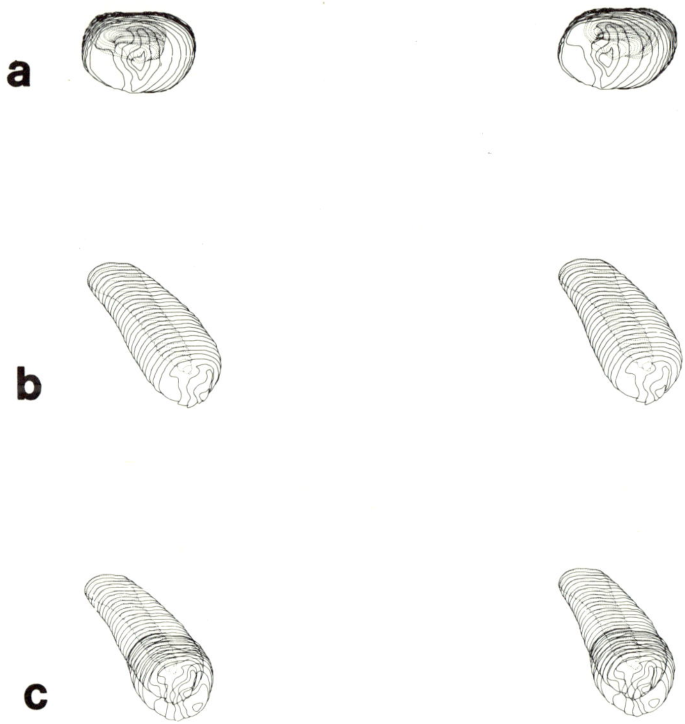

Figure 11.5. Computer-drawn stereo pairs (different scales) of several views of a human lower right first premolar compiled using the x, y, z coordinate data from the serial section tracings. In *a*, the transformed x and y axes lie in the x, y plane. Hidden line removal has been applied to each individual structure, but in order to see the relationship of one structure to another (i.e., relationship of pulp to dentin) hidden line removal has not been applied between structures. *a*. Normal view with no tilting showing dentin and pulp contours only. *b* The same as shown in *a* after a 30° rotation about the x-axis followed by by 30° rotation about the y-axis. *c*. The same view as shown in *b* but with the enamel surface contours included as well.

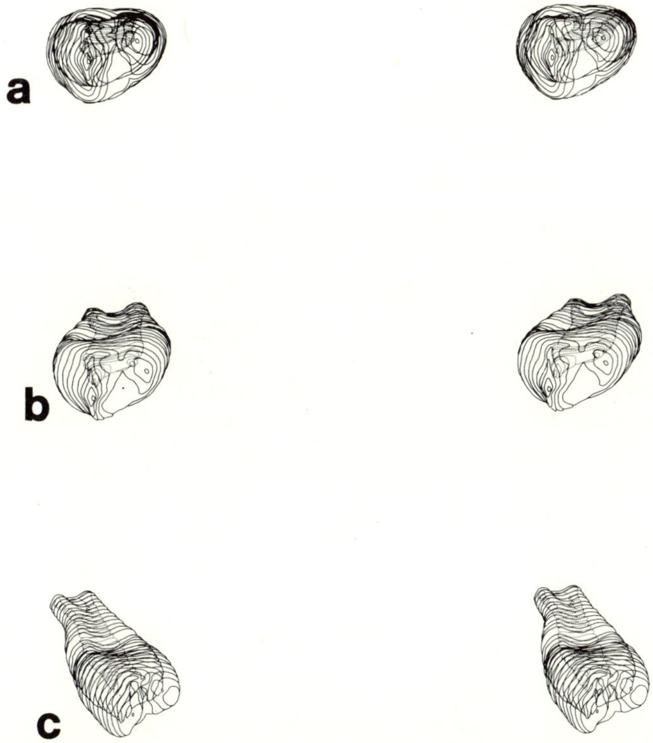

Figure 11.6. Computer-drawn stereo pairs (different scales) of several views of a human upper right first premolar. One section in the series was broken and therefore was not traced. *a.* Normal view with no tilting showing enamel, dentin, and pulp contours. *b.* The enamel contours have not been drawn and the tooth has been tilted 10° about the x-axis. *c.* All the traced contours are shown and the tooth tilted 30° about the x-axis followed by a 30° tilt about the y-axis.

Computer programs to process the data and calculate volumes from the serial section tracings have previously been described (Kimura, Dykes, and Fearnhead 1977).

Results

A typical section and the contours actually traced from that section using the x, y coordinate plotting microscope and subsequently redrawn on microfilm are shown in figure 11.2. Superimposition of all the contours traced from all the sections obtained from one tooth in their correct spatial

Table 11.1. Volumes (cc's).

Tooth	Enamel	Dentin	Pulp
4┘	0.148	0.294	0.027
└4	0.162	0.299	0.027
4┐	0.104	0.289	0.057
┌4	0.109	0.290	0.047

relationship to each other is shown in figure 11.4. Using the z-axis values for each section in addition to the x, y coordinates of the traced outlines, it is possible to draw stereo pairs of any or all the traced outlines, with or without hidden line removal, and from different viewpoints (figs. 11.5 and 11.6). The computed volumes for the enamel, dentin, and pulp for the two pairs of contralateral premolars are given in table 11.1.

Discussion

It is realized that not enough data is presented here to draw any firm conclusions concerning quantitative measurements of volume asymmetry in contralateral premolars. In addition to accurate volume measurements, which the described method is capable of giving, two important facts must be determined before a comparison of volumes of contralateral teeth can usefully be used in describing dental asymmetry quantitatively.

1. Do contralateral teeth start to develop at the same time?
2. Are the growth rates of enamel and dentin the same for contralateral teeth?

The computer-based three-dimensional recording technique described here is capable of supplying an answer to the second question provided there are clearly distinguishable landmarks present in both teeth. These can be either naturally occurring landmarks (e.g., pronounced incremental lines), accidentally deposited therapeutic fluorescent dyes (e.g., tetracycline) or experimentally induced landmarks (e.g., lead citrate, alizarin, etc.). With any of these landmarks common to both teeth it is possible to compute the volume of enamel or dentin laid down between them and comparisons made. If the dates at which these landmarks were induced are known, accurate growth rates in terms of volume of enamel or dentin laid down per unit time can be established.

References

Dykes, E. and J. G. Clement. 1979. The construction and applications of an x, y coordinate plotting microscope. Paper presented at the British Division of the International Association for Dental Research. University of Sheffield.

Kimura, O., E. Dykes, and R. W. Fearnhead. 1977. The relationship between the surface area of the enamel crowns of human teeth and that of the dentino-enamel junction. *Archs. oral biol.* 22:677–682.

Sullivan, P. G. 1972. A method for study of jaw growth using a computer-based three-dimensional recording technique. *J. Anat.* 112:457–470.

12. Basic Principles of Tooth Design

P. W. LUCAS

Most studies of tooth function never make more than incidental references to food. It is commonest in the literature to find use of words such as "crush," "grind," "cut," or "shear" when analyzing tooth design. These words are defined in terms of tooth movements (Kay and Hiiemäe 1974), tooth shape and movement (Rensberger 1973) or a variable mixture of both (Rosenberger and Kinzey 1976; Osborn and Lumsden 1978). They are thus terms describing the masticatory mechanisms of mammals, and represent an attempt to classify them. A similar terminology is used in industry for describing machine types, but in this case it is recognized that the meaning of such terms is often ambiguous (Lowrison 1974). Such descriptions must be kept separate from an analysis of the physical process of food breakdown, which dictates the shape of mammalian cheek teeth.

A model has recently been proposed to explain the design of simple features on teeth by examining the elastic and fracture behavior of foods (Lucas 1979). An attempt to give this model a sounder basis is made here.

Fracture Theory

The purpose of mastication appears to be the particle size reduction, or comminution, of solid foods (Lumsden and Osborn 1977). Thus the fracture of the food is what is important and the achievement of this for as little effort as possible. The simplest mode of loading is to press the teeth onto the food, thus avoiding complicated grip problems. All that is further required of the loading is to make a small contact area at fracture for highest efficiency. The lower limit on such an area arises from the need to

I am extremely grateful to M. A. Khakoo for devising the program and S. D. Walker for checking the mathematics. D. A. Brown gave considerable advice on the food tests. D. A. Luke is thanked for his advice on the manuscript which P. Elson typed. S. P. A. Kariyawasam and S. Metcalfe gave much technical help.

grip the food and not to severely weaken the teeth through high stress concentrations. The latter consideration is a function of the properties of the dental material involved.

A major problem with the theory of fracture as it derives from Griffth (1920) is that it cannot be used for heterogenous materials. While it is known that differences in stiffness inside a material channel the cracks which are formed upon loading (Kendall 1975), it is still difficult to model the behavior of "real" solids. Foods are viscoelastic, displaying both elastic and viscous behavior simultaneously (Mohsenin 1970). However, the viscous element is time-dependent while the elastic behavior, theoretically, is not (Alexander 1968). It might be hoped that the deformation rates in chewing are high enough to limit such viscous behavior and to approximate simple elastic behavior (Peleg 1976). Much permanent deformation would not be expected to be exhibited by structural animal and plant material since the maintenance of shape is important, but non-structural materials—for example, ripe fruits—may show gross yielding.

Elastic Behavior

Hertz (1896) showed that the deformation, D, between the centers of two spheres, A and B, loaded by a normal force, W, could be described by the equation:

$$D = \left[\frac{9\pi^2}{16} \frac{W^2(R_A + R_B)(K_A + K_B)^2}{R_A R_B}\right]^{1/3} \tag{1}$$

where R_A and R_B are the radii of curvature of the spheres A and B respectively and $K_A = 1 - W_A^2/\pi E_A$ and $K_B = 1 - W_B^2/\pi E_B$ where W_A and W_B are the Poisson's ratios and E_A and E_B are the elastic moduli of A and B respectively (see Timoshenko and Goodier (1951).

Simplifications can be introduced. The Poisson's ratios are nearly always fractions of limited variability (though see below). Sphere A will be considered a tooth and B a food. The elastic modulus of A should be much higher than the elastic modulus of B to avoid waste of energy. A more general case of tooth-food contact would be a curved tooth surface against a flat food surface. In this case R_B is infinite and equation (1) then reduces to:

$$D = \left[\frac{9W^2}{16R_A E_B^2}\right]^{1/3} \tag{2}$$

For small contacts where $2A$ is the width of the circle of contact, $D = A^2/2R_B$ (Mohsenin 1970), and substituting in equation (2):

$$A = K\left[\frac{WR_A}{E_B}\right]^{1/3} \qquad (3)$$

where K is a constant. Now supposing R_A to be the radius of curvature of a tooth tip, then this equation might explain the variations in sharpness of mammalian teeth. For given W, matching an increase in the elasticity E_B of the food by an increase in the radius of curvature R_A involves no increase in the contact area. If however, for a given W the condition of $R_A \gg A$, then mismatching E_B and R_A results in a change in the contact area to the power 3/2. For example, doubling R_A increases the actual area of contact by 2.83, and halving E_B achieves the same result.

However, the equation in itself does not predict anything but that teeth possess the smallest R_A possible for the dental material involved. Suppose therefore that the same conditions hold at fracture. The strength of foods may be higher in stiffer foods of higher E_B. Mohsenin and Mittal (1977) present a graph showing an increase in strength with elasticity for man's materials, though this is far from uniform and is particularly variable in metals. However, Tabor (1951) states that harder metals (hardness of metals is linearly related to their strength) generally have higher elastic moduli. If this is the case for foods, then the importance of the stress concentration in the tip of the tooth cusp restricts the sharpness. For tooth tips of semi-elliptical shape with a constant major axis, the stress in the tip is higher than the average stress by a factor inversely proportional to the square root of R_A (Gordon 1978). Thus stiffer stronger foods can be matched to blunter teeth (or larger R_A) to avoid tooth tip fracture, but this does not incur a loss of efficiency (i.e., large contact area) because of equation (3). Thus the simplest case is that in dental-dietary adaptation, the radius of curvature of tooth cusps R_A is matched to the elastic modulus E_B of foods.

Fracture Behavior

I have suggested previously (1979) that a cuspal point contact with hard brittle foods produces "shatter"—the breakage of the foods in smaller pieces—but that softer foods (of low E_B) are generally reluctant to crack, and therefore blades of dimensions greater than those of the food are required.

Savage (1976) mentions that the blades of upper and lower carnassial teeth of carnivores are often concave, showing a sharp inflection in the center of the long dimension of the blade, and gives as one possibility that this facilitates trapping of the food. This seems even more reasonable when the high Poisson's ratios of animal soft tissues are taken into account. Gordon (1978) states that values may be found over unity (1.0), while a table in Mohsenin (1977) records 0.5 for beef muscle, which is similar to soft rubbers and gelatin (Mohsenin 1970). Comminution of such solids is helped by "enclosing" the solids during loading, as otherwise the size of particle that can be divided is very limited.

The equations do not apply to bladed structures, since the area of contact between blade and food is additionally a function of the length of the blade and the linear dimensions of the food surface. Furthermore, because "shatter" does not occur and because the blade is therefore needed to push through the solid, the area of contact will be variable and the energy required to divide the food into two will be in part a function of frictional effects.

Size Limit to Comminution

The particle size of soft foods that can be achieved is probably limited by the ability of the tongue to position such foods under the blades. Hard, brittle, shatterable foods can presumably be broken down until their properties alter at the "brittle-ductile" transition of Kendall (1978a). Kendall predicted that below a critical particle size, hard, brittle materials would become very ductile and be difficult to break down further. He showed that for some materials (Kendall 1978b) this particle size could have a linear dimension of the order of millimeters.

Energy Expenditure

The time which an animal can spend consuming foods is restricted by the power consumption of the body. This is presumably one constraint on the chewing rate of mammals. A lower limit could be the attainment of a strain rate sufficient to get required fracture behavior from the food, since cracks in harder foods increase in speed of growth with strain rate (Kendall 1978b). In uniaxial compressions tests on an Instron testing machine with a carrot, it was found that at very low speeds of testing (< 1 cm/min) considerable flow of the carrot occurred, and the crack did not separate the particles into two. At higher loading speeds (20–25 cm/min) "shatter" occurred and the carrot quickly broke into several pieces.

Alexander (1968) suggests that an increase in loading speed saves energy in the fracture of viscoelastic materials by reducing the energy producing viscous deformation. Uniaxial compression tests on white potato appear to support this statement. The differences between the means of the energy required per unit volume at loading speeds of 0.25 cm/min and 25 cm/min was 30 percent. This is an approximate figure because the stress-strain curve of the potato is slightly non-linear (Mohsenin 1970).

These loading speeds are probably much lower than the chewing speeds

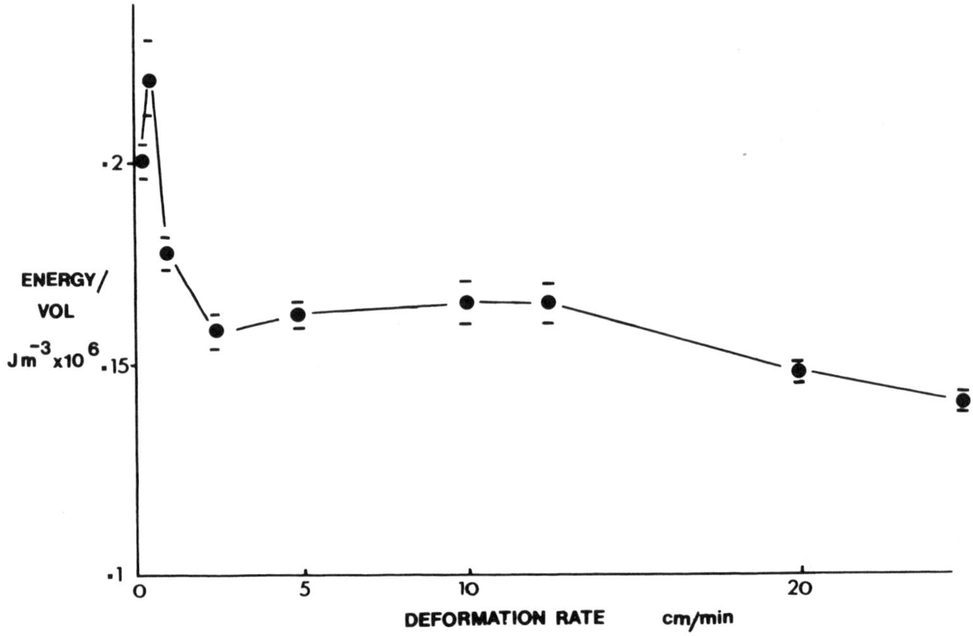

Figure 12.1. The energy required to fracture a unit volume of white potato varies with the speed of loading. The dark circles are the means of 12 to 13 specimen results, and the bars represent the standard deviations. A 100-fold increase in loading speed lowers the energy required by 30 percent.

of mammals. Ahlgren (1976) records the average chewing speed of a man chewing a carrot at 450 cm/min. Byrd, Milberg, and Lutschei (1978) show roughly similar distances of mandibular movement in man and *Macaca fascicularis*, though the latter chews three times as fast as man, as do the miniature pigs studied by Herring (1976). The jaws of the latter traverse much greater distances in this time than the jaws of macaques. The chewing speed for a food positioned between any two teeth depends on the distance of these teeth from the temporo-mandibular joint of course (Bourne 1977).

The constant loading speeds of testing machines do not mimic more than the "first bite" of a mammal. In subsequent chews it has been shown that the loading in man is possibly unilaterally sinusoidal (Sharma and Sherman 1973). However the nearer such tests imitate the chewing situation, the more difficult it is to define and measure the parameters involved.

Beyond the first bite when a "pestle and mortar" dentition is used for shatterable foods, the area of contact will increase, and so will the force and the energy input. The power consumption of mastication, however, can be controlled within certain limits by ensuring that the "mortar" has a larger radius of curvature than the "pestle" (Lucas 1979).

The Radius of Curvature of Tooth Tips

The effective sharpness or bluntness of tooth tips can be measured by fitting curves to drawings of the tooth. If the theory is to have any relevance to mastication, the radius of curvature might be expected to be a constant in a particular species. The unworn cusp tips of the premolar and first two molar teeth of man were measured by magnified camera lucida drawings. Rectilinear coordinates were taken, and these were fed into a computer program (very kindly devised by M.A. Khakoo of University College London) solving a general equation for conic section curves

$$X^2 = AY^2 + BY \tag{4}$$

where B/2 is the radius of curvature at the cusp tip at X = O, Y = O (this equation is similar to that of Biggerstaff 1972). The results in table 12.1 show considerable variation, with a slight tendency toward sharper cusps toward the back of the mouth. Though it may not be related, loading speeds would be smaller at more distal tooth positions.

Table 12.1. The radius of curvature of the cusp tips of unworn human cheek teeth, illustrating their bluntness.

Tooth	Cusp	N	Radius of Curvature Mean (mm)	S.D.
Upper first premolar	Buccal	7	.64	.21
	Lingual	8	.66	.21
Upper second premolar	Buccal	8	.81	.15
	Lingual	8	.64	.23
Lower first premolar	Buccal	8	.69	.25
	Lingual	5	.62	.11
Lower second premolar	Buccal	7	.63	.14
Upper first molar	Metacone	4	.67	.12
	Paracone	4	.57	.16
	Protocone	4	.78	.35
	Hypocone	5	.66	.13
Upper second molar	Metacone	4	.52	.15
	Paracone	4	.67	.20
	Protocone	4	.58	.08
Lower first molar	Metaconid	5	.81	.16
	Entoconid	4	.52	.11
	Protoconid	5	.60	.31
	Hypoconid	3	.44	.08
	Hypoconulid	3	.56	.32
Lower second molar	Metaconid	4	.49	.29
	Entoconid	4	.50	.36
	Protoconid	3	.62	.02
	Hypoconid	2	.51	.08

The equation fitted mesial or distal views of the cusps, but it was found difficult to fit buccal and lingual views on the premolars. When the coefficient A is positive, the cusp tip is elliptical; when A is zero or nearly zero (± 0.1 in this study) the shape was treated as a parabola; and when A is negative, the shape is hyperbolic. Of the 113 cusps measured, 102 were elliptical, 6 were parabolic, and 5 were hyperbolic.

The Direction of Natural Selection

Whether produced experimentally in the laboratory or in an animal adapting to a new habitat, a mismatch between dentition and diet is unlikely to result in a complete failure to masticate the food.

For example, the blunt dentition of man appears maladapted to eating soft but strong foods of low elastic modulus like animal soft tissues. Rather, the teeth seem adapted to hard brittle foods of high modulus. Therefore the energy a man expends obtaining swallowable particles of raw meat with his cheek teeth might be expected to be far higher than a carnivore doing the same with its carnassials. In standardized conditions (for instance a standard particle feed size) this could be measured by the time taken to process equal amounts of food.

Conversely, a carnivore given hard brittle foods to break with its carnassials does not incur an energy "cost" but fractures its sharp blade edge because of the high stress concentration there. It would be perfectly capable of eating the hard brittle food with blunted teeth but would suffer only in an attempt to resume masticating its natural diet. This seems a reasonable explanation of the ferret experiment of Berkovitz and Poole (1977).

This gives some indication of how natural selection may act. An animal adapted to a hard diet wastes considerable energy eating soft, strong foods. Thus it may be expected to limit its feeding around an optimum value of E_B for its tooth shape. In the reverse case an animal adapted to soft food risks a path of no return in eating too hard a food.

Some effect of animal size may be important in this context. Small animals may not have the capability to generate as high forces as large animals. For example Robins (1977) produced loads of 2.630–5.125 kg in the incisors of adult male rats of body weights between 305 and 555 g. Thexton and Hiiemäe (1975) report a force of 2.5 kg in the jaw muscles for the opossum (*Didelphis marsupialis*) of body weight 2.5–3.5 kg. Buckland-Wright (1975) recorded forces of 2.0–23.25 kg at the canines in the cat (*Felis domestica*), and up to 28 kg at the carnassials. The cats had body weights between 1.7–3.8 kg. However, in man, peak forces in male Eskimos have been reported up to 158 kg, though an average for the first molar bite of young people is reported at 45 kg (Jenkins 1978). The biting force is to some extent dependent on the size of the masticatory muscles, which are generally smaller in small animals, and this might bar the latter from eating certain foods. It needs to be pointed out, though, that loads in mastication are reported to be much less than the maximum producible (Anderson 1956).

General Discussion

Most simple analysis of materials assumes failure at very small strains (Gordon 1968), but this is not typical of many biological solids (Gordon 1978). Until alternative concepts of solid behavior are found however, it is necessary to utilize existing methods. The problem with a such general model as described here is its superficiality, but due to the nature of the problem, application of complicated stress analysis (like finite element analysis) will produce results too particular to have general application. For example the loading of muscle parallel to the fibers will produce a different result to a loading perpendicular to them, and this applies to any heterogenous food where the elements in the composite arè oriented. Some element of randomness in the presentation of food to the equipment must be expected then. It is this point that justifies the generality here. Nearly all this study has concentrated on food as a determinant of tooth form, but any theory exists in a wilderness if, as at present, little is known about the physical properties of the dietary solids of mammals.

REFERENCES

Ahlgren, J. 1976. Masticatory movements in man. In D.J. Anderson and B. Matthews, eds., *Mastication*, pp. 119–127. Bristol: Wright.

Alexander, R.M. 1968. *Animal Mechanics*. Seattle: University of Washington Press.

Anderson, D.J. 1956. Measurement of stress in mastication. *J. Dent. Res.* 35:664–674.

Berkovitz, B.K.B. and D.F.G. Poole. 1977. Attrition of the teeth in ferrets. *J. Zool.* 183:411–418.

Biggerstaff, R.H. 1972. Three variations in dental arch form estimated by a quadratic equation. *J. Dent. Res.* 51:1509.

Bourne, M.C. 1977. Compression rates in the mouth. *J. Text. Stud.* 8:373–376.

Buckland-Wright, J.C. 1975. The structure and function of cat skull bones in relation to the transmission of biting forces. Ph.D. dissertation, University of London.

Byrd, K.E.., D.J. Milberg, and E.S. Lutschei. 1978. Human and macaque mastication: a quantitative study. *J. Dent. Res.* 57:834–843.

Gordon, J.E. 1968. *The New Science of Strong Materials*. London: Penguin.

—— 1978. *Structures*. London: Penguin.

Griffith, A.A. 1920. Phenomena of rupture and flow in solids. *Phil. Trans. Roy. Soc.* A221:163–194.

Herring, S.W. 1976. The dynamics of mastication in pigs. *Archs. Oral Biol.* 21:473–480.

Hertz, H.R. 1896. *Miscellaneous Papers.* London: Macmillan.

Jenkins, G.N. 1978. *The Physiology and Biochemistry of the Mouth.* 4th ed. Oxford: Blackwell.

Kay, R.F. and K.M. Hiiemäe. 1974. Jaw movement and tooth use in recent and fossil primates. *Am. J. Phys. Anthrop.* 40:227–256.

Kendall, K. 1975. Control of cracks by interfaces in composites. *Proc. Roy. Soc. Lond.* A341:409–428.

—— 1978a. Complexities of compression failure. *Proc. Roy. Soc. Lond.* A361:245–263.

—— 1978b. The impossibility of comminuting small particles by compression. *Nature* 272:710–711.

Lowrison, G.L. 1974. *Crushing and Grinding.* London: Butterworths.

Lucas, P.W. 1979. The dental-dietary adaptations of mammals. *N. Jb. Geol. Paläont. Mh.* 8:486–516.

Lumsden, A.G.S. and J.W. Osborn. 1977. The evolution of chewing: a dentist's view of palaeontology. *J. Dentistry* 5:269–287.

Mohsenin, N.N. 1970. *Physical Properties of Plant and Animal Materials.* New York: Gordon and Breach.

—— 1977. Characterization and failure in solid foods with particular reference to fruits and vegetables. *J. Text. Stud.* 8:169–193.

Mohsenin, N.N. and J.P. Mittal. 1977. Use of rheological terms and correlation of compatible measurements in food texture research. *J. Text. Stud.* 8:395–408.

Osborn, J.W. and A.G.S. Lumsden. 1978. An alternative to "thegosis" and a re-examination of the ways in which mammalian molars work. *N. Jb. Geol. Paläont. Abh.* 156:371–396.

Peleg, M. 1977. Operational conditions and the stress-strain relationship of solid foods—theoretical evaluation. *J. Text. Stud.* 8:283–295.

Rensberger, J.M. 1973. An occlusion model for mastication and dental wear in herbivorous mammals. *J. Paleontol.* 47:515–528.

Robins, M.W. 1977. Biting loads generated in the laboratory rat. *Archs. Oral Biol.* 22:43–47.

Rosenberger, A.L. and W.G. Kinzey. 1976. Functional patterns of molar occlusion in platyrrhine primates. *Am. J. Phys. Anthrop.* 45:281–298.

Savage, R.J.G. 1977. Evolution in carnivorous mammals. *Palaeontology* 20:237–271.

Sharma, F. and P. Sherman. 1973. Evaluation of some textural properites of foods with the Instron universal testing machine. *J. Text. Stud.* 4:344–353.

Tabor, D. 1951. The Hardness of Metals. London: Oxford University Press.

Thexton, A.J. and K.M. Hiiemäe. 1975. The twitch-contraction characteristics of opossum jaw musculature. *Archs. Oral Biol.* 20:743–748.

Timoshenko, S. and L.N. Goodier. *Theory of Elasticity.* New York: McGraw-Hill.

13. Ecological Aspects of Dental Functional Morphology in the Plio-Pleistocene Rhinoceroses of Europe

MIKAEL FORTELIUS

It has been repeatedly observed that dental function in mammals is still poorly understood. Herbivore teeth particularly have been little studied from a functional viewpoint; yet there is a slowly growing body of information on the subject (Hiiemäe 1967; Crompton and Hiiemäe 1969, 1970; Butler 1972; Rensberger 1973, 1975; Janis 1979; Lucas 1979; and several others). The present study is an attempt to present briefly the application of such information to a specific problem, that of the ecological adaptations of the rhinoceroses of the European Plio-Pleistocene. Only the main results are given here; the methods and data will be presented later in another context, together with a more detialed discussion.

Functional morphology is a subject tinted with a quasi-teleological jargon. In reality this is, it is to be hoped, just a convenient shorthand to avoid tedious repetition of the complex and incompletely understood processes involved in evolution. In this text teleological wording is used in this "practical" sense only. The terminology employed to describe size,

I am very much indebted to Björn Kurtén, University of Helsinki, and John M. Rensberger, University of Washington, for stimulating discussions and valuable suggestions. I would also like to thank the following for generous assistance: Andrew P. Currant, British Museum (Natural History), London; Ann-Marie Forsten, University of Helsinki; Claude Guérin, Université Claude Bernard, Villeurbanne; Eirik Granqvist, Museum of Zoology, Helsinki; Jeremy Hooker, British Museum (Natural History), London; Wighart v. Koenigswald, Hessisches Landesmuseum, Darmstadt; Heintz-Georg Klös, Zoo Berlin; Adrian Lister, University of Cambridge; Hans Loose, Rijksmuseum van Geologie en Mineralogie, Leiden; Peter Lucas, Guy's Hospital Medical School, London; Anthony J. Stuart, University of Cambridge; and Antony J. Sutcliffe, British Museum (Natural History), London.

locomotory adaptations, stratigraphy, etc. is used only in a very general sense, unless otherwise specified. The rhinoceros taxonomy has been subject to some considerable disagreement. This text follows Loose (1975) for living and most Plio-Pleistocene species, otherwise "common usage."

The Plio-Pleistocene rhinoceroses have often been used as paleo-ecological indicators. "The study of the Rhinocerotidae permits pre-historians to both date the deposits and to establish the paleo-ecology of the area" (Guérin 1976:405). Yet the use of large mammals as such indicators is in many ways problematical. A case in point is that of *Dicerorhinus hemitoechus*, commonly known as the "steppe rhino" and frequently described as "hypsodont." As a matter of fact its teeth are essentially brachyodont (see p. 00), as pointed out by Osborn (1900:264)—certainly a diet of mature steppe grasses seems unlikely for this species.

The concept of the "steppe rhino" goes back to Zeuner (1934), who took it from Wüst (1922). Zeuner based his interpretation on skull morphology, not on dental criteria; yet his word is universally invoked in support of the "grazing, hypsodont steppe rhino." If this study will help to lay that unfortunate concept to rest I think it will have justified itself.

The Rhinoceros Dentition

General Features

As those of most herbivores, the cheek teeth of the rhinoceroses form a continuous row and function as a unit. In general there are three premolars (p^2_2–p^4_4) and three molars (M^1_1–M^3_3), which occlude in a fashion indicated in figure 13.1. The upper teeth are roughly square in occlusal view, and consist of a π-shaped arrangement of lophs: a longitudinal ectoloph and trans-versal proto- and metalophs (fig. 13.2a). In M^3 the metalop is reduced and the tooth is triangular in occlusal view (fig. 13.1, fig. 13.11). The lower teeth

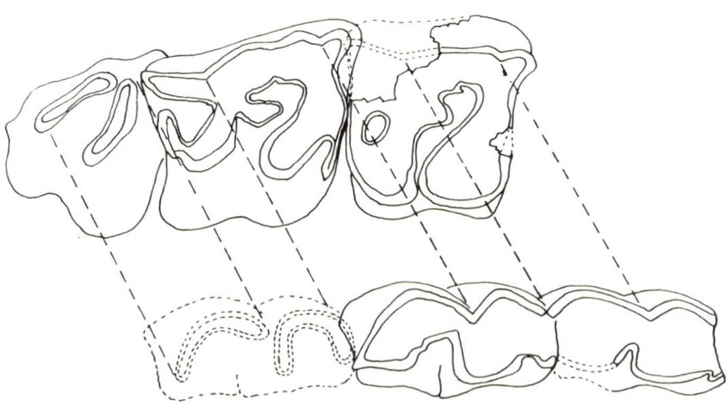

Figure 13.1. Occlusion in the molars of a rhinoceros, buccal up, lingual down.

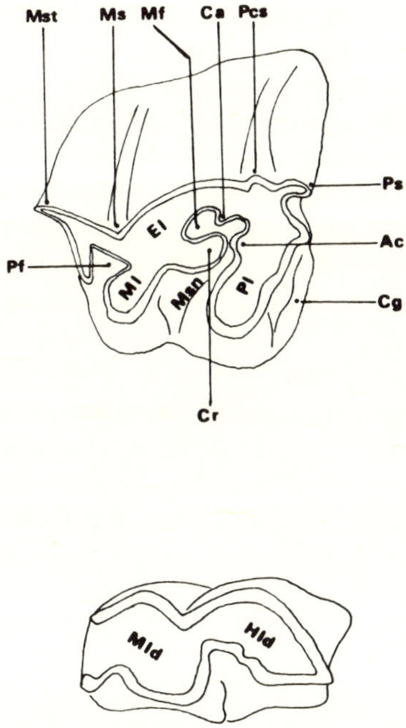

Figure 13.2. Rhinoceros dental terminology, M^2 (above) and M_2 (below).
Key: Acr = antecrochet, Ca = crista, Cg = cingulum, Cr = crochet, El = ectoloph, Hld = hypolophid, Mf = mediofossette, Ml = metaloph, Mld = metalophid, Ms = mesostyle, Msn = medisinus, Mst = metastyle, Pcs = paracone style, Pf = postfossette, Pl = protoloph, Ps = parastyle.

are narrower, and consist of a hypo- and a metalophid behind each other (fig. 13.2b).

The occlusal motion is anteromediad and inclined upwards (fig. 13.1), as is seen from straie in the occlusal surfaces of upper teeth (fig. 13.3). The lophids of the lower teeth wear rather distinct facets into the buccal parts of the occlusal surfaces of the corresponding upper teeth. This indicates that the occlusal motion is at least loosely tracking.

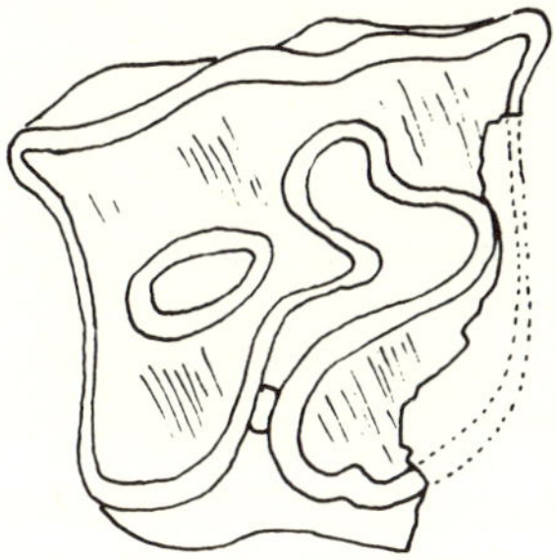

Figure 13.3. Striae in the occlusal surface of a worn M^2 dex. of *Coelodonta antiquitatis*, to show the direction of the occlusal stroke.

As most mammals investigated so far, rhinoceroses chew on one side at a time. The mandibular symphysis is firmly fused, and there is no possibility for independent movement of the two halves. During the occlusal stroke the teeth occlude only on the active side, a fact that is quite apparent even from a dried skull.

The existence of two phases of the occlusal power stroke (the part of the stroke during which the teeth are in occlusion; Crompton and Hiiemäe 1970:27f) has been argued for many primates and herbivores (Mills 1967) and for certain other groups. Kay and Hiiemäe (1974) define the two phases (I and II) as being two parts of a continuous power stroke. Janis (1979) has demonstrated, by cine fluoroscopic studies, that a minor phase II is present in *Procavia*, but argues that, contrary to herbivorous primates, ungulates tend to reduce phase II, which may be lost altogether in grazing forms.

The power stroke of a rhinoceros is also conveniently thought of as consisting of two phases (fig. 13.4). During the stroke the masticatory force vector probably remains essentially constant (Rensberger in litt. 1979),

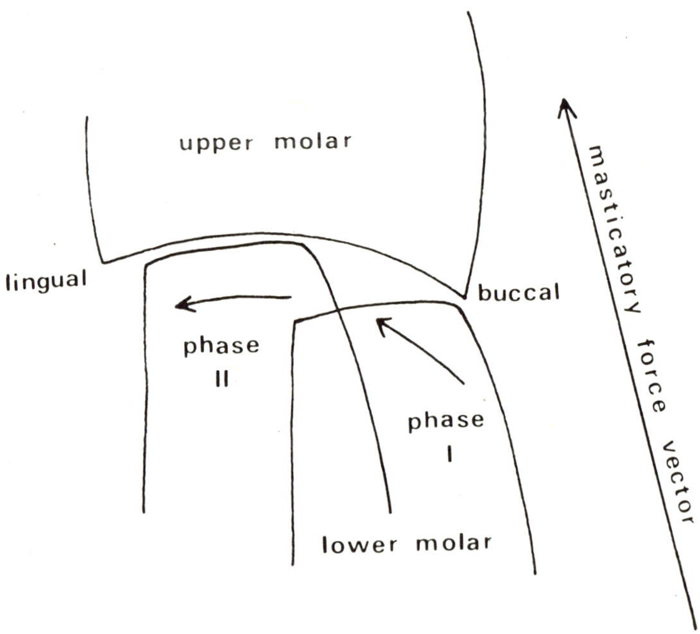

Figure 13.4. Occlusal mechanics of a brachyodont rhinoceros in relation to an imaginary masticatory force vector. The more inclined the plane of contact (and hence the path of motion) is to the force vector, the higher is the occlusal pressure and the lower the speed. As the area of contact is smaller during phase I, pressure is probably relatively uniform throughout, whereas speed is higher during phase I than during phase II. Discussion in text. Modified from sketch by John M. Rensberger.

while the motion is guided chiefly by the shapes of the occlusal surfaces, which in turn are regulated by differential wear due to structural differences, such as inequalities in enamel thickness (see below). (Naturally, the different muscular complexes involved influence the motion in different ways, and the possibility remains that the masticatory force vector varies considerably during the stroke. Rhinoceroses have well developed masseter and pterygoideus complexes, and these are important for lateral jaw movements [Crompton and Hiiemäe 1969:30]. The essentially constant masticatory force vector is used here, however, as a hypothesis with a high explanatory value.)

Initially, during phase I, only the buccal edge of the lower tooth is in contact with the corresponding upper tooth (fig. 13.4). In terms of wear facets this phase is represented by the buccal part of the occlusal surface of the upper tooth, where the ridge-and-valley structure produced by the crescent-shaped buccal edges of the lophids is clearly visible. The motion is closer to the direction of the masticatory force vector than during the following phase, and speed is high. Because of the small area of contact, pressure is probably quite high as well, and it appears that wear is fastest during phase I (see below). Further linguad, contact is made over the entire occlusal surfaces. Movement is mainly perpendicular to the masticatory force vector, and a low speed condition results. As there is a major component of masticatory force normal to the plane of contact, and the actual points of contact are few during each given moment of phase II, pressure is probably high and the comminution of food consequently efficient (Rensberger 1973:518). Phase I may perhaps be said to correspond to the shearing, phase II to the crushing/grinding component of dental function. In brachyodont dentitions both phases are well developed, wherease in hypsodont dentitions phase I is much reduced, resulting in a dominantly high pressure "grinding" type of occlusal stroke.

Words such as "cutting," "shearing," "crushing" and "grinding" have traditionally been used to describe dental function. Rensberger (1973:517) discusses the significance of these terms, and concludes that "the ability of opposing surfaces to divide food. . .seems to be a function of pressure in all four types of processes." Moreover, the function is not linear: too little pressure will only waste energy deforming the food without breaking it, while "overpressure will only needlessly wear and shorten the life of the teeth" (Rensberger 1973:517). Thus maintenance of a constant pressure during wear and consequent change in shape and area of the occlusal contact appears to be a key to understanding the functional morphology of herbivore teeth.

The jaw-joint of the rhinoceroses, as that of most herbivores, is situated above the level of the toothrows, resulting in a more equal distribution of the biting force (Crompton and Hiiemäe 1969:33) and therefore of the

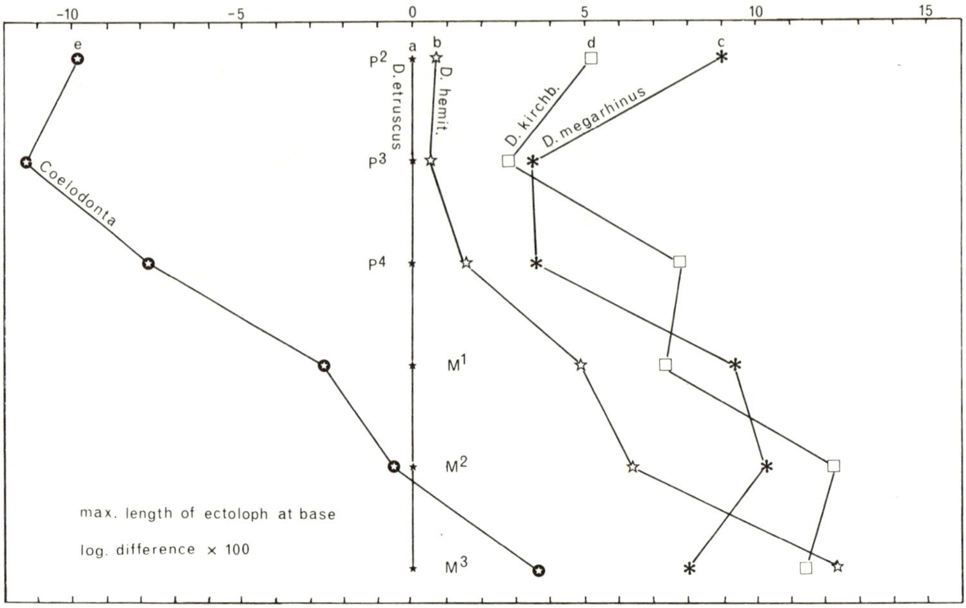

Figure 13.5. Log. ratio diagram to show proportions of the upper toothrow in the Plio-Pleistocene species discussed in the text. Note the relative enlargement of the molars in hypsodont *Coelodonta* and, to a lesser degree, in *Dicerorhinus hemitoechus*. Also note the similarity between *D. megarhinus* and *D. kirchbergensis*. Based on means of complete dentitions, except *Coelodonta* based on means of isolated teeth. From Fortelius (1979), discussion in text.

occlusal pressure. In hypsodont dentitions, where "grinding" function is important, the molars are relatively enlarged (fig. 13.5). This is related to the fact that the biting force is greatest in the posterior part of the toothrow (close to the joint), and it means that this area is relatively more important in forms with a dominantly phase II high pressure occlusal stroke.

The terms brachyodont and hypsodont, conventionally employed to denote differences in crown height, have been used rather indiscriminately in the past, and at present they carry little information to the uninitiated reader. Usually in combinations with more, less, sub-, semi-, pseudo-, etc. they are frequently used in a relative sense only, and may even be outright misleading. (The statement "*Dicerorhinus hemitoechus* is more hypsodont than *D. etruscus* is true in a relative sense, but gives the incorrect notion that the former is hypsodont, which it is not.) The terms are used here in the sense of Osborn (1903:106), i.e., describing functionally distinct types. The morphological and structural differences between the two are summarized in table 13.1.

Table 13.1. Morphological and structural differences between brachyodont and hypsodont cheek teeth of rhinoceroses as defined in the text

Brachyodont	*Hypsodont*
Upper Teeth	Upper Teeth
Occlusal surface concave	Occlusal surface flat
Enamel thickness unequal	Enamel thickness uniform
Buccal and lingual walls inclined, buccal side higher	Buccal and lingual walls vertical, of subequal height
Ectoloph wear profile saw-toothed	Ectoloph wear profile straight
Secondary folds of molars separate, no mediofossette	Secondary folds of molars confluent, forming mediofossette
Styli distinct	Styli indistinct
M^3 triangular, with confluent ecto- and metalophs	M^3 square, with separate ecto- and metalophs
Lower Teeth	Lower Teeth
Lophids with crescent-shaped buccal walls	Lophids with flattened buccal walls
General	General
Crowns often lower	Crowns often higher
Cementum absent or thin[a]	Cementum thick
Enamel smooth, relatively thin	Enamel rough, relatively thick

[a]Note that cementum is very capriciously preserved in fossil teeth. Rough enamel usually indicates former presence of cementum.

Brachyodont Teeth

The vast majority of the rhinoceroses have teeth that may be called brachyodont (fig. 13.6). The upper cheek teeth have characteristically projecting ectoloph edges, which form the main shearing blades of the dentition, corresponding to phase I of the occlusal stroke. The occlusal surface is concave, and the transversal lophs form a "plateau" linqually. The buccal and lingual walls slope outwards towards the base of the teeth, which are apically quite narrow.

In functional terms the key feature of the brachyodont dentition is the concave wear surface of the upper cheek tooth, which produces the shearing ectoloph edge buccally. The main factor regulating the shape of the occlusal surface is differential wear brought about (or at least influenced) by inequalities in enamel thickness: enamel is thin medially and thick buccally and lingually (fig. 13.7). In worn teeth the medial enamel ridges become thicker as the basal parts of the crown are exposed (fig. 13.8) and the relative concavity is thus decreased with wear. The functional

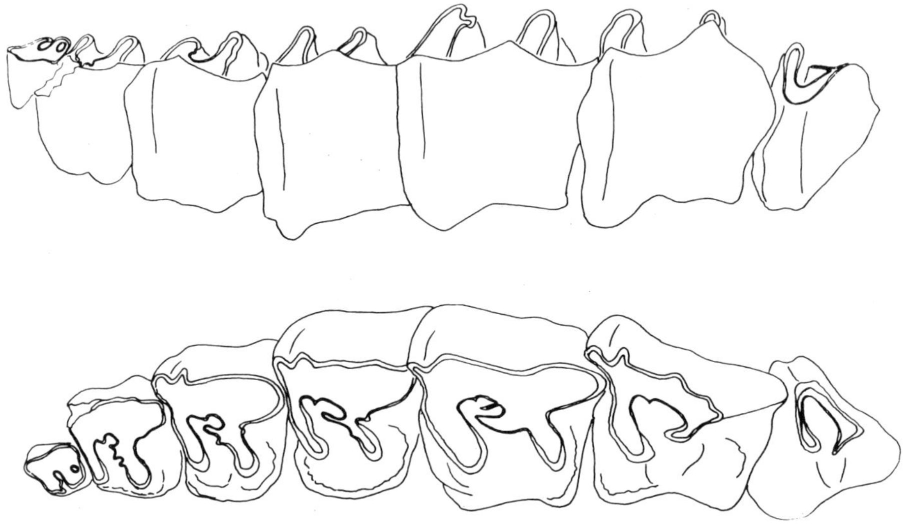

Figure 13.6. Upper cheek teeth of a brachyodont rhinoceros, *Diceros bicornis*. Buccal view (reversed) above, occlusal view below. Note the medially thin enamel ridges, the concave occlusal surface, and the saw-toothed ectoloph wear profile.

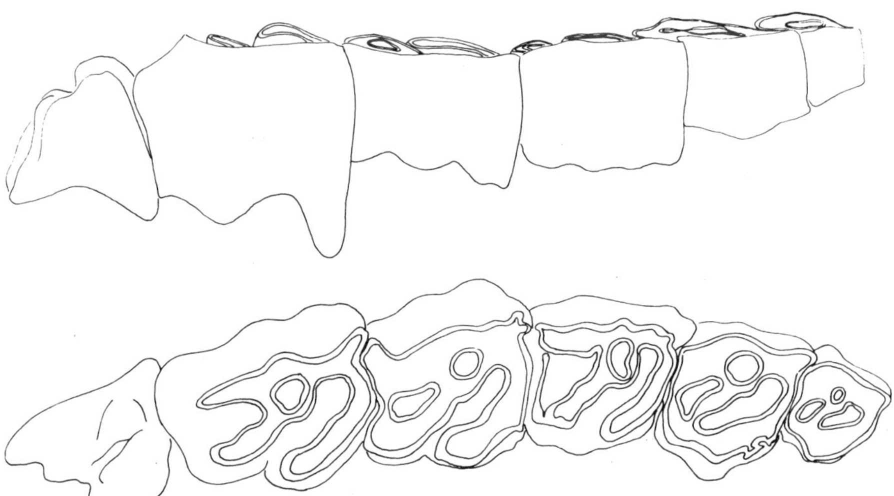

Figure 13.7. Upper cheek teeth of a hypsodont rhinoceros, *Ceratotherium simum*. Buccal view (reversed) above, occlusal view below. Note the uniform thickness of the enamel ridges, the flat occlusal surface, and the straight ectoloph wear profile.

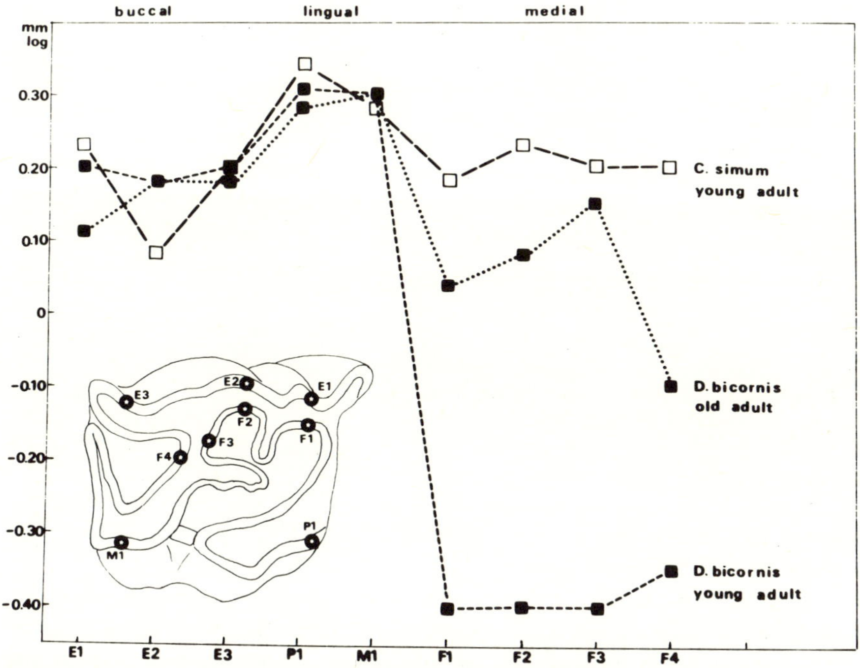

Figure 13.8. Changes in enamel thickness with wear in various parts of the crown in a brachyodont (*Diceros bicornis*) and a hypsodont (*Ceratotherium simum*) rhinoceros. In the former the medial enamel ridges become thicker with wear as the basal, thick parts are exposed, serving to reduce wear medially and to prevent exposure of the pulp cavity as well as to maintain the concavity of the occlusal surface at a functional level (see text). The "young adults" are those depicted in figures 13.6 and 13.7.

implications of this are obvious. The occlusal motion cannot, at least for any greater length, continue against the direction of the masticatory force vector. To remain functional, the upper tooth must not become too strongly concave. Moreover, if the concavity were not reduced in this way, the pulp cavity would become exposed while there was still "tooth left" buccally and lingually, which would mean a waste of dental matter and an unnecessarily reduced life span.

Wear, for some reason, is faster buccally than lingually in upper cheek teeth. Tooth morphology compensates for this. The buccal side is initially higher than the lingual side (particularly in specialized, higher crowned brachyodont species like *Diceros bicornis* and *Dicerorhinus kirchbergensis*), and the tooth is gradually rotated inwards during wear, as the curved ectoloph emerges to replace what has been worn away (fig. 13.9). This mechanism for maintaining shearing efficiency during wear is not unknown among mammals (cf. Mellett 1977:112). In senile rhinoceroses the

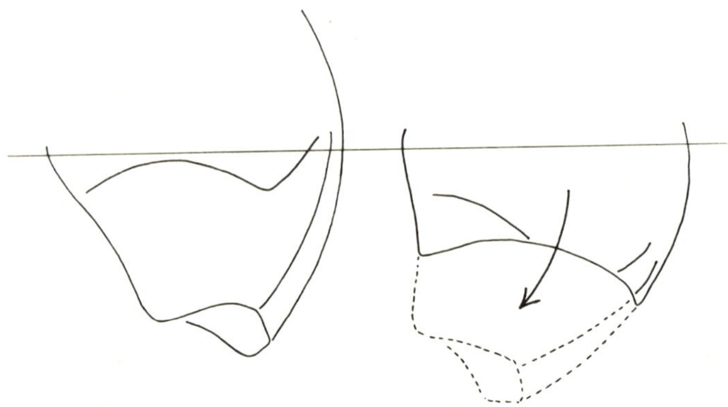

Figure 13.9. Inward rotation of a brachyodont molar to compensate for buccally faster wear and to maintain buccal shearing efficiency, anterior view (in part imaginary). Buccal right, lingual left.

teeth frequently "hang" inwards beneath the palate, showing the final stage of the process.

Brachyodont lower teeth are characterized by the smooth, crescent-shaped buccal walls of the lophids (fig. 13.1). These shear against the ectoloph edge, producing a saw-toothed ectoloph wear profile, which increases the effective length of the blade. Styli are frequently formed on the upper teeth to strengthen the structure at the critical points, and may also serve to trap food (Rensberger in litt. 1979). Cingula are frequently present when crowns are low. This may perhaps be interpreted as an adaptation to protect the gums from injury by thorns and splinters, the need for which would lessen with increaseing crown height and disappear with a shift to a diet free from such components (e.g., grass).

Hypsodont Teeth

Hypsodont rhinoceros teeth differ from brachyodont teeth in having a flat, occlusal surface and a flat octoloph wear profile (fig. 13.7). The occlusal surfaces of opposing teeth (left and right) are inclined toward each other in a tectiform fashion, much as in horses. This allows a one-phase upward-inward occlusal stroke as described below. The teeth have steep walls, and are prism-shaped in outline rather than apically constricted. Buccal and lingual sides are more equally developed than in brachyodont teeth, and the enamel is of a more equal thickness over the entire occlusal surface (fig. 13.7). Accordingly, no concave wear of the "brachyodont" kind is observed. (Interestingly, among geomyoid rodents the situation is reversed: brachyodont forms have enamel of unit thickness while hypsodont forms have strong inequalities, which in this case serve to *maintain* the flat occlusal surface [Rensberger 1975:11f]. Function, not structure, is what differentiates the two types.)

The ectoloph edge does not project as in brachyodont teeth. Instead the wear surface is flat so that pressure is probably more uniformly high over the tooth, and on the occlusal surface there are several parallel blades (enamel ridges), which serve to comminute the food much as in modern horses (cf. Rensberger 1973:519). The molars (literally the "grinders") show the differences between hypsodont and brachyodont more clearly than do the premolars, which are relatively reduced in hypsodont forms (fig. 13.5).

Cementum, thin or absent in brachyodont teeth, is often present as a heavy coating and may fill up the fossettes, producing the characteristic wear-retarding, relief-enhancing, three-component structure found in many hypsodont forms, notably modern horses and elephants. In the Pleistocene giant rhinoceros *Elasmotherium* the cheek teeth had continuous growth, the enamel was plicated, and cementum filled all cavities (fig. 13.10d). In molar development this rhinoceros is rivaled only by a few rodents, being by far the most hypsodont of all large herbivores.

Particularly in the upper molars the effective length (perpendicular to the occlusal motion) of the enamel ridges is increased by the formation of the

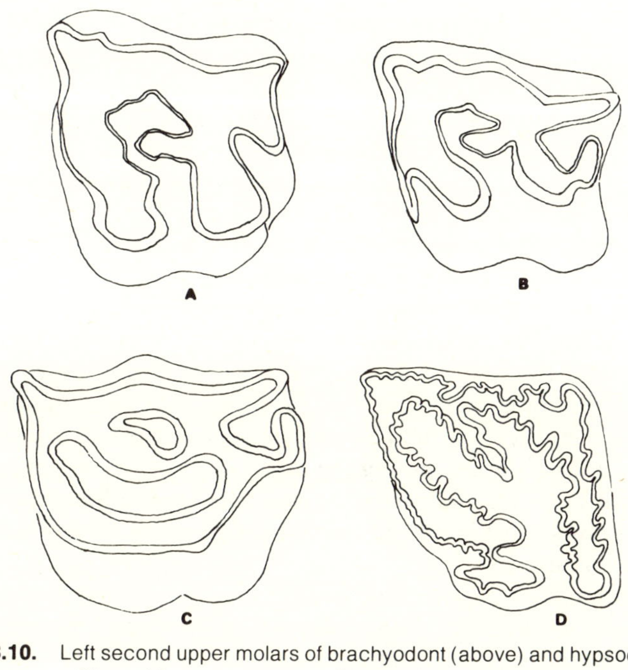

Figure 13.10. Left second upper molars of brachyodont (above) and hypsodont (below) rhinoceroses. A. *Dicerorhinus etruscus*; B. *Dicerorhinus hemitoechus*; C. *Coelodonta antiquitatis*; and D. *Elasmotherium*. Discussion in text. Not to scale.

mediofossette as crochet and crista (the two main secondary folds, fig. 13.2a) merge, and by the posterior (distal) deflection of the trasversal lophs, which may eventually become confluent (fig. 13.10c) M^3 is enlarged, and a separate metaloph is secondarily acquired by all truly hypsodont rhinoceroses, in which M^3 comes to resemble M^1 and M^2 in shape.

The lower teeth of hypsodont rhinoceroses have buccally flattened lophids, which correspond to the flat ectoloph wear profile. The ectoloph, correspondingly, is not as regularly folded as in brachyodont teeth, and the styli are indistinct. The enamel is thicker and has a rough surface, which probably is related to the presence of a cementum coating (the thinner the cementum, it seems, the smoother the enamel). Since buccal and lingual development is subequal and wear rather uniform, no marked inward rotation is observed as the upper tooth is worn down.

Functional and Dietary Considerations

Although the traditional view of hypsodonty is that it is an adaptation to grass eating (or, in functional terms, to increased abrasion), several objections have been raised to this interpretation (e.g., White 1959 and references). The rather confused debate has been much hampered by the fact that "hypsodonty" has not been adequately defined, nor have its functional implications been considered (apart from the fact that higher crowns take longer to wear down), except by a few authors (e.g., Osborn 1903; Rensberger 1973, 1975).

Increased crown height may have several causes, but can be broadly interpreted as a response to increased wear. This in turn may be a result of a need to chew more food per unit time, or food that causes more wear. The need to chew more food may be related to increased size, a more energy-consuming behavior, a colder climate, food of lower nutritive quality or some such factor. Increased wear may result directly from more abrasive food or increased contamination with extraneous abrasive material, or indirectly from food requiring higher occlusal pressures for efficient comminution (Rensberger in litt. 1979). It is argued here that true hypsodonty involves higher total occlusal pressures (a dominantly horizontal power stroke), and that it is primarily a dietary adaptation rather than a general way of increasing the life span of the animal when dental wear is limiting. Thus "indexes of hypsodonty" are of limited value: true hypsodonty is *not* just increased crown height nor yet the appearance of a cementum coating.

The combination of shearing and crushing functions implicit in the brachyodont morphology seems adapted to the comminution of bulky vegetation like twigs, fruits, and soft plants, such as is indeed the main food of living brachyodont rhinoceroses. The high occlusal pressures of the hypsodont dentition, on the other hand, seem to be an adaptation to the rapid comminution of fibrous, thin and tough vegetation such as mature grass, which is the food of the only truly hypsodont living rhinoceros, *Ceratotherium simum*.

Thus the traditional interpretation of hypsodonty as an adaptation to grass eating is supported, but in more than the sense of increased resistance to abrasion. Rather, the high occlusal pressures enable the animals not only to eat young and tender grass occasionally (as do many bears, pigs and deer, for example), but also to specialize on old, tough and fibrous grass. (Obviously, it is more or less impossible to distinguish between the different factors influencing tooth morphology in each separate case, since many grasses, for instance, are *both* abrasive *and* tough and of a low nutritive quality to boot; and since grazing animals are frequently larger than their browsing ancestors. Still, knowledge of the existence of the various factors is of some value in discussing the ecological adaptations of brachyodont and hypsodont animals.)

Vegetation, of course, does not occur in two distinct types, and neither do rhinoceros teeth. There are intermediates between brachyodont and hypsodont, some of which are difficult to place (the living Indian "unicorn," *Rhinoceros unicornis*, for example). Despite this the two categories remain reasonably well defined, and indeed the vast majority of the rhinoceroses are clearly brachyodont, most of the remaining few being equally clearly hypsodont. Brachyodonty and hypsodonty thus probably represent mutually exclusive dietary adaptations rather than "primitive" or "advanced" conditions, although all hypsodont forms are, of course, ultimately derived from brachyodont ancestors.

The Rhinoceroses of Plio-Pleistocene Europe

The classical study concerning rhinoceros dietary adaptations is that of Zeuner (1934). Zeuner was able to demonstrate, species for species, a close relationship between skull shape and the average height of the food vegetation for living rhinoceroses. By applying the same principles to fossil species, Zeuner was able to reconstruct the "mode of life" (Lebensweise) of these. In recent years, Loose (1975) has revised the work and further developed the method. Neither author is primarily concerned with teeth, and their interpretations are based on the inferred height of the food rather than, as in the present study, on its (equally inferred) abrasive, nutritive or structural properties. In no instance do the two different approaches reach contradictory results, but a combination of both gives a more complete picture than either one alone.

In the following the ecology of the rhinoceroses of Plio-Pleistocene Europe is discussed briefly. The data concerning body size and locomotory

Table 13.2. Ecological adaptations of the European Plio-Pleistocene rhinoceroses, summary.

SPECIES	Age	Size[a]	Locomotory Adaptation[a]	Teeth	Height of Food[b]	Diet	Biotope
Elasmotherium	Middle	very large	?	hypsodont	?	mainly	?
Coelodonta antiquitatis	and	medium	graviportal		low	grass	"cold steppe"
Dicerorhinus hemitoechus	Late	medium	mediportal	brachyodont		low growing herbs, grass	temperate open habitats
Dicerorhinus kirchbergensis	Pleistocene	large	graviportal		high	browse	deciduous forest (?)
Dicerorhinus etruscus	Early Pleistocene	small	cursorial				savanna,
Dicerorhinus jeanvireti	Lower Villafranchian	medium	subcursorial				open
Dicerorhinus megarhinus	Pliocene	large	mediportal				woodland

[a] Adapted from Fortelius 1979.
[b] Based on skull shape (Zeuner 1934; Loose 1975) (see text for explanation).

adaptations have been adapted from an unpublished manuscript (Fortelius 1979) and will be presented later elsewhere. The interpretations are summarized in table 13.2.

Dicerorhinus megarhinus (*de Christol 1834*), *Pliocene*

This large, relatively cursorially built rhinoceros had brachyodont teeth with low crowns (note that this is not tautology: functionally brachyodont teeth may be quite high crowned). The upper molars have apically constricted transversal lophs. This probably served to enhance the concavity of the occlusal surface and thus the shearing efficiency by causing rapid wear during the initial stages. Cingula are well developed, and the diet was probably similar to that of present-day browsing rhinoceroses; twigs, leaves, fruits etc. As the skeletal build suggests an open biotope, savanna-like conditions seem indicated. The same is true for the smaller and more cursorial species *D. jeanvireti* Guérin 1972, which in ecological terms appears to be rather intermediate between *D. megarhinus* and *D. etruscus*.

Dicerorhinus etruscus (*Falconer 1868*), *Early Pleistocene**

There seems to be good reason to regard *D. etruscus* as a complex of several subspecies, yet many features appear common. The following is based mainly on material from the "Cromer Forest Bed" of East Anglia, England.

D. etruscus was a very lightly built rhinoceros, with a skull shape indicative of food of intermediate height, comparable to that of *Diceros bicornis* (Zeuner 1934:49; Loose 1975:17). There are several interesting parallels between the two species in dental development as well. Both have brachyodont teeth with strong cingula and a thin cementum coating, and both have steep hypo- and metacones lacking the apical constriction found in *D. megarhinus* and *D. kirchbergensis*. This may indicate a relatively smaller importance of the shearing function (phase I of the occlusal stroke), and thus a functionally less extreme brachyodont condition. Compared with *D. megarhinus*, *D. etruscus* was smaller, more lightly built and probably of a less mesic dietary adaptation, and may represent an adaptation to a more cursorial mode of life and a drier, more open biotope.

Dicerorhinus hemitoechus (*Falconer 1868*), *Middle and Late Pleistocene*

Quite possibly *D. hemitoechus* is part of the same species complex as *D. etruscus*; certainly there are many common features. Two subspecies are

*I have deliberately avoided a more precise stratigraphic terminology, the European Pleistocene stratigraphy being at present subject to extensive revision. The designations "Early," "Middle," and "Late" Pleistocene are quite adequate for the present purpose.

recognized by Azzaroli (1961). The following applies mainly to the Late Pleistocene representatives, and is based primarily on material from England.

Although larger and more graviportal than *D. etruscus*, *D. hemitoechus* was the most cursorial of the Middle and Late Pleistocene rhinoceroses, with a locomotory build similar to that of *D. megarhinus*. The skull shape indicates a diet of low growing vegetation (Zeuner 1934:52, Loose 1975:22). The teeth are rather similar to those of *D. etruscus*, but the crowns are higher, cingula are reduced and there is more cementum. The molars are relatively enlarged (fig. 13.5) and M^3 is very nearly hypsodont, with a rather flat occlusal surface. In the lower teeth the lophids are perhaps more flattened buccally (cf. Staesche 1941:131). All these differences are in the direction of increased hypsodonty, and this has done much to consolidate the concept of a truly hypsodont, grazing "steppe rhino" mentioned in the introduction. The truth is that *D. hemitoechus* had mainly brachyodont teeth with some modifications toward hypsodonty, possibly indicating a shift to a diet containing more mature grasses. Probably this rhinoceros was an inhabitant of various open environments, such as the locally deforested river valley at Barrington, Cambridge (one of the main sites for remains of this species), with its rich flora of low growing vegetation (Gibbard and Stuart 1975:496).

Dicerorhinus kirchbergensis *(Jäger 1839), Middle and Late Pleistocene*

This large rhinoceros was of a more graviportal build than *D. megarhinus*, which is otherwise closely resembled, but still more cursorial than the larger of the living rhinoceroses. The skull shape is indicative of browsing (Zeuner 1934:52), and the species is commonly known as the "forest rhino." The teeth are brachyodont and the crown height similar to that of *D. hemitoechus*, but with a marked constriction of the transversal lophs of the upper molars. As in *D. megarhinus*, this gives a more strongly concave occlusal surface and increased shearing efficiency, indicating perhaps a more mesic dietary adaptation. The unequal wear and the gradual inwards rotation of the upper cheek teeth are very marked, and the ectolophs are characteristically curved around the longitudinal axis of the teeth. Cementum is thin and the enamel smooth, and the molars are not enlarged as in *D. hemitoechus* (fig 13.5). M^3 is low crowned and generally little developed, being no higher than the corresponding tooth in *D. megarhinus*, although there is a marked difference in premolar height between the two species (Fortelius 1979:34, 41, 44). This indicates that the importance of the posterior teeth was relatively small, a view consistent with the fact that M^3 shows none of the "grinding" adaptations found in *D.*

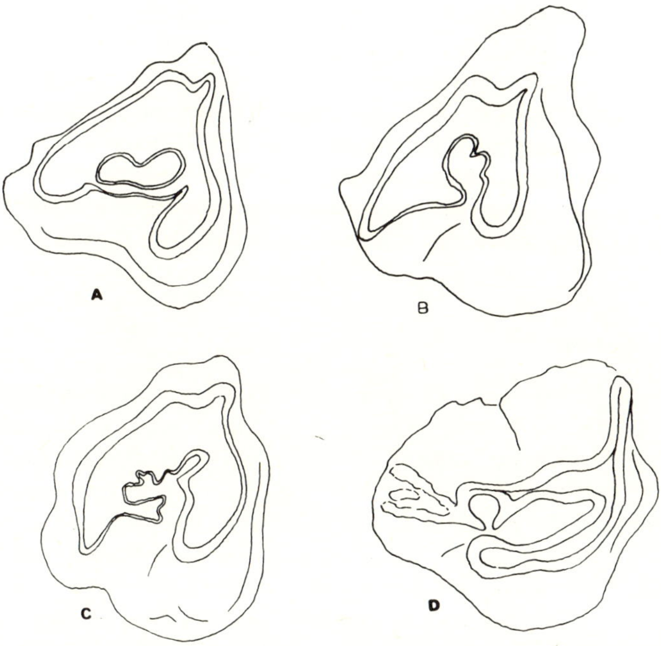

Figure 13.11. Right third upper molars of rhinoceroses. A. *Dicerorhinus* etruscus; B. *Dicerorhinus kirchbergensis*; C. *Dicerorhinus hemitoechus*; and D. *Coelodonta antiquitatis* (discussion in text). Not to scale.

hemitoechus (figs. 13.11b, 13.11c). As far as dietary adaptation is concerned, the deciduous forest appears a plausible biotope for this species. The graviportal build seems consistent with a relatively closed environment.

Coelodonta antiquitatis (*Blumenbach 1799*), *Middle and Late Pleistocene*

The woolly rhinoceros was a medium sized animal of compact, graviportal build, which may well reflect its relatively "cold" adaptation. Although the interpretation of the skull shape of this animal is not wholly clear, it appears to have carried its head low, at least for much of the time (Zeuner 1934:54f; Loose 1975:29). The teeth are extremely similar to those of *Ceratotherium simum* (cf. figs. 13.7, 13.10c and 13.11d), and a similar diet seems indicated. The ecological adaptation has been subject to some disagreement, and this is hardly the place to enter into that discussion. However, the teeth do indicate that the principal food of this animal was grass or some hypothetical, closely similar vegetation. Possibly the animal was seasonal, with a different diet in summer and winter, as was suggested by Loose (1975:24).

Another hypsodont rhinoceros of the Pleistocene, which may have occurred in Europe, was *Elasmotherium*, "(un) grand mammifère le plus mystérieux du Quaternaire européen" (Guérin 1976:407). Although little is known about this animal, it appears unlikely to have been in ecological competition with any of the more "traditional" European rhinoceroses: it was a huge animal with enormous, evergrowing molars (fig. 13.10d), and its remains are mainly found in southern Russia.

Conclusion

Perhaps the most striking fact about the Plio-Pleistocene rhinoceroses of Europe is the abrupt ecological change that seems to take place at the end of the Early Pleistocene (table 13.2). Until then, it appears, Europe was witnessing a succession of species with little temporal overlap. The ecological adaptations of these species were essentially similar, although a transition from large and mediportal to small and cursorial may be observed. With the onset of the Middle Pleistocene that type of rhinoceros disappears, and instead four ecologically different species are found. It is true, of course, that of these four species *Elasmotherium*, if present at all, was probably rare in Europe, and there was a temporal alternation between *Coelodonta antiquitatis* and the two *Dicerorhinus* species, the former being mainly glacial, the latter mainly interglacial elements. Yet *Dicerorhinus kirchbergensis*, *D. hemitoechus*, and *Coelodonta antiquitatis* are known to have occurred in the same general area at the same general time (e.g., Guerin 1973), and in any case the contrast to the Pliocene and early Pleistocene is dramatic. It is very tempting to regard the increased ecological diversity as a result of the increased environmental complexity in time and space, due to the climatic "deterioration" of the Ice Age.

References

Azzaroli, A. 1961. Validitá della specie *Rhinoceros hemitoechus* Falconeri. *Paleontographia Italica* 16:21–34.

Butler, P. M. 1972. Some functional aspects of molar evolution. *Evolution* 26(3):474–483.

Crompton, A. W. and K. Hiiemae. 1969. How mammalian molar teeth work. *Discovery* 5(1):23–34.

——. 1970. Molar occlusion and mandibular movements during occlusion in the American opossum, *Didelphis marsupialis* L. *Zool. J. Lin. Soc.* 49:21–47.

Fortelius, M. 1979. Ekologisk-funktionell morfologi hos Europas plio och pleistocena noshörningar. Master's thesis, Dept. of Zoology, University of Helsinki.

Gibbard, P. L. and A. J. Stuart. 1975. Flora and vertebrate fauna of the Barrington Beds. *Geol. Magazine* 112(5):493–501.

Gúerin, C. 1972. Une nouvelle espèce de rhinocéros (Mammalia, Perissodactyla) à Vialette (Haute-Loire, France) et dans d'autres gisements du Villafranchien inferieur européen: *Dicerorhinus jeanvireti* n. sp. *Docum. Lab. Geol. Sci. Lyon* 49:53–150.

Gúerin, C. 1973. Les trois espèces de rhinocéros (Mammalia, Perissodactyla) du gisement pléistocène moyen des abîmes de La Fage à Noailles (Correze). *Nouv. Arch. Mus. Hist. Nat.* 2:55–84.

—— 1976. Les périssodactyles: Rhinocerotidés. In H. de Lumley, ed., *La Prehistoire Française*, 1:405–408.

Hiiemäe, K. 1967. Masticatory function in mammals. *J. Dent. Res.* 46 5:1 suppl.:883–893.

Janis, Ch. 1979. Mastication in the hyrax and its relevance to ungulate dental evolution. *Paleobiology* 5(1):50–59.

Kay, R. F. and K. Hiiemäe. 1974. Jaw movements and tooth use in recent and fossil primates. *Am. J. Phys. Anthrop.* 40:227–256.

Loose, H. K. 1975. Pleistocene Rhinocerotidae of W. Europe with reference to the recent two-horned species of Africa and S.E. Asia. *Scripta Geologica* 33:1–59.

Lucas, P. W. 1979. The dental-dietary adaptations of mammals. *N. Jb. Paläont. Mh.* 1979(8):486–512.

Mellett, J. S. 1976. Paleobiology of North American *Hyaenodon* (Mammalia, Creodonta) *Vertebr. Evolution* 1:1–134.

Mills, J. R. E. 1967. A comparison of lateral jaw movements in some mammals from wear facets on the teeth. *Archs. Oral Biol.* 12:645–661.

Osborn, H. F., 1900. Phylogeny of the rhinoceroses of Europe. *Bull. Amer. Mus. Nat. Hist.* 13:229–267.

—— 1903. The extinct rhinoceroses. *Mem. Amer. Mus. Nat. Hist.* 1(3):75–164.

Rensberger, J. M. 1973. An occlusion model for mastication and dental wear in herbivorous mammals. *J. Paleont.* 47(3):515–528.

Rensberger, J. M. 1975. Function in the cheek tooth evolution of some hypsodont geomyoid rodents. *J. Paleont.* 49(1):10–22.

Staesche, K. 1941. Nashörner der Gattung *Dicerorhinus* aus dem Diluvium Württembergs. *Abh. Reichsst. Bodenforsch.* NF 200:1–148.

White, T. E. 1959. The endocrine glands and evolution, no. 3: os cementum, hypsodonty, and diet. *Univ. Michigan Mus. Paleont. Contr.* 13:211–265.

Wüst, E. 1922. Beiträge zur Kenntnis der diluvialen Nashörner Europas. *Cbl. Min. Geol. Paläont.* 1922:641–656.

Zeuner, F. E. 1934. Die Beziehungen zwischen Schädelform und Lebensweise bei den rezenten und fossilen Nashörnern. *Ber. naturforsch. Ges. Freiburg i. Br.* 34:21–80.

14. The Effects of Attrition on Tooth Dimensions of San (Bushmen)

J. F. VAN REENEN

ABSTRACT

Teeth show evidence of wear soon after they erupt into the mouth. In this study an attempt was made to determine the dimensional changes which take place in teeth as the result of attrition. The investigation was undertaken on the teeth of a group of San (Bushmen) skulls. Each tooth was first classified for degree of attrition and then the three dimensions; crown height, mesiodistal length, and labiolingual diameter were determined. Crown height diminished the most and the labiolingual diameters the least as wear of the teeth progressed, while the mesiodistal lengths occupied an intermediate position. More wear tended to occur in the teeth of males than in females. The maxillary first incisors tended to show more reduction of all three dimensions than any of the other maxillary teeth. This was also the case for crown height in the mandibular first incisors, but not for the other two dimensions where the first molars tended to show most reduction in size. The tooth material values were similar for all three dimensions when the mean values for the first incisor to first molar were added together. The problems of measuring teeth and the need to define techniques used in metrical studies are discussed. The view is expressed that measurements of teeth undertaken on dentitions which show evidence of attrition are likely to be low estimates of the original dimensions of the teeth.

Attrition has been defined as the wearing away of the substance of a tooth under the stress of mastication. It may be either occlusal, interproximal, or a combination of both (van Reenen 1964). When it is occlusal, the crown height of the tooth is reduced, and when it is interproximal, the mesiodistal length of the tooth is shortened. Since the buccal or labial and the lingual or palatal surfaces of the teeth are convex, especially in the molars and

I wish to acknowledge with thanks a grant received from Elida Gibbs (Pty) Ltd., which made it possible to undertake some of the statistical analyses. Thanks are also due to P. V. Tobias, Department of Anatomy, University of the Witwatersrand, Johannesburg, for encouragement and guidance.

premolars, with the widest points of bulging well below the occlusal surface, the labiolingual dimension of teeth are not affected even by a moderate degree of occlusal wear. When substantial occlusal attrition takes place the maximum labiolingual diameters of the teeth are also reduced.

The crowns of teeth begin to show evidence of attrition shortly after they erupt into the mouth. This wear continues throughout the lifespan of the individual, and its rate is related to the nature of the diet and other factors (van Reenen 1964). It is more marked in primitive peoples than civilized men, is found in the majority of ancient skulls, and is a prominent feature of the dentition of Stone Age man (Begg, 1954).

Thus the small mammalons on the incisal edges of the front teeth are clearly discernible when the incisors erupt but usually wear away soon after the teeth come into occlusion, leaving them with entire and often straight incisal edges. Similarly, the occlusal surfaces of the first molars begin to show facets of wear after they come into occlusion, particularly in primitive peoples where the cusps are lost and the crowns are worn flat soon after they reach adulthood (Campbell 1925; Davies and Pedersen 1955; van Reenen 1964).

Facets of occlusal wear are evident even in modern man existing on a refined diet. By the time he reaches adulthood, and certainly by the time he reaches middle age, these can clearly be seen. Wear facets are used by some clinicians (Lewin 1979) as a guide in determining the chewing pattern of a subject. For as Pleasure and Friedman (1938) put it, "teeth are worn differently because they were used differently." The wear patterns, therefore, reveal details about tooth usage and the motions and forces applied to the dentition. Not all people wear their teeth in the same way, but a particular pattern of wear is often a characteristic of a certain population or group (Molnar 1971). Soon after eruption each tooth usually makes point contact with the adjacent teeth on their mesial and distal sides. Interproximal wear increases this from a point to a surface contact. Wolpoft (1971) points out that this wear, first, is the result of lateral or side-to-side movement of the teeth and second, is dependent on the mesial force vector, or how hard the teeth push on each other when lateral movement of the teeth occurs under the force of mastication.

Qualitative and quantitative methods have been used to study attrition (Lysell 1958). In the case of the former some workers have made a detailed study of the actual facets of occlusal wear on each tooth (Murphy 1968). Others have simply observed the tissues exposed by the loss of tooth substance, and this has formed the basis of several classifications of attrition which have been put forward over the last century (Broca 1879; Martin 1928; Goldstein 1932; Davies and Pedersen 1955).

A review of the literature (Barker 1972) shows that few quantitative studies on tooth attrition have been undertaken. This is surprising in view of the fact that measurements of teeth are so frequently undertaken in anthropological studies. Barker (1972) in fact believes that Lysell (1958) has been the only worker who has attempted to recompute the original dimensions of worn teeth. Lysell (1958) showed that the mesiodistal diameters of the teeth decrease with increasing age. He did not observe sex differences in the degree of attrition, which was most marked in the first incisors and least pronounced in the third molars. The mandibular teeth wore more than the maxillary teeth except for the premolars, where the opposite was the case.

Mehta and Evans (1966) investigated the amount of attrition of teeth in Arkansas Indian skulls by measuring the mesiodistal diameters of all the teeth to determine the proximal wear, and by measuring crown height to determine the occlusal wear. They showed that extensive attrition of teeth in these people compensated for an inherent excessive tooth material over the basal bone.

The purpose of the present study was to attempt to show some of the dimensional changes which take place in teeth as a result of attrition. The study was carried out on San (Bushmen) skulls housed in museums in South Africa and Europe. Bushmen, as these people used to be called (Jenkins and Tobias 1977), are a yellow-skinned people, short in stature, who possess a number of very distinctive morphological characteristics which have been well described (Dart 1937, 1938; Tobias 1955, 1956). Their teeth are said to be microdont (van Reenen 1961), and this is another of the many features which distinguish them from the other indigenous peoples of southern Africa. The main reasons for choosing the San were that it had been well established by most anthropologists working in this field that their teeth showed both occlusal and interproximal attrition (Galloway 1937). The degree of wear is usually marked by the time they reach a mature age, and in those who survive into old age, the crowns of most of the teeth are reduced to the root surfaces (van Reenen 1964). The San population is unique in this regard, because attrition of the teeth is seldom as marked in the other population groups of southern Africa.

Materials and Methods

In only a few instances was the identity of the person known before death. The majority of skulls had to be classified as San on the possession of certain well described morphological features. Almost all the material had been examined, however, by well known physical anthropologists such as Drennan, Dart, Broom, Wells, Trevor-Jones, Tobias, and others, and their comments on this point were usually placed in the museum records, which were available for reference. These records, the opinion of other anthropologists working in the museum, and a physical examination of each adult skull and pelvic bones when available, made it possible to sex the material studied. In this regard reference was made to the differential sex characteristics listed by Brothwell (1963). The skulls which could not be classified as male or female with a fair degree of certainty were not used.

The dentitions of 140 male and 102 female skulls were included in the study. The degree of attrition for each tooth was recorded using the classification described by Davies and Pedersen (1955). According to this classification a tooth is placed in stage 0 when no or only slight wear of enamel has occurred; in stage 1 when there is clear evidence of enamel wear; in stage 2 when dentine is exposed, and in stage 3 when secondary dentine is exposed. A further stage was added to the classification, stage 4, where the pulp canal has become exposed and the crowns have been worn down to the necks of the teeth. Smith (1977) used a similar classification but added yet another category, stage 5, where the crown of the tooth had worn away completely. Observations on teeth without crowns were not made in this study. In view of the small number of teeth placed in stage 4, the observations made for this category were also discarded. Since this classification of tooth wear gives only a qualitative indication of the changes taking place on the occlusal surfaces of the teeth, the quantitative changes had to be assessed by tooth measurements.

Each tooth was measured with a Vernier sliding caliper, and the measurements were recorded to the nearest 0.1 mm. The beaks of the caliper were sharpened, to enable them to be placed between two adjacent teeth when this was necessary. Three measurements were made, the first being crown height (CH). This was taken as the distance between the tip of the mesiobuccal cusp and the deepest point of the cemento-enamel junction on the buccal or labial side of the tooth and measured parallel to the long axis of the tooth. This measurement was used by Black (1902), Weidenreich (1937), and Moorrees (1957). When the tip of the mesiobuccal cusp had been worn away, the one point of the caliper was placed on the occlusal or incisal surface, as near to the buccal or labial surface as possible.

The second measurement was the mesiodistal crown diameter (MD). This was taken as the greatest dimension of the tooth crown, measured in a mesiodistal direction, with the caliper held in a horizontal plane parallel to the occlusal, labial, or buccal surface of the tooth. In the case of teeth rotated in the jaw, the measuring points were taken as those which would have existed had the tooth been in a normal position. This measurement was used by Martin (1928), Pedersen (1949), Korenhof (1960), Tobias (1967) and others.

The third was the labiolingual or buccolingual crown diameter (LL). This was taken as the greatest distance between the labial and lingual surfaces of the tooth crown measured in a plane parallel to the projected mesiodistal diameter. The technique used to take this measurement is described by Tobias (1967). Each tooth was first classified for degree of attrition and then measured. Because of the small sample size the metrical data for right and left sides for each class of tooth were pooled. The mean dimensions of each class of tooth for each stage of attrition were determined and the data then subjected to statistical analysis. Thus, the mean, range, standard deviation, standard error of mean, and coefficient of variation were calculated for each of the three sets of measurements. Because of the small number of teeth falling into stages 0 and 1 degrees of attrition, the values of these two classes were also combined into one class.

Results

The mean values of the three diameters measured for each class of tooth at each of the various stages of attrition are set out in tables 14.1–14.16. Crown height diminished the most and the labiolingual diameters the least as wear of the teeth progressed, while the mesiodistal diameters occupied an intermediate position.

Table 14.1. Metrical characters of maxillary first incisor at three stages of attrition.

Metrical Character	Stage of Attrition	Sex	Number	Mean	Range	S.D.	S.E.	C.V.
CH	0 + 1	M	2	10.6	–	–	–	–
		F	4	11.2	11.5–11.00	0.26	0.13	2.34
	2	M	26	8.0	10.2–5.5	1.30	0.26	16.36
		F	14	8.6	10.0–6.5	1.17	0.31	13.62
	3	M	5	4.4	6.4–1.7	–	–	–
		F	8	4.7	6.4–3.2	1.15	0.41	24.81
MD	0 + 1	M	2	9.3	–	–	–	–
		F	8	8.0	9.0–6.0	1.11	0.39	13.97
	2	M	48	8.0	9.8–6.5	0.62	0.09	7.67
		F	21	7.8	8.3–7.0	0.47	0.10	6.06
	3	M	8	6.7	8.0–5.5	–	–	–
		F	7	6.9	7.0–6.0	0.85	0.32	12.26
LL	0 + 1	M	2	7.3	–	–	–	–
		F	8	6.9	7.5–6.0	0.63	0.22	9.16
	2	M	49	6.9	8.3–5.9	0.49	0.07	7.10
		F	23	6.7	7.3–6.0	0.36	0.07	5.35
	3	M	8	6.7	–	–	–	–
		F	10	6.7	7.8–5.2	0.78	0.25	11.57

S.D. = standard deviation. CH = Crown height.
S.E. = standard error of mean. MD = Mesiodistal length.
C.V. = coefficient of variation. LL = Labiolingual diameter.

Table 14.2. Metrical characters of maxillary second incisor at three stages of attrition.

Metrical Character	Stage of Attrition	Sex	Number	Mean	Range	S.D.	S.E.	C.V.
CH	0 + 1	M	11	8.8	9.9–7.0	0.93	0.28	10.54
		F	4	9.0	9.0–8.0	–	–	–
	2	M	24	7.3	9.6–3.4	1.35	0.28	18.52
		F	15	7.0	8.8–5.2	1.18	0.30	16.82
	3	M	6	4.0	5.4–2.0	1.30	0.53	32.80
		F	7	4.3	7.5–2.7	1.74	0.66	40.14
MD	0 + 1	M	17	6.6	8.0–5.5	0.67	0.16	10.17
		F	6	6.7	7.5–6.0	0.61	0.25	9.08
	2	M	39	6.4	7.3–5.2	0.54	0.09	8.45
		F	19	6.4	7.5–5.3	0.61	0.14	9.56
	3	M	12	5.4	7.2–4.3	0.82	0.24	15.29
		F	12	5.2	6.2–4.0	0.61	0.18	11.64
LL	0 + 1	M	17	6.2	7.0–5.2	0.57	0.14	9.17
		F	7	6.3	7.4–5.6	0.63	0.24	10.05
	2	M	39	6.3	7.8–5.0	0.54	0.09	8.64
		F	24	6.3	8.4–5.5	0.63	0.13	9.98
	3	M	12	6.3	7.0–5.5	0.36	0.10	5.71
		F	12	5.5	6.5–4.6	0.60	0.17	10.96

Table 14.3. Metrical characters of maxillary canine at three stages of attrition.

Metrical Character	Stage of Attrition	Sex	Number	Mean	Range	S.D.	S.E.	C.V.
CH	0 + 1	M	16	8.8	11.4–5.5	1.57	0.39	17.91
		F	5	8.9	11.0–7.5	1.30	0.58	14.67
	2	M	42	7.5	9.5–3.4	1.19	0.18	15.91
		F	25	7.1	9.6–5.0	1.31	0.26	18.63
	3	M	9	5.3	6.0–4.0	0.67	0.22	12.69
		F	10	5.0	8.2–2.1	2.07	0.66	41.15
MD	0 + 1	M	22	7.5	8.7–6.5	0.64	0.14	8.51
		F	10	7.3	7.8–6.5	0.42	0.13	5.85
	2	M	60	7.4	8.3–6.0	0.48	0.06	6.47
		F	34	7.2	8.0–5.5	0.56	0.10	7.87
	3	M	11	6.6	8.0–5.0	0.87	0.26	13.23
		F	13	6.3	7.5–5.3	0.70	0.20	11.16
LL	0 + 1	M	22	7.6	8.8–6.5	0.57	0.12	7.42
		F	9	7.4	8.0–6.7	0.39	0.13	5.33
	2	M	62	7.7	9.2–6.5	0.61	0.08	7.89
		F	34	7.4	8.3–6.2	0.57	0.10	7.67
	3	M	10	7.8	9.0–7.3	0.56	0.18	7.15
		F	13	7.3	8.0–6.0	0.64	0.18	8.74

Table 14.4. Metrical characters of maxillary first premolar at three stages of attrition.

Metrical Character	Stage Attrition	Sex	Number	Mean	Range	S.D.	S.E.	C.V.
CH	0 + 1	M	21	6.8	8.1–5.4	0.86	0.19	12.61
		F	10	7.5	9.7–6.8	0.90	0.28	6.21
	2	M	50	6.1	8.0–3.7	1.08	0.15	17.88
		F	34	5.8	7.4–3.8	0.94	0.16	16.26
	3	M	8	5.2	10.0–4.0	2.04	0.72	38.88
		F	11	3.4	5.2–2.0	1.00	0.30	28.93
MD	0 + 1	M	31	6.5	7.5–5.4	0.53	0.10	8.13
		F	18	6.9	7.3–6.0	0.36	0.09	5.30
	2	M	72	6.7	8.0–5.4	0.52	0.06	7.73
		F	47	6.4	7.5–5.2	0.49	0.07	7.68
	3	M	11	5.8	6.5–5.0	0.55	0.17	9.47
		F	14	5.6	8.0–4.8	0.79	0.21	14.23
LL	0 + 1	M	31	8.4	9.5–7.0	0.55	0.10	6.56
		F	17	8.2	9.1–3.5	1.25	0.30	15.31
	2	M	70	8.7	10.0–7.0	0.53	0.06	6.11
		F	45	8.4	10.0–4.2	0.89	0.13	10.63
	3	M	11	8.4	9.0–7.5	0.48	0.14	5.71
		F	15	8.3	9.2–7.3	0.48	0.13	5.82

Table 14.5. Metrical characters of maxillary second premolar at three stages of attrition.

Metrical Character	Stage of Attrition	Sex	Number	Mean	Range	S.D.	S.E.	C.V.
CH	0 + 1	M	21	6.2	7.5–4.6	0.76	0.17	12.25
		F	18	6.5	8.0–5.5	0.72	0.17	11.15
	2	M	48	5.4	7.0–3.3	0.88	0.13	16.14
		F	31	5.5	9.9–3.5	1.27	0.23	23.12
	3	M	3	3.6	4.4–2.7	–	–	–
		F	3	4.5	6.4–4.8	–	–	–
MD	0 + 1	M	32	6.5	7.3–5.4	0.51	0.09	7.81
		F	24	6.5	7.5–5.8	0.49	0.10	7.53
	2	M	75	6.4	7.5–5.3	0.54	0.06	8.41
		F	45	6.5	7.5–5.2	0.52	0.08	8.06
	3	M	3	5.4	5.7–5.2	–	–	–
		F	4	5.9	6.8–4.5	0.97	0.48	16.45
LL	0 + 1	M	32	8.5	9.3–7.3	0.55	0.10	6.49
		F	24	8.4	9.2–7.3	0.47	0.10	5.66
	2	M	74	8.7	10.0–6.5	0.67	0.08	7.72
		F	43	8.5	9.7–7.0	0.55	0.08	6.44
	3	M	3	8.4	9.3–7.5	–	–	–
		F	3	8.2	8.8–8.2	–	–	–

Table 14.6. Metrical characters of maxillary first molar at three stages of attrition.

Metrical Character	Stage of Attrition	Sex	Number	Mean	Range	S.D.	S.E.	C.V.
CH	0 + 1	M	19	6.8	8.2–5.0	0.81	0.19	12.01
		F	17	6.8	8.8–5.7	0.70	0.17	10.38
	2	M	106	5.7	8.9–3.5	0.86	0.08	15.12
		F	73	5.7	8.0–3.0	0.87	0.10	15.24
	3	M	15	4.5	6.3–3.0	1.16	0.30	25.92
		F	7	4.6	7.5–2.2	1.79	0.68	39.31
MD	0 + 1	M	24	10.7	12.0–9.9	0.56	0.12	5.24
		F	18	10.6	12.0–9.9	0.57	0.14	5.37
	2	M	140	9.9	11.5–8.0	1.24	0.11	12.55
		F	94	10.0	11.7–6.4	0.87	0.09	8.77
	3	M	22	9.0	10.7–6.8	0.98	0.21	10.86
		F	11	9.1	10.0–8.0	0.65	0.20	7.10
LL	0 + 1	M	24	11.1	11.9–10.0	0.55	0.11	4.99
		F	19	11.0	12.5–9.8	0.80	0.18	7.23
	2	M	140	10.8	12.6–8.4	0.65	0.06	6.04
		F	91	10.7	12.0–7.7	0.63	0.07	5.89
	3	M	21	10.8	12.1–9.8	0.67	0.15	6.22
		F	12	10.2	11.3–8.1	0.85	0.25	8.34

Table 14.7. Metrical characters of maxillary second molar at three stages of attrition.

Metrical Character	Stage of Attrition	Sex	Number	Mean	Range	S.D.	S.E.	C.V.
CH	0 + 1	M	59	6.5	8.1–5.0	0.70	0.09	10.69
		F	33	6.6	8.0–5.0	0.78	0.14	11.82
	2	M	67	6.0	10.0–4.0	1.11	0.14	18.38
		F	34	5.6	7.7–3.8	0.79	0.14	14.11
	3	M	3	3.7	5.0–1.5	–	–	–
		F	2	6.4	–	–	–	–
MD	0 + 1	M	78	9.9	11.1–8.2	0.66	0.08	6.66
		F	46	9.9	11.3–8.6	0.67	0.10	6.78
	2	M	85	9.8	11.0–8.45	0.74	0.08	7.57
		F	45	9.7	11.2–8.0	0.87	0.13	8.97
	3	M	3	8.7	9.0–8.5	–	–	–
		F	1	8.7	–	–	–	–
LL	0 + 1	M	78	11.0	12.5–9.6	0.64	0.07	5.86
		F	46	10.7	12.0–9.4	0.57	0.08	5.33
	2	M	87	11.0	12.3–10.0	0.55	0.06	4.99
		F	41	10.9	12.0–10.0	0.51	0.08	4.64
	3	M	3	10.7	11.0–10.5	–	–	–
		F	1	11.0	–	–	–	–

Table 14.8. Metrical characters of maxillary third molar at three stages of attrition.

Metrical Character	Stage of Attrition	Sex	Number	Mean	Range	S.D.	S.E.	C.V.
CH	0 + 1	M	55	6.2	9.4–3.6	0.87	0.12	14.01
		F	23	5.8	7.0–5.0	0.59	0.12	10.20
	2	M	25	5.4	8.5–4.0	0.97	0.19	17.89
		F	9	5.0	6.0–3.8	0.73	0.24	14.70
	3	M	–	–	–	–	–	–
		F	–	–	–	–	–	–
MD	0 + 1	M	76	8.6	11.0–5.0	0.92	0.11	10.72
		F	31	8.4	9.5–6.5	0.82	0.15	9.73
	2	M	32	8.3	9.6–6.5	0.76	0.13	9.16
		F	12	7.7	9.3–5.5	1.95	0.56	25.36
	3	M	–	–	–	–	–	–
		F	–	–	–	–	–	–
LL	0 + 1	M	78	10.5	13.0–7.0	1.04	0.12	9.83
		F	31	10.2	11.3–8.2	0.76	0.14	7.38
	2	M	32	10.5	12.3–9.6	0.66	0.12	6.29
		F	11	9.9	11.3–6.4	1.31	0.39	13.16
	3	M	–	–	–	–	–	–
		F	–	–	–	–	–	–

Table 14.9. Metrical characters of mandibular first incisor at three stages of attrition.

Metrical Character	Stage of Attrition	Sex	Number	Mean	Range	S.D.	S.E.	C.V.
CH	0 + 1	M	6	8.0	8.4–7.3	0.41	0.17	5.15
		F	5	8.2	9.0–6.0	1.23	0.55	15.02
	2	M	30	6.0	10.5–4.0	1.41	0.26	23.31
		F	22	6.1	8.7–3.5	1.43	0.30	23.31
	3	M	4	3.2	4.8–1.0	–	–	–
		F	9	3.3	4.5–2.3	0.82	0.28	24.73
MD	0 +1	M	6	5.4	5.4–5.0	0.47	0.15	6.94
		F	5	5.4	6.3–4.3	0.84	0.38	15.51
	2	M	45	4.7	5.5–3.0	0.60	0.09	12.15
		F	29	4.8	5.5–4.0	0.51	0.09	10.47
	3	M	8	3.9	4.5–3.4	–	–	–
		F	12	4.0	4.5–3.2	0.41	0.12	10.11
LL	0 + 1	M	6	5.3	5.5–4.8	0.27	0.11	5.13
		F	5	5.6	6.2–5.0	0.61	0.27	10.91
	2	M	46	5.3	6.2–4.5	0.42	0.06	7.81
		F	29	5.4	6.5–4.8	0.47	0.08	7.88
	3	M	8	5.2	6.0–4.5	–	–	–
		F	12	5.3	6.0–4.5	0.40	0.12	7.47

Table 14.10. Metrical characters of mandibular second incisor at three stages of attrition.

Metrical Character	Stage of Attrition	Sex	Number	Mean	Range	S.D.	S.E.	C.V.
CH	0 + 1	M	6	7.5	9.0–5.0	1.72	0.70	22.78
		F	6	8.5	9.8–7.0	0.94	0.38	10.96
	2	M	36	6.4	8.8–3.2	1.23	0.20	19.05
		F	24	6.9	9.0–4.5	1.05	0.22	15.35
	3	M	5	3.3	4.5–2.0	1.01	0.45	30.35
		F	12	3.9	5.1–2.0	1.00	0.29	25.68
MD	0 + 1	M	8	5.9	6.5–5.3	0.50	0.18	8.49
		F	7	5.7	6.0–5.5	0.25	0.10	4.36
	2	M	56	5.3	6.4–4.0	0.55	0.07	10.31
		F	34	5.4	6.5–4.5	0.54	0.09	9.90
	3	M	13	4.2	5.5–3.5	0.65	0.18	15.26
		F	12	4.4	5.4–3.8	0.53	0.15	12.09
LL	0 + 1	M	8	5.6	6.0–5.1	0.32	0.11	5.65
		F	7	5.7	6.6–5.3	0.54	0.20	9.49
	2	M	57	5.7	6.6–5.0	0.43	0.06	7.53
		F	36	5.6	6.0–4.0	0.45	0.08	8.14
	3	M	13	5.9	6.5–5.3	0.37	0.10	6.20
		F	15	5.7	6.2–4.9	0.34	0.09	5.95

Table 14.11. Metrical characters of mandibular canine at three stages of attrition.

Metrical Character	Stage of Attrition	Sex	Number	Mean	Range	S.D.	S.E.	C.V.
CH	0 + 1	M	12	9.1	11.4–6.7	1.36	0.39	14.93
		F	10	9.5	11.0–8.0	1.00	0.32	10.50
	2	M	54	7.8	11.0–4.0	1.55	0.21	19.80
		F	30	7.4	11.0–3.0	1.83	0.33	24.59
	3	M	4	3.8	4.6–2.8	–	–	–
		F	9	5.7	7.3–3.6	1.16	0.39	20.32
MD	0 + 1	M	15	6.9	7.8–5.5	0.64	0.17	9.30
		F	13	6.9	8.0–6.0	0.55	0.15	7.95
	2	M	74	6.7	8.7–4.7	0.55	0.06	8.22
		F	44	6.5	8.0–5.0	0.59	0.09	8.96
	3	M	10	6.0	8.0–4.4	1.22	0.39	20.24
		F	11	5.9	7.0–4.6	0.71	0.21	12.10
LL	0 +1	M	14	6.7	8.0–6.0	0.53	0.14	8.02
		F	13	6.6	7.0–6.0	0.38	0.11	5.75
	2	M	76	7.2	9.0–5.3	1.30	0.15	18.12
		F	44	6.7	7.6–5.5	0.47	0.07	6.98
	3	M	10	7.3	8.0–6.4	0.47	0.15	6.44
		F	12	7.0	7.5–6.2	0.42	0.12	6.02

Table 14.12. Metrical characters of mandibular first premolar at three stages of attrition.

Metrical Character	Stage of Attrition	Sex	Number	Mean	Range	S.D.	S.E.	C.V.
CH	0 + 1	M	18	7.0	8.0–6.0	0.66	0.16	9.52
		F	17	7.2	8.5–5.5	0.76	0.18	10.61
	2	M	51	5.7	10.4–1.0	1.58	0.22	27.70
		F	37	5.5	7.5–3.0	1.16	0.19	21.04
	3	M	1	3.0	–	–	–	–
		F	1	3.5	–	–	–	–
MD	0 + 1	M	23	6.9	8.0–6.0	0.65	0.14	9.44
		F	25	6.7	7.5–6.0	0.46	0.09	6.95
	2	M	61	6.0	8.0–5.0	0.98	0.13	16.27
		F	42	6.5	7.7–5.3	0.56	0.09	8.59
	3	M	5	5.4	7.0–4.4	–	–	–
		F	4	5.6	7.0–4.8	1.01	0.51	18.28
LL	0 + 1	M	27	7.5	8.3–6.9	0.44	0.08	5.90
		F	25	7.2	8.0–6.3	0.54	0.11	7.46
	2	M	69	7.5	9.0–6.2	0.55	0.07	7.25
		F	42	7.3	8.5–5.5	0.63	0.10	8.57
	3	M	8	7.3	8.5–6.5	0.70	0.25	9.59
		F	4	7.3	7.4–7.0	0.17	0.09	2.39

Table 14.13. Metrical characters of mandibular second premolar at three stages of attrition.

Metrical Character	Stage of Attrition	Sex	Number	Mean	Range	S.D.	S.E.	C.V.
CH	0 + 1	M	20	6.3	8.0–4.8	0.70	0.16	11.09
		F	15	6.5	8.0–5.3	0.83	0.21	12.67
	2	M	47	5.02	7.3–1.5	1.13	0.17	22.49
		F	33	4.6	7.0–1.2	1.17	0.20	25.37
	3	M	4	2.6	3.0–2.5	–	–	–
		F	–	–	–	–	–	–
MD	0 + 1	M	25	6.5	7.5–5.4	0.54	0.11	8.26
		F	23	6.9	8.0–6.0	0.58	0.12	8.33
	2	M	55	6.0	8.3–5.2	1.42	0.19	23.76
		F	42	6.6	7.4–5.5	0.46	0.07	7.05
	3	M	5	4.9	7.0–2.5	–	–	–
		F	1	5.0	–	–	–	–
LL	0 + 1	M	28	7.9	8.5–7.3	0.38	0.07	4.88
		F	22	7.7	8.3–6.8	0.43	0.09	5.66
	2	M	62	8.0	9.3–6.0	0.60	0.08	7.49
		F	41	7.9	8.6–6.6	0.54	0.08	6.88
	3	M	10	7.6	9.0–5.5	1.08	0.34	14.24
		F	1	7.5	–	–	–	–

Table 14.14. Metrical characters of mandibular first molar at three stages of attrition.

Metrical Character	Stage of Attrition	Sex	Number	Mean	Range	S.D.	S.E.	C.V.
CH	0 + 1	M	11	6.6	7.8–5.0	1.00	0.30	15.11
		F	7	5.9	7.0–5.3	0.59	0.22	9.85
	2	M	88	4.8	6.2–2.5	0.86	0.09	18.00
		F	4	5.0	7.3–1.2	1.14	0.16	22.80
	3	M	9	2.7	6.0–1.0	1.93	0.64	72.01
		F	6	3.1	4.5–2.5	0.74	0.30	23.60
MD	0 + 1	M	13	9.0	12.0–5.0	2.76	0.77	30.68
		F	6	10.9	12.0–9.8	0.83	0.34	7.58
	2	M	101	8.1	12.3–8.5	3.04	0.30	37.61
		F	65	10.4	11.9–8.0	0.79	0.10	7.62
	3	M	12	6.8	10.9–8.0	3.93	1.13	58.00
		F	10	9.4	10.9–8.2	0.83	0.26	8.79
LL	0 + 1	M	14	10.5	11.7–9.0	0.77	0.21	7.37
		F	7	10.8	11.8–10.2	0.63	0.24	5.85
	2	M	105	10.4	11.9–9.0	0.60	0.06	5.76
		F	69	10.2	11.3–9.2	0.52	0.06	5.17
	3	M	13	10.2	11.7–9.4	0.70	0.19	6.82
		F	11	10.2	10.8–9.2	0.57	0.17	5.64

Table 14.15. Metrical characters of mandibular second molar at three stages of attrition.

Metrical Character	Stage of Attrition	Sex	Number	Mean	Range	S.D.	S.E.	C.V.
CH	0 + 1	M	43	6.1	7.3–4.5	0.65	0.10	10.74
		F	31	6.1	7.7–5.0	0.71	0.13	11.58
	2	M	59	4.7	6.4–2.8	1.00	0.13	21.18
		F	37	4.8	6.9–3.0	0.84	0.14	17.52
	3	M	2	2.1	2.8–1.8	–	–	–
		F	–	–	–	–	–	–
MD	0 + 1	M	55	10.7	12.7–9.1	0.67	0.09	6.30
		F	34	10.5	12.5–9.5	0.74	0.13	7.00
	2	M	80	10.3	12.0–8.6	0.68	0.08	6.58
		F	46	10.3	12.4–9.0	0.79	0.12	7.70
	3	M	4	9.8	10.8–8.0	–	–	–
		F	–	–	–	–	–	–
LL	0 + 1	M	58	10.3	12.0–9.2	0.57	0.07	5.51
		F	35	10.1	11.4–9.3	0.57	0.10	5.58
	2	M	77	10.5	11.6–9.1	0.63	0.07	6.00
		F	46	10.3	11.4–9.3	0.52	0.08	5.02
	3	M	4	10.1	10.5–9.5	–	–	–
		F	–	–	–	–	–	–

Table 14.16. Metrical characters of mandibular third molar at three stages of attrition.

Metrical Character	Stage of Attrition	Sex	Number	Mean	Range	S.D.	S.E.	C.V.
CH	0 + 1	M	54	5.6	7.0–4.0	0.66	0.09	11.78
		F	29	5.5	7.3–3.5	0.81	0.15	14.80
	2	M	26	4.4	7.0–2.0	1.06	0.21	23.94
		F	9	4.4	6.0–3.0	0.86	0.29	19.43
	3	M	–	–	–	–	–	–
		F	1	4.2	–	–	–	–
MD	0 + 1	M	75	10.1	13.0–7.0	0.79	0.09	7.82
		F	38	9.4	11.5–7.5	0.98	0.16	10.42
	2	M	34	9.6	11.1–7.0	0.93	0.16	9.70
		F	14	9.6	11.0–7.2	1.01	0.27	10.50
	3	M	–	–	–	–	–	–
		F	1	11.5	–	–	–	–
LL	0 + 1	M	75	10.0	12.6–8.3	0.73	0.08	7.28
		F	38	9.6	11.4–8.0	0.78	0.13	8.20
	2	M	34	9.9	11.5–8.0	0.76	0.13	7.62
		F	14	9.7	10.5–8.8	0.60	0.16	6.23
	3	M	–	–	–	–	–	–
		F	1	10.0	–	–	–	–

The differences in mean values between the first and second and the first and third stages of attrition for each class of tooth are set out in table 14.17. This table shows that, with some exceptions, more wear occurred in the teeth of males than females.

As far as the different classes of teeth are concerned, the maxillary first incisors tended to show more reduction of all three dimensions than any of the other maxillary teeth. This was also the case for crown height in the mandibular teeth, but not for the other two dimensions, where the first molar and then the first incisor tended to show the most reduction.

There appeared to be a distinct tendency for crown height to reduce most in first incisors, then in second incisors, and third in the canine teeth. This pattern was more typical in the maxillary than mandibular teeth. The third and second molars tended to be the most conservative of the teeth in this regard, showing less reduction in tooth dimensions than the other teeth. Table 14.17 shows the sum of the mean values for the first incisor to first molar (6 teeth) and for first incisor to second molar (7 teeth), for each of the dimensions measured. This is called the tooth material values, a concept Ashley Howes (1954) first introduced for the sum of the mesiodistal lengths of the teeth. In this study this concept has been extended to include the sum of the crown heights and the sum of the labiolingual diameters.

Table 14.18 shows that when the tooth material values exclude the second molars, they are very similar for all three dimensions, the differences amounting to not more than 4 mm. When the second molars are included, then the labiolingual values for both the maxillary and mandibular teeth are the highest. The tooth material values also show that the greatest overall reduction is in the crown height dimensions, then in the mesiodistal dimensions, and last in the labiolingual dimensions.

Table 14.19 shows the amount of reduction in tooth material values which occurs with attrition. These values show that more overall reduction of the crown height and mesiodistal dimensions occur in the mandibular teeth. At the third stage of attrition this reduction amounts to more than 50

Table 14.17. Reduction in mean tooth dimensions by attrition (mm).

Metrical Character	Stage of Attrition	Sex	I^1	I^2	C^1	P^1	P^2	M^1	M^2	M^3
CH	(0 + 1) −2	M	2.6	1.5	1.3	0.7	0.8	1.1	0.5	0.8
		F	2.6	2.0	1.8	1.7	1.0	1.1	1.0	0.8
	(0 + 1) −3	M	6.2	4.8	3.5	1.6	2.6	2.3	2.8	–
		F	6.6	4.7	3.9	4.1	2.0	2.2	2.2	–
MD	(0 + 1) −2	M	1.3	0.2	0.1	0.2	0.1	0.8	0.1	0.2
		F	0.2	0.3	0.1	0.5	0.0	0.6	0.1	0.7
	(0 + 1) −3	M	2.6	1.2	0.9	1.1	1.1	1.7	1.2	–
		F	1.1	1.5	1.0	1.3	0.6	1.5	1.2	–
LL	(0 + 1) −2	M	0.4	+0.1	+0.1	+0.3	0.2	0.3	0.0	0.0
		F	0.2	0.0	0.0	+0.1	+0.1	0.4	+0.2	0.0
	(0 + 1) −3	M	0.6	+0.1	+0.2	0.0	0.1	0.3	0.3	–
		F	0.2	0.2	0.1	+0.1	0.2	0.8	+0.3	–
			I_1	I_2	C_1	P_1	P_2	M_1	M_2	M_3
CH	(0 + 1) −2	M	2.0	1.1	1.3	1.3	1.3	1.8	1.3	1.2
		F	2.1	1.6	2.1	1.6	1.9	0.9	1.3	1.1
	(0 + 1) −3	M	4.8	4.2	5.3	4.0	3.7	3.9	3.5	–
		F	4.9	4.6	3.8	3.6	3.5[a]	2.8	3.5[a]	1.3
MD	(0 + 1) −2	M	0.7	0.6	0.2	0.9	0.5	0.9	0.4	0.5
		F	0.6	0.3	0.4	0.2	0.3	0.6	0.2	+0.2
	(0 + 1) −3	M	1.5	1.7	0.9	1.5	1.6	2.2	0.9	–
		F	1.4	1.3	1.0	1.1	1.9	1.6	0.8[a]	+2.1
LL	(0 + 1) −2	M	0.0	+0.1	+0.5	0	+0.1	0.1	+0.3	0.1
		F	0.2	0.1	+0.1	+0.1	+0.2	0.4	+0.2	0.0
	(0 + 1) −3	M	0.2	+0.3	+0.6	0.2	0.3	0.3	0.1	–
		F	0.3	0.0	+0.4	+0.1	0.2	0.4	–	+0.4

[a]Estimated values.

Table 14.18. Tooth material values at three stages of attrition.

Tooth Material	Stage of Attrition	Sex	CH	MD	LL
I^1–M^1	0 + 1	M	48.0	47.1	49.1
(Maxilla)		F	49.9	46.0	48.2
I^1–M^2		M	54.5	57.0	60.1
(Maxilla)		F	56.5	55.8	58.9
I^1–M^1	2	M	4.0	44.8	49.1
(Maxilla)		F	39.7	44.3	47.8
I^1–M^2		M	46.0	54.6	60.1
(Maxilla)		F	45.3	54.0	58.7
I^1–M^1	3	M	27.0	38.9	48.4
(Maxilla)		F	26.5	39.0	46.2
I^1–M^2		M	30.7	47.6	59.1
(Maxilla)		F	32.9[a]	47.7	57.2
I_1–M_1	0 + 1	M	44.5	40.6	43.5
(Mandible)		F	45.7	42.6	43.4
I_1–M_2		M	50.5	51.3	53.7
(Mandible)		F	51.8	53.1	53.5
I_1–M_1	2	M	35.7	36.8	44.1
(Mandible)		F	35.5	40.2	43.1
I_1–M_2		M	40.4	47.1[a]	54.6
(Mandible)		F	40.3	50.5	53.4
I_1–M_1	3	M	18.6	31.2	43.4
(Mandible)		F	22.5	34.3	43.0
I_1–M_2		M	21.1	41.0[a]	53.5
(Mandible)		F	25.1	44.0	53.0[a]

[a] Estimated values

percent of the original crown height tooth material values and to a 20 percent reduction in the mesiodistal tooth material values.

Table 14.19 also shows that the crown height values reduce more in females than males between the first two stages of attrition. The reverse, however, holds when the difference in values between the first and third stages of attrition is considered. On the other hand the reduction in mesiodistal values is always greater in males than in females.

There appears to be a slight increase in male tooth material values for the labiolingual dimensions during the initial stages of attrition. This apparent increase is also seen for this dimension in some of the individual tooth values (table 14.17).

Dental attrition may be regarded as a measure of the destructive effects of oral function on the dentition. The magnitude of these effects as indicated by the degree of attrition will depend in part upon the intrinsic resistance offered by individual teeth. This resistance is related to factors such as the size, shape, form, hardness, and position of the tooth in the mouth.

An incisor tooth offers less resistance to occlusal wear than a molar by virtue of its blade-like shape and relatively small occlusal surface. Moreover, its CH dimension is considerably greater than that of a molar, and consequently a 2 mm loss of this dimension is functionally less significant in an incisor than in a molar.

In an attempt to equate the differences in resistance offered by each tooth to attrition, reduction in crown height at the third stage of attrition was taken to represent 10 units of wear, and the units of wear for the other dimensions (MD, LL) were then calculated in these units. In practice, this amounted to determining the mathematical factor which, when multiplied by the reduction in CH value at the third stage of attrition, would increase it to 10 mm. The values of the reduction in the MD and LL dimensions at the third stage of attrition were then multiplied by this factor. The same exercise was repeated for each category of tooth (see table 14.20).

Table 14.19. Reduction in tooth material values by attrition.

			Reduction in mm					
			CH		MD		LL	
Tooth Material	Stage of Attrition	Sex		%		%		%
I^1–M^1	(0 + 1) −2	M	8.0	16.7	2.3	4.9	0.0	0.0
(Maxilla)		F	10.2	20.4	1.7	3.7	0.4	0.8
	(0 + 1) −3	M	21.0	43.8	8.2	17.4	0.7	1.4
		F	23.4	46.9	7.0	15.2	2.0	4.1
I^1–M^2	(0 + 1) −2	M	8.5	15.6	2.4	4.2	0.0	1.5
(Maxilla)		F	11.2	19.8	1.8	3.2	0.2	2.9
	(0 + 1) −3	M	23.8	43.7	9.4	16.5	1.0	1.7
		F	23.6[a]	41.8	8.1	14.5	1.7	2.9
I_1–M_1	(0 + 1) −2	M	8.8	19.8	3.8	9.4	+0.6	+1.4
(Mandible)		F	10.2	22.3	2.4	5.6	0.3	0.7
	(0 + 1) −3	M	25.9	58.2	9.4	23.2	0.1	0.2
		F	23.2	50.8	8.3	19.5	0.4	0.9
I_1–M_2	(0 + 1) −2	M	10.1	20.0	4.2[a]	8.2	+0.9	+1.7
(Mandible)		F	11.5	22.2	2.6	4.9	0.1	0.2
	(0 + 1) −3	M	29.4	58.2	10.3[a]	20.1	0.2	0.4
		F	26.7	51.5	9.1[a]	17.1	0.5[a]	0.9

[a] Estimated values

Table 14.20. Calculated reduction in tooth dimensions when mean CH loss at third stage of attrition is increased to 10 mm.

Jaw	Tooth Character	Sex	Calculated Reduction in mm						
			I^1	I^2	C	P^1	P^2	M^1	M^2
Max	MD	M	4.19	2.49	2.57	6.87	4.23	7.39	4.28
		F	1.64	3.19	2.56	3.17	3.00	6.81	5.45
Mand		M	3.12	4.04	1.69	3.75	4.32	5.64	2.57
		F.	2.85	2.82	2.63	3.05	5.42	5.71	2.28
Max	LL	M	0.96	0.20	0.00	0.00	0.38	1.30	1.07
		F	0.29	1.70	0.25	0.00	1.00	3.63	1.36
Mand		M	0.41	+0.71	+1.13	0.50	0.81	0.76	0.28
		F	0.61	0.00	+1.05	+0.27	0.57	1.42	0.85
Ranking order			5	6	7	4	2	1	3

The first molars showed the greatest relative loss in mesiodistal and labiolingual crown lengths for males and females with one exception (mandibular labiolingual dimension in males). Generally this was followed by the second premolars, second molars, and first premolars. The canines were the most resistant teeth to mesiodistal and labiolingual wear, and generally this was followed by the second and then the first incisors. This seems to indicate that most of the chewing in this population was done in the first molar region.

Discussion

This study shows that all tooth dimensions are influenced by attrition. Crown height is the dimension most influenced by wear, and the labiolingual lengths the least affected. This is not an entirely new observation, because most anthropologists have for this very reason avoided recording crown height. For example, Pedersen (1949) did not include measurements of crown height "because occlusal attrition is a definite obstacle to accurate measurement of that dimension." In the light of the findings of the present study, investigators might reconsider the value of recording crown height. If a reduction in this dimension is suspected either because of reference to previously established mean values or because of visible signs of attrition, then the crown height length could give some indication of the magnitude of reduction of the mesiodistal and even labiolingual diameters of the teeth being studied. Crown height could almost be considered as the most useful indicator of the magnitude of reduction of the other tooth dimensions. In this regard, the finding that the tooth material values (excluding the second molars) are similar for all three dimensions may be a useful method for recomputing the original dimensions of the teeth, at least in San. This may, of course, not hold for other groups.

The definition of the crown height measurement requires attention. Pedersen (1949) pointed out that it is "an ill-defined measurement because it has to be taken from the gingival enamel border which is curved in various fashions." The wear produced by attrition affects the shape of the incisal or occlusal edge of the buccal surface, which usually becomes irregular and thereby adds to the difficulties of deciding on the exact points of measurements. Moreover, the sliding calipers usually employed for these measurements are difficult to use because the upper and lower points of measurement are often not parallel to the long axis of the tooth. If the upper arm of the calipers had an extension at right angles to the lower arm, which could rotate through an angle of 90°, it might be considerably easier to undertake this measurement. This would also help to increase the accuracy in recording this dimension.

There is little doubt that it is difficult to measure teeth. Hrdlička (1924) was skeptical about the value of tooth measurements and recommended that the points of measurement and the technique of measurement needs to be clearly defined. Goose (1963) defines some of the difficulties of measuring teeth and stresses that "unless the basic data are sound," statistical treatment of the findings is of little value. Recently Taylor (1978) goes so far as to conclude that "individual teeth or groups of teeth *in situ* are unsuitable for measurement, first because the morphology of a tooth is lacking in precise anatomical landmarks, and second because of the nature of variation. This makes it impossible for any particular tooth, or hypothetical average, to be acceptably representative of its kind or of any ethnic group." Although there is a great deal of truth in these statements, the view expressed might be regarded as being rather skeptical. The problems of tooth measurement may yet be solved and further research may well put forward proposals to deal with the question of variations in tooth morphology.

Some of the means, particularly the labiolingual diameters of the teeth, increased with attrition. This is clearly incorrect and is probably due to the small number of teeth falling into the first category of attrition. The mean values in cases of a small sample size are probably poor estimates of the true values. The high coefficient of variation values in these cases seem to support this contention. It is also possible that in some cases the measurements may well have been inaccurately recorded. This may be due not only to the difficulty of undertaking the actual measurements but also to the fact that the teeth probably moved or tipped in the mouth as a result of attrition during the lifetime of the individual.

There is considerable support for the view expressed by many workers (Klatsky 1939; Sicher 1953; Begg 1954; Neiburger 1977) that attrition should take place to allow teeth to adapt to the changing oral environment of man during his lifespan. Occlusal and interproximal wear allow the teeth to move into new positions of neutrality. This does not imply that heavy attrition is required, nor that the degree of attrition should be the same for every person. It is likely, therefore, that many of the teeth measured in this study tipped or moved as the result of wear.

It is difficult to decide with accuracy the long axis of a tooth when the occlusal surface has worn considerably and when the tooth is still in its bony socket. Assuming that buccal or lingual tipping had taken place in the teeth measured, the amount of tilt may not have been accurately estimated; this would provide an explanation for the finding that the labiolingual lengths of the teeth measured tended to increase rather than decrease with attrition in certain instances.

Measurements of teeth ought ideally to be undertaken soon after eruption of the teeth. More metrical studies ought to be undertaken on adolescents, and in this regard the use of orthodontic record casts may be useful. Measurements of teeth undertaken on dentitions which show evidence of attrition are likely to be low estimates of the original dimensions of the teeth.

It is clear from the present study that the true dimensions of Bushmen teeth have as yet not been determined by those who have attempted to do this, including myself. In making this statment it is assumed that it is possible to record these measurements with a fair degree of accuracy. The dimensions of San teeth would presumably be greater than those recorded for the first degree of attrition. The latter were found to be greater than the mean values obtained when the selection of teeth was not based on degree of attrition but on the exclusion of teeth showing marked wear. When this approach was used and measurements of teeth showing little, one, and two degrees of attrition were pooled in order to obtain mean values, these values were often smaller than those recorded for the first degree of attrition in the case of crown height and the mesiodistal diameters (unpublished data). Surprisingly enough, the mean values obtained using the latter approach were very close to those obtained by Drennan (1929) over fifty years ago (unpublished data).

The finding that more tooth wear occurred in male than in female Bushmen is not unexpected. The male physique and particularly the jaws are more robust than those of the female. This is likely to be related to genetic factors and also to the fact that males have always had to do the hunting, an arduous task requiring great strength and stamina. The females, on the other hand, have always been the food gatherers. They collect roots, fruits, berries and other food which requires far less effort than hunting. It is assumed that robust males exert more chewing force than the more delicately built females, an assumption which is not based on observation.

The tendency for crown height to be reduced more in the incisor and canine regions than in the molars is likely to be one of the characteristics of this particular group. This, however, has also been observed for Australian aborigines (Murphy 1968). As hunters and food-gatherers living a Stone Age existence, living San have been observed to frequently use their teeth—and especially the anterior teeth—as tools (van Reenen 1964). In some skulls where severe attrition of the teeth had taken place, the anterior teeth were rounded and out of occlusion, indicating that the loss of tooth substance had not been caused by the grinding of one tooth surface on another, but rather by the abrasive action of foods and other substances held between the anterior teeth during practices such as rope-making and the preparation of leather goods.

An unexpected finding was the observation that some of the mean values for males were smaller than those for females during the initial stages of attrition. For example, table 14.18 shows that the tooth material values for crown height and mesiodistal lengths in the maxilla and mandible were greater for females than for males during the initial stages of attrition. Studies on living Bushmen (van Reenen 1970) showed that males have larger teeth than females, and the findings of the present study therefore probably indicate that more tooth wear occurred at an earlier age in males than in females. The finding that, particularly in males, the maxillary first incisor was more worn than any of the other teeth, suggests that in these people it was the tooth more frequently used in function than the other teeth. It is known that proprioception of this tooth is good, and moreover, it is a more robust tooth and in a more stable position than the opposing anterior mandibular teeth, which are located in a mobile jaw. The observation that the canines were the most resistant teeth to mesiodistal and labiolingual wear is probably related to the fact that this class of tooth has the longest root and offers great resistance to bucco-lingual movement. They would therefore tend to wear less on their mesial and distal surfaces than the other teeth.

References

Barker, B.C.W. 1972. Concerning the possibility of estimating the original dimensions of worn aboriginal teeth. *Austral. Dent. J.* 17:448–453.

Begg, P. R. 1954. Stone Age man's dentition with reference to anatomically correct occlusion, aeitology of malocclusion and a technique for its treatment. Part I, *Am. J. of Orthod.* 40:298–312.

Black, G. V. 1902. *Descriptive Anatomy of the Human Teeth.* 4th ed. Philadelphia: S. S. White.

Bothwell, D. R. 1963. *Digging up Bones.* London: British Museum.

Broca, P. 1879. Instructions relatives a l'étude anthropologique du systemè dentaire. *Bull. de la Soc. d' Anthrop. de Paris.* Ser 3, T.Z. p. 128.

Campbell, T. D. 1925. *Dentition and Palate of the Australian Aboriginal.* Adelaide: Hassel Press.

Dart, R. A. 1937. The physical character of the Bushmen. *Bantu studies.* 11:175–246.

Dart, R. A. 1938. Fundamental human facial types in Africa. *S. Afr. J. Sci.* 35:341–348.

Davies, T. G. H. and P. O. Pedersen. 1955. The degree of attrition of the deciduous teeth and first permanent molars of primitive and urbanized Greenland Natives. *Br. Dent. J.* 99:35–43.

Drennan, M. R. 1929. The dentition of a Bushman tribe. *Ann. S. Afr. Mus.* 24:61–88.

Galloway, A. 1937. *The Skeletal Remains of Mapungubwe.* Cambridge University Press.

Goldstein, M. 1932. Caries and attrition in the molar teeth of the Eskimo mandible. *Am. J. Phys. Anthrop.* 16:421–430.

Goose, D. H. 1963. Dental measurement: an assessment of its value in anthropological studies. In D. R. Bothwell, ed., *Dental Anthropology*, pp. 125–148. Oxford: Pergamon Press: New York: MacMillan.

Howes, A. E. 1954. A polygon portrayal of coronal and basal arch dimensions in the horizontal plane. *Am. J. Orthod.* 40:811–831.

Hrdlicka, A. 1924. New data on the teeth of early man and certain fossil European Apes. *Am. J. Phys. Anthrop.* 7:109–132.

Jenkins, T. and P. V. Tobias. 1977. Nomenclature of population groups in Southern Africa. *Afr. Stud.* 36:49–55.

Klatsky, M. 1939. Dental attrition. *J. Am. Dent. Assoc.* 26:73–84.

Korenhof, C. A. W. 1960. *Morphogenetical Aspects of the Human Upper Molar.* Utrecht: Uitgeversmaatshaapij Neerlandia.

Lewin, A. 1979. Personal communication.

Lysell, L. 1958. Qualitative and quantitative determination of attrition and the ensuing tooth migration. *Acta Odond. Scand.* 16:278–292.

Martin, R. 1928. *Lehrbuch der Anthropologie*, 2nd ed. Jena: Gustav Fischer.

Mehta, J. D. and C. C. Evans. 1966. A study of attrition of teeth in the Arkansas Indian skulls. *Angle Orthod.* 36:248–257.

Molnar, S. 1971. Human tooth wear, tooth function and cultural variability. *Am. J. Phys. Anthrop.*, 34:175–189.

Moorrees, C. F. A. 1957. *The Aleut Dentition*, Cambridge: Harvard University Press.

Murphy, T. R. 1968. The progressive reduction of tooth cusps as it occurs in natural attrition. *Dent. Prac.* 19:8–14.

Neiburger, E. J. 1977. Flat-plane occlusion in the development of man. *J. of Prosth. Dent.* 38:459–469.

Pedersen, P. O. 1949. The East Greenland Eskimo Dentition. *Meddel. om Gronland.* 142:1–256.

Pleasure, M. A. and S. W. Friedman. 1938. Practical full denture occlusion. *J. Am. Dent. Assoc.*, 25:1606–1617.

Sicher, H. 1953. The biology of attrition. *Oral Surg., Oral Med., Oral Pathol.*, 6:406–412.

Smith, P. 1977. Diet and attrition in the Natufians. *Am. J. Phys. Anthrop.*, 37:233–238.

Taylor, R. M. S. 1978. *Variation in Morphology of teeth. Anthropologic and Forensic aspects.* Springfield, Ill: C. Thomas.

Tobias, P. V. 1955/56. Les Bochimans Auen et Naron de Ghanzi. Contribution à l' étude des "Anciens Jaunes" Sud-Africains. *L'Anthropologie*, 59:235–252;429–-461; 60:22–52;268–289.

Tobias, P. V. 1967. *Olduvai Gorge*, vol. 2. Cambridge: Cambridge University Press.

van Reenen, J. F. 1961. The use of the concept of tooth material as an indication of tooth size in a group of Kalahari Bushmen. *S. Afr. J. Sci.*, 57:347–352.

—— 1964. Dentition, jaws, and palate of the Kalahari Bushman, Part II. *J. Dent. Assoc. S. Afri.*, 19:38–44.

—— 1970. Sexual dimorphism in tooth size in Bushmen. *J. Dent. Assoc. S. Afr.*, 25:376–381.

Weidenreich, F. 1937. The dentition of Sinanthropus Pekinensis. *Palaeontol. Sinica* (whole series 101; new series D-I).

Wolpoft, M. H. 1971. Interstitial wear. *Am. J. Phys. Anthrop.* 34:205–208.

III

POPULATIONS

15. Dental Asymmetry as an Indicator of Developmental Stress in a Free-Ranging Troop of *Macaca Fuscata*

G. Gisela Nass

Mammals are characterized by morphological symmetry, yet directional asymmetry of certain organs is the norm. Lundström (1961) refers to these characteristics as "fundamental asymmetries." Additionally, qualitative and quantitative asymmetries exist to varying degrees in features that evolved homologously in both halves of the body. In fact, complete morphological symmetry is an ideal phenotypic expression never completely attained (Bader 1965; Biggerstaff 1979; Lundström 1961).

Optimal phenotypic realization of a given genotype requires a genetically well buffered and environmentally stress-free ontogenetic development. This concept has been expounded by Waddington (1957) with his term "developmental canalization," which denotes an ontogenetic pathway through which phenotypic values come to their expression inspite of environmental and/or genetic noise. According to Waddington, the degree and intensity of canalization may differ for individuals, causing adult features to deviate from the mean by different degrees. In sum, developmental stresses that interfere with optimal phenotypic expressions bring about a "breakdown in a polygenic system that controls the development of diploid organisms and buffers them against environmental stresses" (Baylit, Doykos, and Swanson 1968).

I wish to thank Sam Weinstein and Nicholas Piesco for reading the manuscript and suggesting several improvements. Edward Harris did most of the computational analyses, which is gratefully acknowledged. This work was funded, in part, by NIDR grant No. 1-T32-DE07047-02.

As the only unalterable body tissue once formed, the mammalian dentition is an excellent indicator of the quality of canalization during early development. In spite of the implied assumption that bilaterally antimeric teeth are under identical genetic control, metric and morphological dental asymmetries are not unusual. According to Kollar (1972), and Kollar and Baird (1970b), form and size of teeth originate in the local specificity of mesenchyme derived from neural crest and pre-otic ectomesenchymal cells. The genetic potential of proliferation, differentiation, and migration of embryonic mesenchymal tissue ideally is present in bilaterally equal dimensions to insure development of mirror images in the maxillary and mandibular teeth for proper masticatory function. However, past studies on dental metrics and morphology revealed the not uncommon existence of tooth asymmetries in humans as well as in other mammals (Bader 1965; Fischer 1954; Haataja, Rantala, and Ranta 1972; Harris and Nweeia 1980; Lundström 1961).

These findings not only evoke questions of causality, but also lead to the exploration of differences in expression. Applying both univariate and multivariate analyses to tooth dimension in twins, previous workers found evidence for the genetic basis of tooth crown size dimensions, but not for its expressed asymmetry (Potter and Nance 1976; Potter, Nance, Yu and Davis 1976). Further, it has been suggested that developmental instabilities are primarily responsible for asymmetric expressions in the teeth (Adams and Niswander 1976; Bailit, Workman, Niswander, and McClean 1970).

Accepting the contention that dental metric asymmetry is an expression of reduced canalization, an investigation of its potentially variable expression in mesiodistal and buccolingual dimensions, sex, arcade, and tooth type is warranted. Previous studies on dental asymmetry have been concerned with some of these variables, but have been confined to extant human groups (Adams and Niswander 1976; Bailit, Workman, Niswander, and McClean 1970; Doyle and Johnston 1977; Garn et al. 1966; Harris and Nweeia, in press), prehistoric humans (Sciulli 1978; Siegel and Doyle 1975b), and laboratory animals (Sciulli, Doyle, Kelly, P. Siegel, and M. I. Siegel 1979) to determine the relative genetic and environmental causes of this particular dental development. However, none of these studies was concerned with mensurational dental asymmetry of a population living in a relatively stress-free environment. The subjects of the study on which this paper reports, a troop of Japanese macaques, very closely met the criterion of such an environment, where abundant nutritional supplies, and a minimum of climatic, predatory, and social stresses were available to them. This troop provided excellent opportunity to quantify the natural expression of dental asymmetry under minimal development stress, and provided baseline information of the quality of developmental canalization in the deciduous and permanent teeth of male and female non-human primates.

Materials and Methods

In 1972, a troop of 138 *Macaca fuscata fuscata* that had previously undergone a natural fission from a larger troop was transplanted from Arashiyama near Kyoto, Japan to an enclosed range near Laredo, Texas. These macaques are native to Japan, and this particular troop came from a temperate, lush forest environment with optimal nutritional conditions and absence of predators (Clark 1974). Since the early 1950s, they had been provisioned with additional food items and medical care by researchers of the University of Kyoto. In short, their environmental imposed a minimum of stress and a maximum opportunity for a well-canalized development.

In 1973, the troop was subjected to its first physical and dental examination since its arrival in Texas. Of the total troop, 128 animals or 93 percent were caught and processed. The unequal sex distribution of 85 females and 43 males was a natural product of their social system.

Full-mouth alginate impressions were taken on the anesthesized animals, and casts were poured in dental stone. Measurements were taken from the casts with Helios dial caliphers measuring to 0.05 mm accuracy. Maximum mesiodistal and buccolingual tooth crown dimensions were measured in the standard manner (Leutenegger 1971; Leutenegger 1976; Moorrees 1957; Schuman and Brace 1954). For both dimensions, an adequate number of deciduous and permanent tooth pairs was available (table 15.1).

Pairing design tests (Woolf 1968) were used in assessing intra-observer repeatability by tooth type. F-ratios, ranging between 0.0 and 3.2, demonstrated a high degree of repeatability, clearly eliminating the contribution of measurement error to the expressed variability.

Analysis of the data was based on the differences between the left and right mesiodistal and buccolingual measurements of antimeres. To eliminate size differences among the teeth, transformation of the data was performed by dividing the side differences—d—by the mean dimension of

Table 15.1. Measurable deciduous and permanent pairs of teeth.

	Deciduous			Permanent		
	MD	BL	Total	MD	BL	Total
Male	149	105	254	168	124	292
Female	197	130	327	555	429	984
Total	346	235	381	723	553	1,276

the left and right antimeres, resulting in the transformed difference, *d* (Harris and Nweeia, 1980). Testing of the side differences disclosed no cases of directional asymmetry, ascertaining that all side variations are due to fluctuating asymmetry for all teeth in both arcades. Consequently, the measure of interest was the magnitude of asymmetry rather than the side difference. In order to prevent positive and negative values canceling each other out, the absolute values of the left-right side differences were taken.

$$d = \frac{|L - R|}{\frac{L + R}{2}}$$

The rationale of this approach was based on the premise that a certain quantity of asymmetry in a small tooth pair is of greater biological importance than the same quantitative asymmetry in a large tooth pair. Further, other sources of variation in tooth crown size, i.e., individual and sex differences, are thereby eliminated from the variance.

The mesiodistal and buccolingual data from the deciduous and permanent teeth were analyzed separately via a three-way factorial analysis of variance (Model I) (Sokal and Rohlf, 1969; Steele and Torrie 1960), testing for significant sources of variation of main effects, and first and second order interactions of sex, arcade, and tooth type. To examine sources of variation of the deciduous and permanent dentitions combined, main effects and first and second order interactions were subsequently computed for significance of dentition, arcade, and tooth type.

Results and Discussion

Morphological asymmetry is a manifestation of an organism's inability to attain identical development in bilateral structures. As a peculiarity of quantitatively variable characters, it is a strong indicator of the genotype's ability or lack thereof to effectively control ontogenetic development. This concept of developmental canalization, a genetic system that buffers ontogeny against developmental "noise," has been discussed by Waddington (1957, 1960), Bader (1965), Adams and Niswander (1976), Bailit et al. (1970), and Doyle and Johnston (1977).

Once calcified, tooth size can no longer respond to environmental or genetic influences. Since the deciduous teeth and first permanent molars complete morphodifferentiation prenatally in macaques (Booth 1972), and the remaining permanent teeth do so during the first few postnatal years, developmental noise has limited temporal opportunity to evoke impaired morphological responses, such as bilaterial metric asymmetries.

Both the deciduous and permanent dentitions evidenced asymmetries of

Table 15.2. Results of a three-way factorial analysis of variance for sex, arcade, and tooth type differences in bilateral asymmetry in mesiodistal and buccolingual dimensions of the deciduous and permanent dentitions.

Source of Variation	Deciduous				Permanent			
	Mesiodistal		Buccolingual		Mesiodistal		Buccolingual	
	D.F.	F-ratio	D.F.	F-ratio	D.F.	F-ratio	D.F.	F-ratio
Main effects								
sex	1	0.12	1	4.95*	1	0.35	1	0.20
arcade	1	11.61*	1	0.01	1	0.29	1	7.19*
tooth type	9	3.41*	7	1.83	13	2.47*	13	10.02*
First order interactions								
sex–arcade	1	4.54*	1	0.88	1	0.02	1	4.03*
sex–tooth type	9	0.45	7	0.20	13	0.59	13	0.06
arcade–tooth type	9	2.78*	7	0.27	13	1.03	13	0.17
Second order interactions								
sex–arcade–tooth type	9	1.77	7	0.14	13	0.20	13	0.49

*$P < 0.05$

the "fluctuating" type described by Van Valen (1957). In consideration of their ontogenetic differences, the results were discussed separately for the two dentitions, and for the analysis combining both.

Deciduous Teeth

The results of a three-way factorial analysis for mesiodistal and buccolingual dimensions are given in table 15.2. The significant main effects contributing to variation are arcade and tooth type for mesiodistal dimensions, and sex for buccolingual dimensions. Significant first and second order interactions were found only for the mesiodistal dimension between sex and arcade and for arcade and tooth type. There were no significant second order interactions in either dimension.

Main Effects. From these analyses, several important observations surface. To begin with, greater overall asymmetry is found in male teeth, significantly so in buccolingual direction (fig. 15.1). Since the deciduous teeth are formed prenatally, and one may assume that the mothers experienced similar pregnancies for male and female fetuses, this sex difference could possibly be interpreted in terms of reduced developmental canalization in males. Previous studies suggest this phenomenon in human children (Greulich 1947; Greulich, Crimson, and Turner 1953; Stini 1969, 1971).

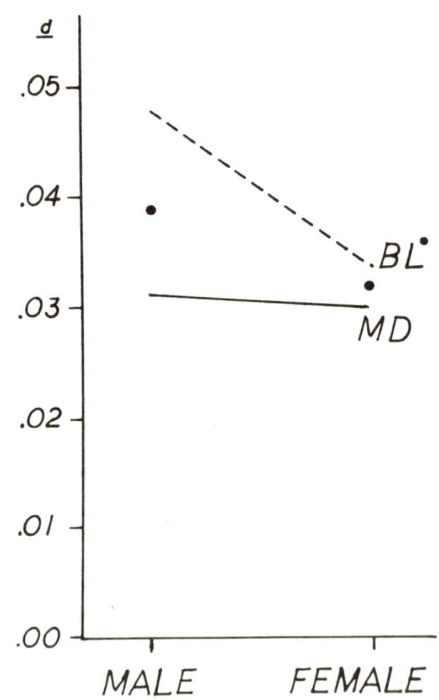

Figure 15.1. Deciduous teeth—mediodistal and buccolingual asymmetries in males and females.

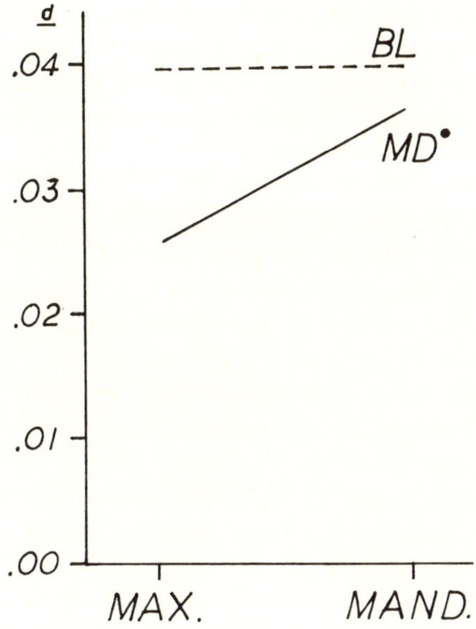

Figure 15.2. Deciduous teeth—mesiodistal and buccolingual asymmetries in the maxilla and mandible, with sexes pooled.

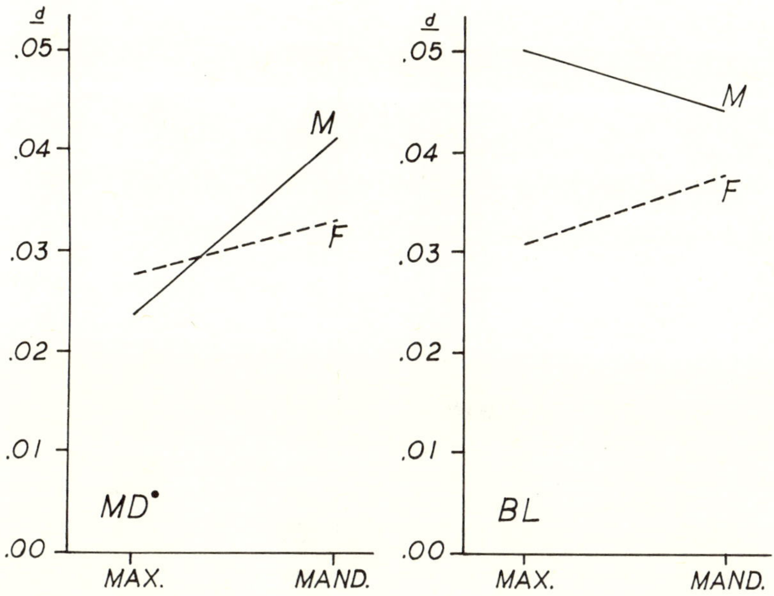

Figure 15.3. Deciduous teeth—mesiodistal and buccolingual asymmetries in the maxilla and mandible for males and females.

The significant arcade difference originates in greater asymmetry on the mandible (fig. 15.2), for which no ready explanation exists. Likewise, when comparing the coefficients of variation in a dental metric study of Colobine monkeys, it also becomes apparent that greater variability occurs in the mandibular teeth, particularly in mesiodistal dimension (Leutenegger 1971). Although these results are not directly comparable, they evoke speculations on possibly different relationships between early tissue differentiation and tooth calcification in the maxilla and mandible.

First Order Interactions. Figure 15.3 represents first order interactions between sex and arcade in mesiodistal and buccolingual directions, the former being significant. This sex-arcade interaction simply means that in mesiodistal dimension, males exhibit significantly greater asymmetries than do females, in the mandible. The overall greater asymmetries in males, and in buccolingual dimension, can again be seen.

The other significant first order interaction in mesiodistal dimension, arcade and tooth type, is shown in figure 15.4. When sexes are pooled, asymmetries in anterior teeth, i.e., incisors and canines, are significantly greater in the mandible, while those of the molars are actually greater in the maxilla. Although the pattern for buccolingual dimensions is essentially

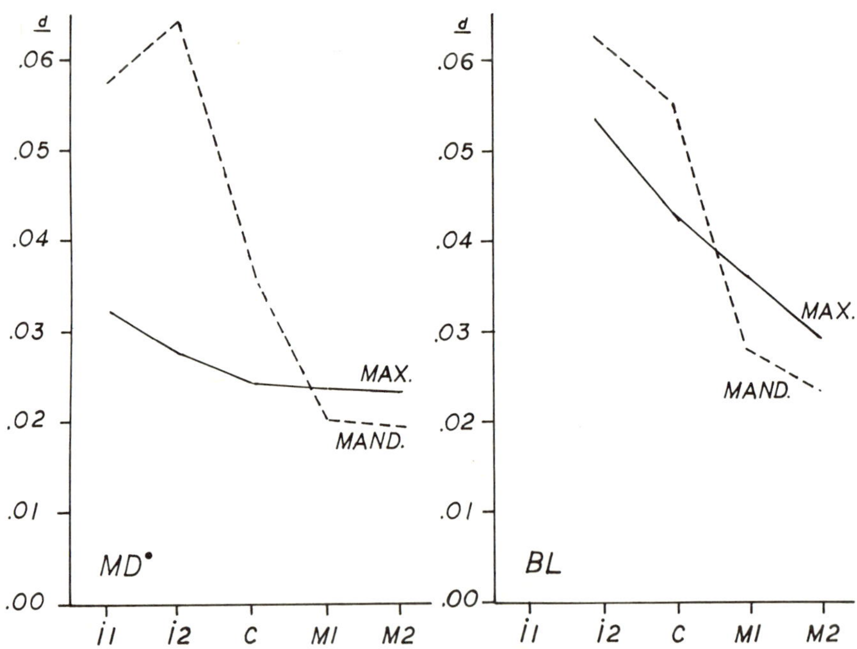

Figure 15.4. Deciduous teeth—mesiodistal and buccolingual asymmetries in the maxilla and mandible by tooth type, with sexes pooled.

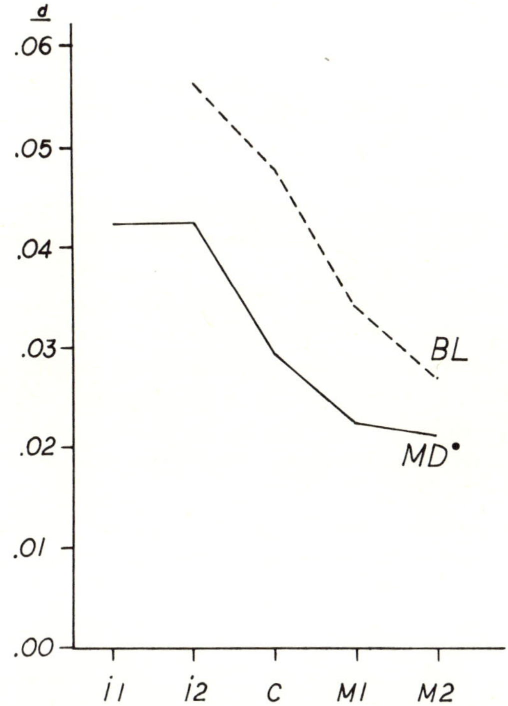

Figure 15.5. Deciduous teeth—mesiodistal and buccolingual asymmetries by tooth type, with sexes and arcades pooled.

similar, the interaction remains statistically non-significant. Since buccolingual data for mandibular deciduous first incisors in males is not available, arcade comparison could not be made for those teeth. This rather specific pattern of asymmetric expression in deciduous teeth begs the question of genetic control of development in incisors, canines, and molars. Potter and her co-workers (Potter, Nance, Yu, and Davis 1976) determined from multivariate analyses of twins that tooth size is primarily genetic in origin, and additionally, that the lower anterior teeth are influenced by an independent genetic system in both dimensions. Furthermore, they state: "Observed tooth size asymmetry would seem to be primarily environmental" (Potter et al. 1976). From these results it may be assumed that the intensity of developmental canalization is not the same for anterior and posterior teeth in the mandible.

Tooth Dimension. Comparison of buccolingual and mesiodistal dimensions reveals an overall greater expression of asymmetry in the former (fig. 15.5). Selection pressures for asymmetry are postulated to be greater in mesiodistal direction, not only because of its particularly important role in proper occlusion, but also because of the limited space availability in anterior-posterior direction. Evidence of separate influences by environmental sources of variation on mesiodistal and buccolingual dimensions support this view (Potter et al. 1976).

Permanent Teeth

Table 15.2 presents the results of the three-way factorial analysis of variance for sex, arcade, and tooth type differences in bilateral asymmetry in mesiodistal and buccolingual dimensions. Significant main effects are by arcade buccolingually, and by tooth type in both directions. Only one significant first order interaction occurs between sex and arcade buccolingually. None of the other interactions are significant in either direction. The overall impression gained from the permanent dentition is that the degree and some patterns of asymmetry diverge from those shown by the deciduous teeth.

Main Effects. Contrary to the asymmetric sex distribution in the deciduous teeth, no statistically significant differences appear in the

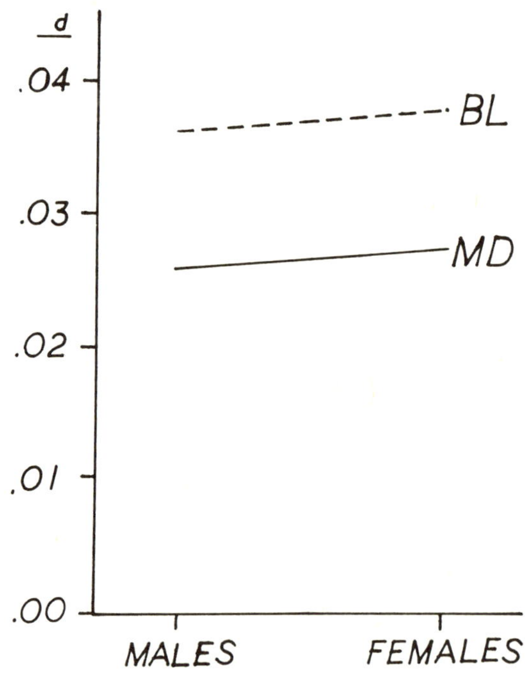

Figure 15.6. Permanent teeth—mesiodistal and buccolingual asymmetries in males and females.

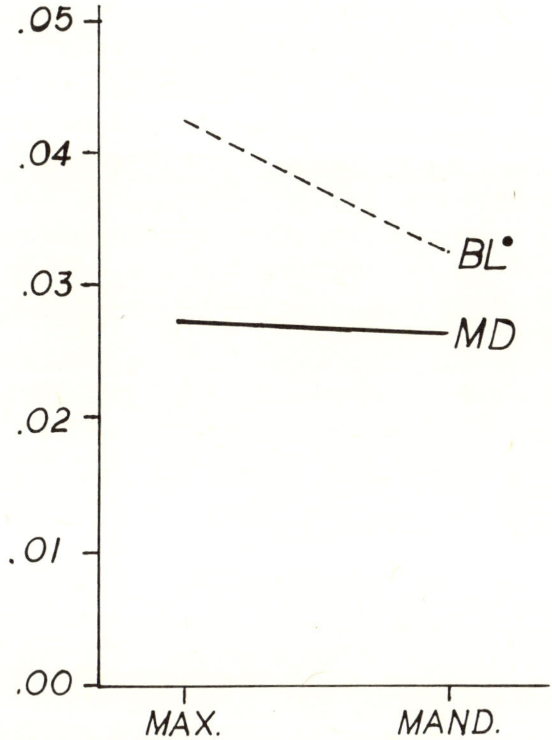

Figure 15.7. Permanent teeth—mesiodistal and buccolingual asymmetries in the maxilla and mandible, with sexes pooled.

permanent dentition, although female asymmetries are slightly greater (fig. 15.6). This corresponds to similar findings in human teeth by Garn et al. (Garn, Lewis, and Keresky 1966, 1967a), Bailit et al. (1970), and Leutenegger (1971), but not by Harris and Nweeia (1980). The Harris study clearly revealed greater overall asymmetries in female teeth, and the authors interpeted this as a partial manifestation of greater environmental pressures. The contention that developmental canalization favors human and possibly non-human primates females either does not bear out for the permanent dentition, or indicates that the results must be interpreted in a different light. It may be argued that selection pressures for dental symmetry may be greater for both females *and* males in the permanent teeth, as their proper function is of greater survival value than that of the deciduous teeth. Another explanation for greater female asymmetry could be found in the Lyon hypothesis, which states that females are mosaics for genes on the X-chromosome because only one, of either maternal or paternal origin, is genetically active in a given cell (Lyon 1961; Witkop 1967). This interpretation seems particularly attractive, since there are strong indications of X-chromosomal involvement in coding for tooth crown size (Alvesalo 1971). However, this does not explain the sex differences in the deciduous dentition.

A second difference between the two dentitions is found in the significantly greater maxillary asymmetry, particularly in buccolingual dimension (fig. 15.7). Garn and his co-workers (1966) and Harris and Nweeia (1980) presented similar results from their human data. Why there should be different developmental canalization for the deciduous and permanent teeth in the maxilla and mandible must for the time being remain an enigma.

Figure 15.8 depicts degrees of asymmetry by tooth type in mesiodistal and buccolingual dimensions, with sexes and arcades pooled. There is a general anterior-posterior decrease of asymmetry along the tooth row

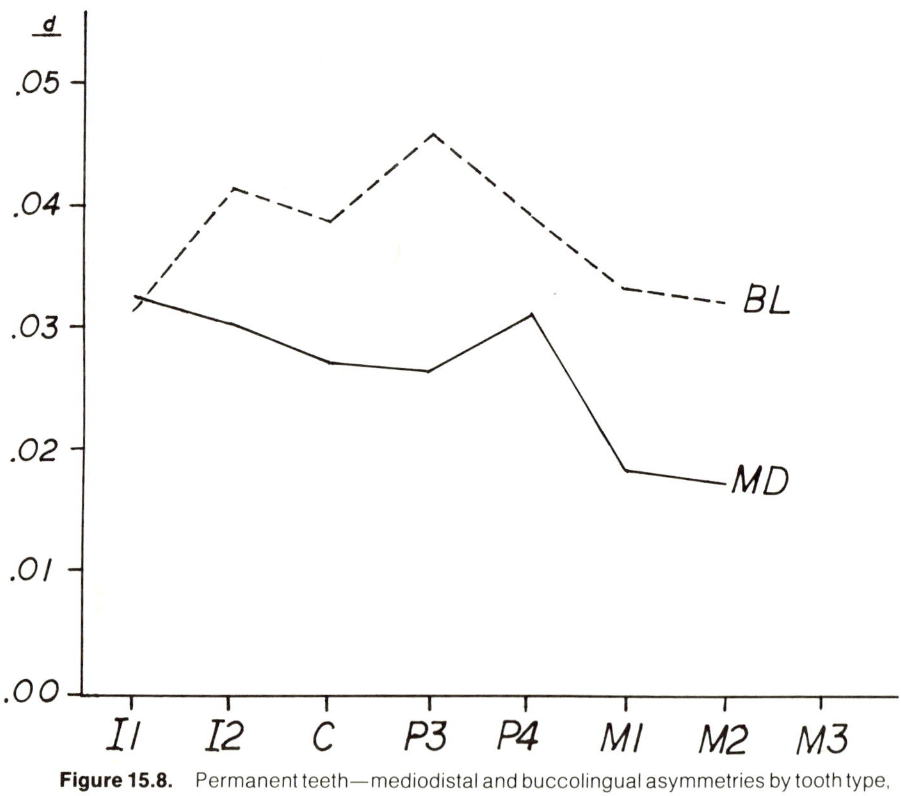

Figure 15.8. Permanent teeth—mediodistal and buccolingual asymmetries by tooth type, with sexes pooled.

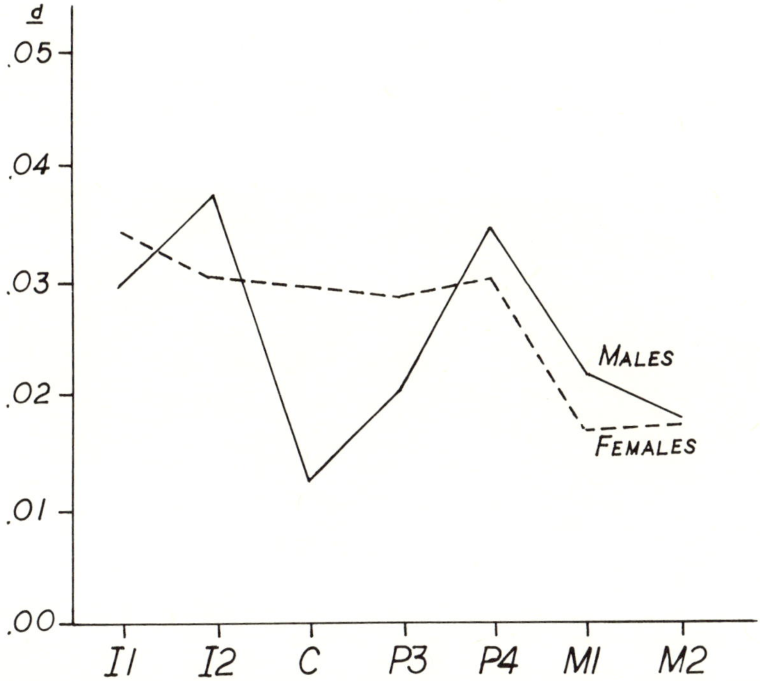

Figure 15.9. Permanent teeth—mesiodistal asymmetries by tooth type in males and females.

which crystallizes into greater tooth-specificity when tooth dimensions and sex are plotted out separately (figs. 15.9, 15.10). The significant contribution to variance by tooth type is clearly the property of both dentitions.

It was Butler (1939) who first put forward the idea of three morphogenetic fields in the heterodont dentition. This theory was expaned further by Dahlberg (1945), who suggested that each "tooth field" contains a stable, "polar" tooth that emanates developmental influences to adjacent teeth with decreasing power over distance. Thus, the most remote tooth in this "polarity" field exhibits greatest variance.

The tooth pattern of asymmetry in these macaques reflects quite literally the field polarity, particularly in mesiodistal direction of the male teeth (figs. 15.9, 15.10). The central incisors, canines, and second molars emerge as "stable" teeth, while the lateral incisors, premolars, and first molars express asymmetries of varying greater intensities. The incisor and canine polarities compare favorably to human studies (Garn et al. 1965, 1966). Further, a particularly non-human primate quality emerges in mesiodistal asymmetry of the canine (fig. 15.9). Male canines exhibit considerably less

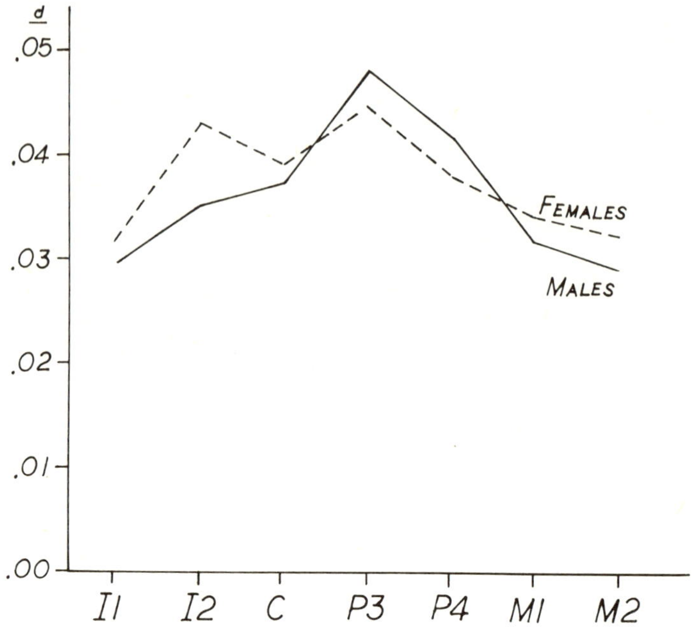

Figure 15.10. Permanent teeth—buccolingual asymmetries by tooth type in males and females.

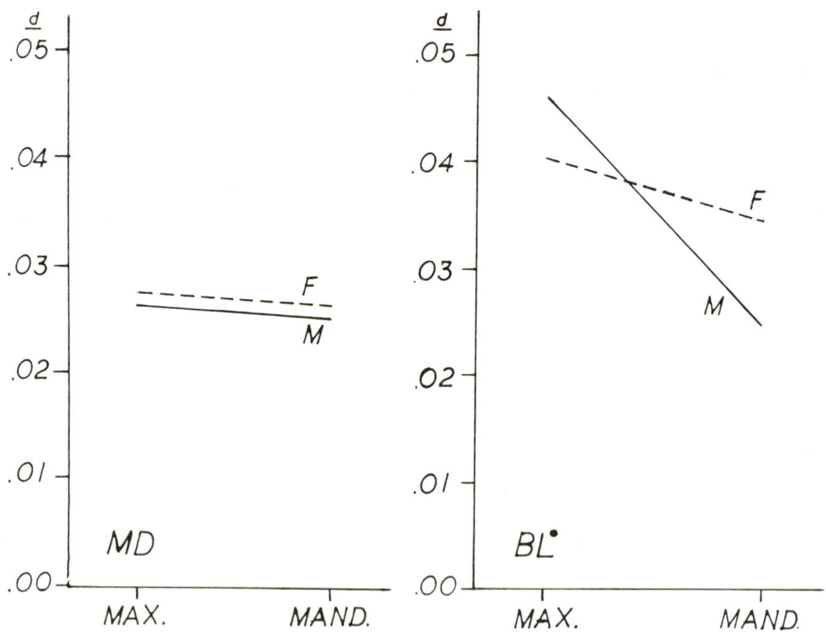

Figure 15.11. Permanent teeth—mesiodistal and buccolingual asymmetries in the maxilla and mandible for males and females.

mean asymmetry than the female counterpart; that is, they seem better canalized, exemplifying an important property of the male behavioral repertoire. In contrast, selection pressures show almost negligible sex dimorphism for canine symmetry in buccolingual direction.

The molar field is characterized by low asymmetry in both dimensions, and both sexes. The second molar seems to be the stable tooth that radiates developmental influence to the first molars and premolars. Leutenegger's (1971) contention that the premolars and molars in Colobine monkeys represent a single morphogenetic field is completely borne out in the buccolingual and partially in the mesiodistal dimension. Deviation of the latter is almost certainly due to the low asymmetry mesiodistally in the third premolar, particularly in males. The sectorial third premolar functions as a honing surface for the maxillary canines, and in this capacity enjoys a particularly selective advantage. The second molar field polarity is not an unusual property in monkeys, and in fact, has been labeled a "simian" feature (Hooijer 1948).

First Order Interactions. The only first order interaction to contribute significantly to variation is found between sex and arcade in buccolingual dimension (fig. 15.11). It means that males are more asymmetric than females in the maxilla, and less so in the mandible. Additionally, the arcade difference for males exceeds the inter-sex difference within each arcade.

Tooth Dimension. As in the deciduous dentition, the buccolingual dimensions are more asymmetric in both sexes, and in both arcades (figs.

Table 15.3. Results of a three-way factorial analysis of variance for dentition, arcade, and tooth type differences in bilateral asymmetry in mediodistal and buccolingual dimensions of the deciduous and permanent dentitions.

Source of Variation	Mesiodistal D.F.	Mesiodistal F-Ratio	Buccolingual D.F.	Buccolingual F-Ratio
Main effects				
dentition	1	5.0*	1	1.57
arcade	1	6.58*	1	2.59
tooth type	9	5.81*	9	2.65*
First order interactions				
dentition–arcade	1	5.97*	1	0.99
dentition–tooth type	9	0.99	9	0.61
arcade–tooth type	9	1.84*	9	0.25
Second order interactions				
dentition–arcade–tooth type	9	1.91*	9	0.28

*$P < 0.05$

15.6, 15.7). This finding, in conjunction with the previously seen presence of a sex-by-tooth difference in the mesiodistal dimension *and* the absence of a difference in the buccolingual dimension (figs. 15.9, 15.10), augments the contention that developmental canalization is more intense mesiodistally.

Deciduous and Permanent Dentitions

To elicit some measure of comparison between the degrees of canalization in the two dentitions, both were subjected simultaneously to a three-way analysis of variance, pooling sex (table 15.3). The observed variation, particularly in mesiodistal dimension, is due to the individual as well as the interactive effects of the factors taken in combination.

Since the differences between the two dentitions have already been discussed, only the most important aspects will be reiterated. The magnitude of asymmetry in both dimensions is systematically greater in the deciduous dentition by tooth type and arcade (figs. 15.12, 15.13), with proportionately greater expressions in the mandible.

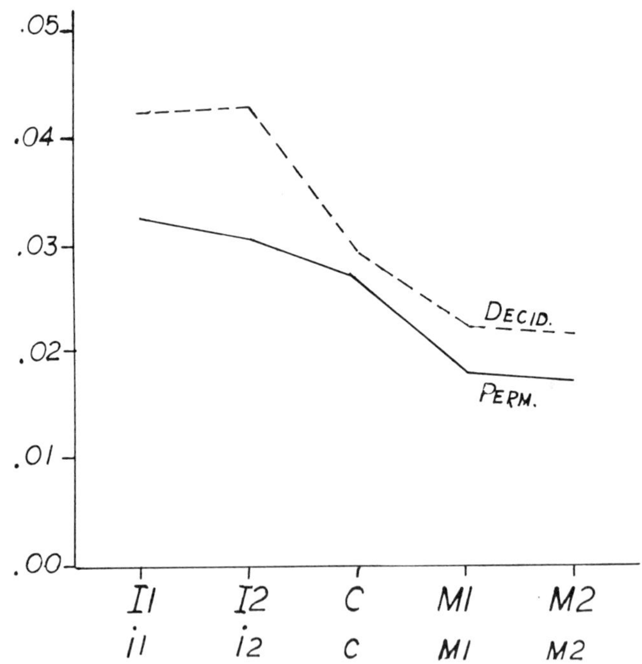

Figure 15.12. Asymmetries of the deciduous and permanent dentitions by tooth type, with sexes and arcades pooled.

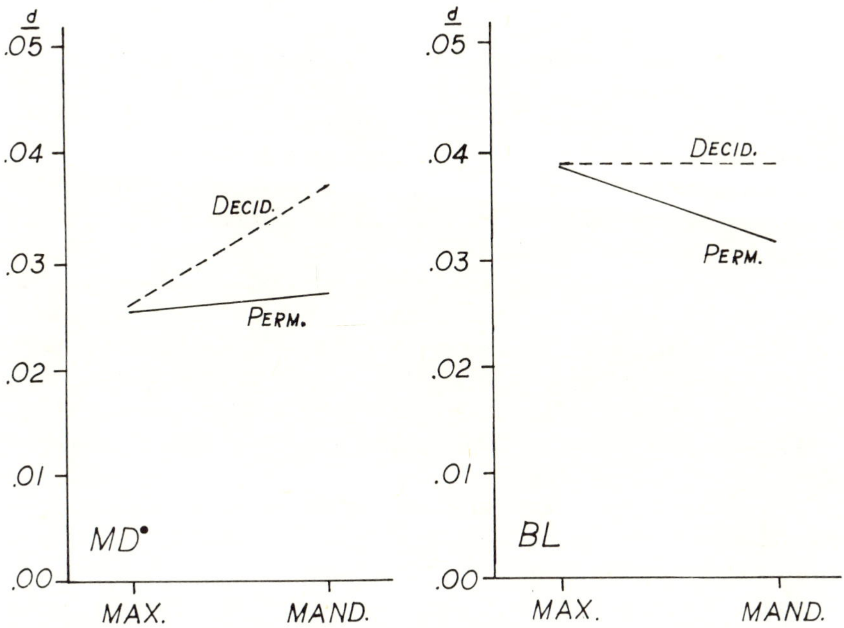

Figure 15.13. Mesiodistal and buccolingual asymmetries in the deciduous and permanent dentitions in the maxilla and mandible, with sexes pooled.

Two basic similarities are inherent in both dentitions: asymmetries are more intense buccolingually (fig. 15.13), and the gradient within and among morphogenetic fields is statistically identical (fig. 15.12). There also appear to be behavioral similarities between the deciduous and permanent first and second molars. In both dimensions they experience less asymmetry than the incisors and canines. It is postulated that they are better canalized, resulting in more efficient genetic buffering against variation. Since primates are endowed with an opposable thumb and, consequently, precision grip, the grasping and slicing functions of mammalian incisors lose in importance. In fact, several macaques of this troop do not suffer detrimental nutritional consequences upon the loss of their anterior teeth.

Summary and Conclusions

Dental asymmetry was assessed in a free-ranging, minimally stressed, non-human primate troop, *Macaca fuscata*. Asymmetric intensity was employed as a measure of developmental canalization in the deciduous and permanent teeth of males and females. The following results, obtained by multivariate analyses, provide quantitative baseline information of dental asymmetry in a normal, healthy macaque troop.

1. Antimeric teeth of the deciduous dentition in *M. fuscata* manifest considerably greater asymmetry than those of the permanent teeth. It is suggested that better developmental canalization of the latter is of functional importance, since those teeth need to serve the animals' nutritional requirements a great deal longer than do the primary teeth.

2. Sex differences in dental asymmetry are expressed differently in deciduous and permanent dentitions. In the former, male teeth are significantly more asymmetric, while in the latter, asymmetries are slightly but not significantly greater in females. The expected lower developmental canalization in males materializes only in the deciduous teeth. It is suggested that similar expressions of permanent tooth asymmetry in males and females may indicate the importance of selection pressures on well-canalized, permanent teeth, regardless of sex. Further, the Lyon hypothesis is implicated in trying to partially explain the sex difference in permanent teeth.

3. Mesiodistal tooth dimensions are uniformly better canalized than buccolingual dimensions in both dentitions. This may be a response to more intense selection for mesiodistal homogeneity, necessary for a space-limited, and occlusionally more important, tooth dimension.

4. While the deciduous dentition is more asymmetric in the mandible, the permanent teeth exhibit less uniformity in the maxilla. An interpretation of these results is difficult to make, and their biological origin must remain an enigma for the time being.

5. Morphogenetic field effects are clearly discernible in deciduous and permanent dentitions, with lower asymmetric expression in the polar teeth. The permanent teeth in particular exemplify a pattern of incisor, canine, and molar fields, the latter encompassing the premolars as well.

The sort of analysis used here is helpful not only in comparing among sexes, arcades, tooth types, and dentitions, but also in defining morphogenetic fields. It will be of interest to see how results from other non-human primates compare to and contrast with this species of macaques.

References

Adams, M.S. and J.D. Niswander. 1976. Developmental 'noise' and congenital malformation. *Gent. Res.* 10:313–317.

Alvesalo, L.J. 1971. The influence of sex chromosome genes on tooth size in man. (academic dissertation) *Proc. Finn. Dent. Soc.* 67:3–54.

Bader, R. S. 1965. Fluctuating asymmetry in the dentition of the house mouse. *Growth* 29:291–300.

Bailit, H. L., J. D. Doykos III, and L. J. Swanson. 1968. Dental development in children with cleft palates. *J. Dent. Res.* 47:664.

Bailit, H. L., P. L. Workman, J. D. Niswander, and C. J. McClean. 1970. Dental asymmetry as an indicator of genetic and environmental conditions in human populations. *Hum. Biol.* 42:626–637.

Biggerstaff, R. H. 1979. The biology of dental genetics. *Yearbook Phys. Anthrop.* 22:215–227.

Booth, S. N. 1972. Prenatal development of the teeth of the Rhesus monkey. Ph.D. dissertation. University of Wisconsin, Madison.

Butler, P. M. 1939. Studies of the mammalian dentition. Differentiation of the postcanine dentition. *Proc. Zool. Soc. Londong* 109:1–36.

Clark, T. W. 1974. Transplantation and adaptation of a troop of Japanese macaques to a Texas bushland habitat. University of Wisconsin, Madison, unpublished.

Dahlberg, A. A. 1945. The changing dentition of man. *J. Am. Dent. Assoc.* 32:676–690.

Doyle, W. J. and O. Johnston, 1977. On the meaning of increased fluctuating dental asymmetry: A cross-population study. *Am. J. Phys. Anthrop.* 46:127–134.

Fischer, B. 1954. Asymmetries of the dentofacial complex. *Angle Orthod.* 24:179–192.

Garn, S. M., A. B. Lewis, and R. S. Keresky. 1965. Size interrelationships of the mesial and distal teeth. *J. Dent. Res.* 44:350–355.

——1966. The meaning of bilaterial asymmetry in the permanent dentition. *Angle Orthod.* 36:55–62.

——1967a. Buccolingual size asymmetry and its developmental meaning. *Angle Orthod.* 37:186–193.

Greulich, W. W. 1951. The growth and developmental status of Guamanian school children in 1947. *Am. J. Phys. Anthrop.* 9:55–70.

Greulich, W. W., C. S. Crimson, and M. L. Turner. 1953. The physical growth and development of children who survived the atomic bombing of Hiroshima and Nagasaki. *J. Pediatr.* 43:121–145.

Haataja, J., A. Rintala, and R. Ranta. 1972. On asymmetric development in the first and second permanent molars in children with cranio-facial anomalies. An Orthopantomographic study. *Proc. Finn. Dent. Soc.* 68:15–19.

Harris, E. F. and M. T. Nweeia. 1980. Dental asymmetry as a measure of environmental stress in the Ticuna Indians of Columbia. *Am. J. Phys. Anthrop.* 53:133–42.

Hooijer, D. A. 1948. Prehistoric teeth in man and of the Orangutan from Central Sumatra, with notes on the fossil Orangutan from Java and Southern China. *Zool. Mededeelingen.* Quoted in D. R. Swindler, J. A. Gavan, and W. M. Turner. 1963. Molar tooth size variability in African monkeys. *Hum. Biol.* 35:104–122.

Kollar, E. J. 1972. Histogenetic aspects of dermal-epidermal interactions. In H.

Slavkin, ed., *Developmental Aspects of Oral Biology*, pp. 125–149. New York: Academic Press.

Kollar, E. J. and G. R. Baird. 1970b. Tissue interactions in embryonic mouse tooth germs. II. Inductive role of the dental papilla. *J. Embryol. Exp. Morph.* 24:173–186.

Leutenegger, W. 1971. Metric variability of the postcanine dentition in *Colobus* monkeys. *Am. J. Phys. Anthrop.* 35:91–100.

——1976. Metric variability in the anterior dentition of African Colobines. *Am. J. Phys. Anthrop.* 45:45–51.

Lundström, A. 1961. Some asymmetries of the dental arches, jaws, and skull, and their etiological significance. *Am. J. Orthod.* 47:81–106.

Lyon, M. F. 1961. Gene action in the X-chromosome of the mouse (*Mus musculus L.*). *Nature* 190:372–373.

Moorrees, C. F. A. 1957. *The Aleut Dentition.* Cambridge: Harvard University Press.

Osborne, R. H., S. L. Horowitz, and F. V. DeGeorge. 1958. Genetic variation in tooth dimensions: A twin study of the permanent anterior teeth. *Am. J. Hum. Genet.* 10:350–356.

Potter, R. H. and W. E. Nance. 1976. A twin study of dental dimension. I. Discordance, asymmetry, and mirror imagery. *Am. J. Phys. Anthrop.* 44:391–395.

Potter, R. H., W. H. Nance, P.-L. Yu, and W. B. Davis. 1976. A twin study of dental dimension. II. Independent genetic determinants. *Am. J. Phys. Anthrop.* 44:397–412.

Schuman, E. L. and C. L. Brace. 1954. Metric and morphologic variations in the dentition of the Liberian chimpanzee, comparisons with anthropoid and human dentitions. *Hum. Biol.* 26:77–93.

Sciulli, P. W. 1978. Developmental abnormalties in the permanent dentition in prehistoric Ohio Valley Amerindians. *Am. J. Phys. Anthrop.* 48:193–198.

Sciuli, P. W., W. J. Doyle, C. Kelley, P. Siegel, and M. I. Siegel. 1979. The interaction of stressors in the induction of increased levels of fluctuating asymmetry in the laboratory rat. *Am. J. Phys. Anthrop.*, 50:279–284.

Siegel, M. I. and W. J. Doyle. 1975b. The effect of cold stress on fluctuating asymmetry in the dentition of the mouse. *J. Exp. Zool.* 193:385–389.

Sokal, R. R. and F. J. Rohlf. 1969. *Biometry. The Principles and Practice of Statistics in Biological Research.* San Francisco: W. H. Freeman.

Steel, R. G. D. and J. H. Torrie. 1960. *Principles and Procedures of Statistics.* New York: McGraw-Hill.

Stini, W. A. 1969. Nutritional stress and growth: Sex differences in adaptive response. *Am. J. Phys. Anthrop.* 31:417–426.

——1971. Evolutionary implications of changing nutritional patterns in human populations. *Am. Anthrop.* 73:1019–1030.

Suarez, B. K. 1974. Neanderthal dental asymmetry and the probable mutation effect. *Am. J. Phys. Anthrop.* 41:411–416.

Van Valen, L. 1962. A study of fluctuating asymmetry. *Evolution* 16:125–142.

Waddington, C. H. 1957. *The Strategy of the Genes.* London: Ruskin House, Allen and Unwin, 1957.

——1960. Experiments on canalizing selection. *Genet. Res.* 1:140–150.

Witkop, Jr., C. J. 1967. Partial expression of sex-linked recessive amelogenesis imperfecta in females compatible with the Lyon hypothesis. *Oral Surg., Oral Med., Oral Path.* 23:174–182.

Woolf, C. M. 1968. *Principles of Biometry.* New York: Van Nostrand.

16. Size and Morphology of Children's Teeth in North Wales

DENYS H. GOOSE AND ERIC E. ROBERTS

The Welsh have always regarded themselves as rather different from the English, as was evident when Welsh Devolution was recently considered in the United Kingdom. Ultimately, the Welsh turned it down, but this does not alter the fact that they regard themselves as a nation. Particularly in North Wales, many prefer to speak Welsh rather than English.

In 1973 a national survey of children in England and Wales was carried out and published in 1975 (Todd). About 13,000 children from ages 5 to 15 were examined dentally and also by questionnaire to assess attitudes towards dental health. One of the intriguing findings was that Welsh children appeared to have a lower incidence of overcrowding, and it seemed worthwhile to examine such children to see what the possible causes could be. For example, were the jaws larger, teeth smaller? Several differences in occlusion were found, and these have been reported elsewhere (Roberts and Goose 1979). From what meager information was available for English children, however, on the whole there seemed no great difference in overall tooth size.

Materials and Methods

Two populations were looked at. The first was made up of all available 12-, 13- and 14-year-old children who were in attendance at secondary schools on the coastal strip of North Wales from Bangor to Abergele, who had lived there all their lives, and whose parents were also natives of that area (fig. 16.1). There were 239 boys and 260 girls. The second group consisted of families living within the island of Anglesey in an area which has had fluoridation since 1956 (1964 for whole island). For this study the parents and children needed to have lived all their lives in Anglesey, the

Figure 16.1. Map of the United Kingdom showing the island of Anglesey (a) and the coastal strip opposite (b).

parents needed to have a reasonable number of their own teeth, and the children should have received fluoride in the water supply since birth (or indeed, since conception). Only 50 families were able or willing to participate. In both studies, upper and lower alginate impressions were taken of teeth, of both thc parents and children.

Fluoridation and Tooth Size

An earlier study (Lovius and Goose 1969) had suggested that when Anglesey had fluoridation in one half of the island with the other acting as a control, children's teeth in the fluoridated area were found to be smaller. However, it was not possible to show significant differences. In a family study it should be-possible to contrast parents with children with regard to their tooth size, since only the children have had fluoride all their lives. One would then expect the children to have smaller teeth. However, a complicating factor enters here, since it is well established that a secular change is occurring in Western populations toward increased tooth size in children today (Hanna et al. 1963; Goose 1967; Garn et al. 1968).

Tables 16.1 and 16.2 show the finding with regard to buccolingual measurements, and it will be observed in only 8 out of 24 were the teeth of the children larger (two significantly). Thus there is substantial evidence for assuming the normal secular trend toward increased tooth size has been interfered with. Somewhat more were larger mesiodistally, (14 out of 28 comparatively and 8 significantly so) but it is difficult to assess the relevance of proximal attrition in the parents in this dimension, and so it is not as revealing as the buccolingual comparisons. That a secular trend does exist in Wales has not been actually proved, but Bowden and Goose (1969) in Liverpool (which is not many miles away from the area measured) showed this was the case for dimensions in the upper teeth, and Lavelle (1972) has shown similar findings in the nearby Midlands. Thus it does seem as though the secular trend for increased tooth size has been interfered with, and this would be consistent with fluoride reducing dimensions.

Table 16.1. Differences in buccolingual dimensions between parents and offspring in mm.

Upper	Father/Son	Mother/Daughter
I^1	0.17	0.10
I^2	0.20	0.04
C	0.32	0.07
PM^1	0.14	−0.24*
PM^2	−0.09	−0.14
M^1	−0.02	−0.04

– Means offspring larger than parent
*At 0.05 level

Table 16.2 Differences in buccolingual dimensions between parents and offspring in mm.

Lower	Father/Son	Mother/Daughter
I_1	0.16	−0.01
I_2	0.10	0
C	0.52	0.15
PM_1	0.29	0.05
PM_2	0.04	−0.25*
M_1	0.09	−0.06

− Means offspring larger than parent
*At 0.05 level

Additional evidence of the effect of fluoride in reducing tooth dimension might be adduced by comparing the Anglesey children with those from our other group from the mainland. In 34 out of 56 comparisons they were smaller.

Correlation coefficients were calculated between parents and offspring, having first ascertained there was no assortative mating. The father/daughter and father/son relationships are the least affected by environmental factors, and they showed that although there were a few individually low values, there was generally no great divergence from 0.5, the theoretical value in accordance with multifactorial inheritance without dominance ($\bar{r} = 0.47$ for father/son and 0.40 for father/daughter). Thus it appears that even if fluoride is acting environmentally on tooth size, it seems to be acting uniformly and does not disturb the heritability unduly.

Shoveling of the Maxillary Incisors

In the Liverpool study referred to earlier it had been noticed that a certain number of individuals had shovel-shaped incisors, and that some of these same individuals had Welsh names. When it became possible to study our Welsh children, it seemed therefore desirable to look into this more carefully.

Models from the previously mentioned children (492 in all) were examined and four categories of incisors noted for shovel, semi or moderate shovel, trace, or none.

The findings for laterals (table 16.3) are shown here along with American whites (Hrdlička 1920) and Chinese living in Liverpool (Goose 1977). The Welsh have substantially more semi and trace shoveling than the whites but of course nothing as much like the Chinese (incidentally 1.2 percent were anomalous and could not be used in the Welsh group).

Table 16.4 shows the similar findings for the central incisors. Again, the Welsh have considerably more shoveling than North American whites, in fact about double in both semi and trace groups. It may be said that

Table 16.3 Percent of shoveling of lateral incisors.

	Shovel	Semi-Shovel	Trace	None
Welsh	0.2	14.3	50.9	33.4
American whites[a]	1.2	8.1	33.2	54.8
Chinese[b]	70.1	29.1	0.8	0

[a] Hrdlička (1920).
[b] Goose (1977).

Hrdlička's figures are not typical, but Lasker's (1957) more recent combined figures for semi and trace shoveling in centrals were 45 percent, still considerably lower than the present findings. Unfortunately there are no figures from whites with which to compare these results in the United Kingdom, although we know populations of other Europeans like the Finns also seem to have relatively high incidences of the minor forms of shoveling (Koski and Hautala 1952).

What can we make of these figures? A first point is that although the figures have been combined for reference to others in the literature there is actually a sex difference (table 16.5) with the female showing more, supporting earlier evidence of Hrdlička (1920) and Morse (1937). Unfortunately, there is not much other material with which to compare this finding, and therefore it appears premature to comment on possible reasons.

It seems to be accepted that shoveling is genetically determined, with Lee and Goose (1972) suggesting a multifactorial inheritance and Blanco and Chakraborty (1976) calculating about 68 percent heritability. Clearly it is part of a Mongoloid dental complex as suggested by Hanihara (1963), but it may be as well to heed the warning by Palomino et al. (1977) that "the value of tooth morphology as an indicator of genetic difference between populations should only be regarded with cautious optimism." However, it seemed worthwhile to study the following possible reasons for these results in Welsh children, who were showing more shoveling than would be expected.

First, the criteria used might not be the same as other investigations. Second, there might be a lingering on of a so-called primitive trait in North

Table 16.4. Percent of shoveling of central incisors.

	Shovel	Semi-Shovel	Trace	None
Welsh	0	15.4	47.8	36.8
American whites[a]	2.0	6.4	23.2	68.4
Chinese[b]	77.5	21.7	0.8	0

[a] Hrdlička (1920).
[b] Goose (1977).

Table 16.5. Shoveling of central incisors.

	Semi-Shovel	Trace	None	Total
Male	24	97	113	234
Female	52	138	68	258
	76	235	181	492

$P < 0.001$.

Wales. Third, in some way some Mongoloid genes might have found their way into Wales, presumably at an early stage in pre-history.

It is hoped the first suggestion can be countered, since plaques produced by Dahlberg were used for standardization, and although actual measurements were not made, it seems likely there was not too much departure from the truth.

As to the second suggestion, it is true that earlier hominid forms show shoveling of their incisors. For example, Robinson (1956) describes this in connection with his classical study of the Australopithecines. Weidenreich (1937) found it in *Homo erectus*, which he concluded was related to the modern Mongoloid population.

Neanderthal man also often showed the trait, but as with the others, it is impossible to be sure of any frequencies, owing to the sporadic nature of fossil discoveries. The most interesting period would be the Upper Paleolithic, but unfortunately there are very few specimens which can be studied in the United Kingdom. In those that could be examined, too much attrition was present to show shoveling.

In discussing this trait in the Belgians, Brabant and Ketelbant (1975) show a decrease since Neolithic times through Megalithic and later periods. Carbonell (1963), however, found a lower proportion in Romano-British than in Saxon, a later population in Britain, although Berry (1976) shows that the ancient inhabitants of the Orkneys have a higher incidence than modern day inhabitants.

The evidence is scanty, but it could be postulated that early man in North Wales retained the "primitive trait" of shoveling from earlier ancestors. In support of this Fleure (1951) discerned Paleolithic characters in modern populations in the nearby Denbighshire moorlands. On the whole, it seems unlikely that the question can be resolved without more suitable specimens to examine, particularly from the Upper Paleolithic in Europe.

The final alternative is that Mongoloid genes found their way by early immigrations into North Wales which on the surface seems unlikely. However, there are pointers to the fact that this might be the case aside from the question of shoveling.

In general, Wales has a higher prevalence of B and O groups of the ABO blood group system. For example, Wales shows gene frequencies of 7.17 percent for B and 69.65 percent for O compared to 5.75 percent and 66.98 percent respectively for the adjacent part of England (Mourant et al. 1968).

Watkin (1956) has shown that in certain areas in North Wales still higher frequencies may be obtained. In a recent investigation, Caernarvon—which is also in the area of our study—showed a B gene frequency of 9.43 percent (Kopec 1970).

Wales also shows differences in Rh blood group frequencies, and the percentages of D negative in Wales is only 16.5 percent compared to 18.4 percent elsewhere in the United Kingdom (Mourant et al. 1968).

In studying shovel-shaped incisors in a northwestern Argentine population, Devoto et al. (1968) noted that there was a very high prevalence of shoveling, and that group O made up 91 percent of the population with only 1 percent Rhesus negatives; and commented on the possible Mongoloid affinities.

Turning to Asia itself, figures in China approximate 20 percent B genes and 60 percent O genes with somewhat lower figures for B in Asian Soviet populations generally. In addition, Mongoloid populations seem to have very low figures for D negatives.

This all seems to suggest that the Welsh blood groups are tending in values towards those of Mongoloid populations.

Like other countries, the early history of Wales is full of conjecture, but there is little doubt of the effects of the Celts who moved into Wales in the centuries just before the Roman invasion. For example Brown (1965) says that the high percentage of blood group O frequencies in North Wales is associated with Celtic speaking peoples. From this period onwards few other invasions of Britain had much effect on Wales and Offa's Dyke, for example, shows in a way the cut-off nature of Wales. It can be presumed that the genetic make-up of the Welsh was not altered substantially by later Viking raids except perhaps in Pembrokeshire in the south.

What could be the relevance of the Celts with respect to Mongoloid genes? In the fifth century B.C. the Celts, of course, were associated with the Hallstadt culture in Austria, but prior to that their origins are obscure. However, there is now archaeological evidence showing their precursors may have come from farther east. An early Indo-European culture of the Copper Age has been shown to have developed in the Russian Steppes. This group, to whom Gimbutas (1965) referred as the Kurgan people, were much influenced by the horse. Being nomadic, they surged out westwards and moved into Europe at different periods, and only later, after they had mixed with local populations, did new cultures arise. Thus perhaps, much later on, the historical Celts came into being. If the general idea of migration from the Russian Steppes becomes accepted, it is easy to see that the close proximity to Mongoloid populations in Central Asia could have caused gene flows to occur. Possibly these included those for shoveling of incisors. It is interesting to note that Axelsson and Kirveskari (1977), when finding the deflecting wrinkle common among Icelanders, suggested that from lack of such proximity this characteristic could not, therefore, be part of the Mongoloid dental complex. Perhaps they were wrong to assume this since the Icelanders had their origin, at least in part, from the Celts!

References

Axelsson, G. and P. Kirveskari. 1977. *Am. J. Phys. Anthrop.* 47:321.

Berry, C. 1976. *Am. J. Phys. Anthrop.* 45:257.

Blanco, R. and R. Chakraborty. 1976. *Am. J. Phys. Antrhop.* 44:233.

Bowden, D. E. J. and D. H. Goose. 1969. *J. Med. Genet.* 6:55.

Brabant, H. and C. Ketelbant. 1975. *Bull. Group Europ Rech. Stomat et Odont.* 18:121.

Brown, E. S. 1965. *Heredity* 20:289.

Carbonell, V. M. 1963. In D. R. Brothwell, ed., *Dental Anthropology*. Oxford: Pergamon Press.

Devoto, F. C. H., N. H. Arias, S. Ringuelet, and N. H. Palma. 1968. *J. Dent. Res.* 47:820.

Fleure, H. J. 1951. *A Natural History of Man in Britain.* London: Collins.

Garn, S. M., A. B. Lewis, and A. Walenga, 1968. *J. Dent. Res.* 47:503.

Gimbutas, M. 1965. *Bronze Age Cultures in Eastern Europe.* The Hague: Mouton.

Goose, D. H. 1967. *J. Dent. Res.* 46:959.

Goose, D. H. 1977. In A. A. Dahlberg and T. M. Graber, *Orofacial Growth and Development*, The Hague: Mouton.

Goose, D. H. and E. E. Roberts. 1979. Possible influence of fluoridation on tooth crown size. *J. Dent. Res.* 58(6):1562–1563.

Hanihara, K. 1963. In D. R. Brothwell, ed., *Dental Anthropology*, Oxford: Pergamon Press.

Hanna, B. L., M. E. Turner, and R. D. Hughes. 1963. *J. Dent. Res.* 42:1322.

Hrdlicka, A. 1920. *Am. J. Phys. Anthrop.* 3:429.

Kopec, A. C. 1970. *The Distribution of the Blood Groups in the U. K.* London: Oxford University Press.

Koski, K. and E. Hautala. 1952. *Am. J. Phys. Anthrop.* 10:127.

Lasker, G. 1957. *J. Forens. Sci.* 2:401.

Lavelle, C. L. B. 1972. *Am. J. Orthod.* 61:29.

Lee, G. T. R. and D. H. Goose. 1972. *J. Med. Genet.* 9:336.

Lovius, B. B. J. and D. H. Goose, 1969. *Brit. Dent. J.* 127:322.

Morse, W. R. 1937. *J. W. China Border Res. Soc.*, Supplement 8 (quoted by Lasker. 1945). *Am. J. Phys. Anthrop.* 3:129.

Mourant, A. E., A. C. Kopec, and K. Domaniewska-Sobczak. 1968 *The Distribution of the Human Blood Groups and Other Polymorphisms.* London: Oxford University Press.

Palomino, H., R. Chakraborty and F. Rothhammer. 1977. *Hum. Biol.* 49:61.

Roberts, E. E. and D. H. Goose. 1979. *Brit. Dent. J.* 146:17.

Robinson, J. T. 1956. *The dentition of the Australopithecinae. Mem. 9.* Pretoria: Transvaal Museum.

Todd, J. E. 1975. *Children's Dental Health in England and Wales.* 1973. London: H. M. S. O.

Watkin, I. M. 1956. *Heredity*, 10:161.

Weidenreich, F. 1937. *The Dentition of Sinanthropus pekinensis. Palaentol. Sinica.*

17. Correlations Between Lower Molar Occlusal Traits in Icelanders

Gudjón Axelsson and
Pentti Kirveskari

ABSTRACT

The groove pattern and the presence of the hypoconulid on the first and second permanent mandibular molars, and the presence of the sixth and the seventh cusps, and the deflecting wrinkle on the second deciduous and first permanent mandibular molars of 1,010 Icelandic children are determined from dental casts. The intertrait correlations are computed after converting the trait expressions into dichotomies. The majority of the correlations are very low. Cusp number and groove pattern prove unrelated, but a significant positive correlation exists between the deflecting wrinkle and groove pattern. An unexpected association is present between the absence of the hypoconulid and the presence of the sixth cusp on M_1. It probably reflects the difficulty of identification of a single distal marginal cusp on M_1. Although some of the correlations are significant they are not believed to be of major importance.

The occlusal surface of human lower molars exhibits several traits that are useful in population studies of the dentition. Cusp number and groove pattern are perhaps the most widely studied traits, but the accessory sixth (c6) and seventh (c7) cusps, and the deflecting wrinkle of the metaconid, also show considerable variation between populations. One of the assumptions made while comparing trait frequencies is that the traits occur independently of each other. When tested, the assumption has mostly proven to be true, but there are some conflicting results. Jørgensen (1955) argued for the independence of cusp number and groove pattern, an argument that was corroborated by others (Garn et al. 1966a). However, Garn et al. (1966b) also noticed a relationship between cusp number and groove pattern among recent human populations, so that populations showing least reduction in cusp number tend to retain the ancestral Y pattern most often. Sofaer et al. (1972) report a significant correlation between cusp number and fissure pattern in two Melanesian populations.

The question of intertrait correlations becomes even more complicated when the possible association of the deflecting wrinkle with the groove pattern is considered. According to Morris (1970) the broad metaconid-hypoconid contact of the Y pattern is often the result of the deflecting wrinkle joining these cusps, not the result of a given cusp size relationship. In other words, the frequency of the Y pattern may well reflect that of the deflecting wrinkle. However, Morris presented no statistical evidence for his argument.

Assuming that extra cusps and enamel ridges have been advantageous by giving added efficiency to chewing, some evidence of a positive correlation between such traits could still be expected in population studies. Some anterior dental traits have been reported to be associated (Berry 1976; Scott 1977). The Carabelli's trait shows a significant correlation with the cingular nodule of the upper lateral incisor (Berry 1976), and with the protostylid of lower molars (Scott 1975). Dental traits show a positive correlation to tooth size, which probably explains why the inter-trait correlations are mostly positive (Sofaer et al. 1972).

Material and Method

We examined the dental casts of 1,010 Icelandic children (see Axelsson and Kirveskari 1979). The following observations were made:

1. cusp number on M_1 and M_2
2. fissure pattern on M_1 and M_2
3. c6 on dm_2 and M_1
4. c7 on dm_2 and M_1
5. deflecting wrinkle on dm_2 and M_1

The cusp number was either 5 or 4 depending on the presence or absence of the hypoconulid. Distinct furrows over the distal-distobuccal border of the occlusal surface were considered indicative of the presence of the hypoconulid.

We counted cusps at the midline or buccal to the midline as hypoconulids, and cusps lingual to it as c6. We followed Jørgensen's (1955) classification of fissure patterns. However, no measurements of the width of the basal cusp contact were made. When the distinction between metaconid-hypoconid contact (Y pattern) and protoconid-entoconid con-

Table 17.1. Correlations of cusp number (CN), fissure pattern (FP), accessory cusps (C6, C7), and deflecting wrinkle (DW) on the lower molar teeth of Icelanders. (Number of cases in parentheses.)

	M_1CN	M_2CN	M_1FP	M_2FP	dm_2C6	M_1C6	dm_2C7	M_1C7	dm_2DW	M_1DW
M_1CN		0.1441* (269)	0.0003 (303)	−0.0920 (192)	0.0105 (202)	−0.1634*** (572)	0.0224 (222)	−0.0145 (716)	0.0166 (119)	0.0800 (291)
M_2CN			−0.1466 (51)	−0.0014 (198)	(3)	−0.1246 (210)	(5)	−0.0032 (243)	(5)	0.0896 (56)
M_1FP				0.0220 (55)	0.0742 (149)	−0.0840 (210)	0.0303 (145)	0.0398 (293)	0.1339 (99)	0.2104*** (300)
M_2FP					(4)	−0.1231 (156)	(5)	0.3145*** (164)	−0.2582 (7)	0.0357 (58)
dm_2C6						0.2305* (102)	0.0317 (204)	0.0049 (196)	0.0668 (121)	0.1228 (140)
M_1C6							0.0805 (108)	0.1285** (544)	−0.1407 (55)	−0.0187 (212)
dm_2C7								0.0656 (210)	0.2306** (135)	0.1291 (135)
M_1C7									0.0124 (118)	0.0762 (290)
dm_2DW										0.5560*** (95)
M_1DW										

*Probably significant
**Significant
***Highly significant

Table 17.2. Cusp number and fissure pattern of lower first and second permanent molars in Icelanders.

		Y-Pattern	+ or X-Pattern	Total
	4 cusps	24 (70.6%)	10 (29.4%)	34 (11.2%)
M_1	5 cusps	190 (70.6)	79 (29.4)	269 (88.8)
	Total	214 (70.6)	89 (29.4)	303 (100.0)
	4 cusps	17 (10.1)	151 (89.9)	168 (84.8)
M_2	5 cusps	3 (10.0)	27 (90.0)	30 (15.2)
	Total	20 (10.1)	178 (89.9)	198 (100.0)

tact (x pattern) was difficult to make by naked eye, a point contact was assumed (+ pattern). For the calculations + and x patterns were combined. c6 and c7 were originally scored according to the classification criteria by Turner et al. (1969, 1970), but the distribution was dichotomized for the present purpose (Axelsson and Kirveskari 1978). The deflecting wrinkle was scored according to Hanihara et al. (1964), but also dichotomized to become a present-absent classification (Axelsson and Kirveskari 1977).

The scorings were made on both sides, but only one side was used in the calculations, so that the traits compared in each individual always came from the same tooth or the same side. The dichotomies of trait expressions were given the numerical values 1 or 0 as follows: cusp number 4 = 0, 5 = 1; fissure pattern x or + = 0, Y = 1; c6, c7, and deflecting wrinkle absent = 0, present = 1. Thereafter a correlation matrix was computed.

Results

The Pearsonian coefficients of correlation and the number of pairs they are based on are given in table 17.1. Cusp number and fissure pattern proved unrelated on both M_1 and M_2 (table 17.2). Cusp number and c6 on M_1 showed a highly significant correlation, c6 being more common on 4-cusped than on 5-cusped M_1 (table 17.3). The fissure pattern and the

Table 17.3. Occurrence of the accessory cusp c6 on 4 and 5-cusped lower first permanent molars in Icelanders.

	4-Cusped	5-Cusped	Total
c6 absent	56 (68.3%)	420 (85.7%)	476 (83.2%)
c6 present	26 (31.7)	70 (14.3)	96 (16.8)
Total	82 (14.3)	490 (85.7)	572 (100.0)

$\chi^2 = 15.27$, 1 df, $p < 0.001$.

Table 17.4. Occurrence of the deflecting wrinkle on lower first permanent molars of Icelanders according to the fissure pattern.

	Deflecting wrinkle		
	Absent	Present	Total
+ or X-pattern	76 (84.4%)	14 (15.6%)	90 (30.0%)
Y-pattern	133 (63.3)	77 (36.7)	210 (70.0)
Total	209 (69.7)	91 (30.3)	300 (100.0)

$\chi^2 = 13.29$, 1 df, $p < 0.001$

deflecting wrinkle on M_1 were significantly associated, the latter being more common on teeth with the Y pattern than on teeth with + or X pattern (table 17.4). The majority of the correlations were positive, but for the most part statistically not significant.

Discussion

Studies of intertrait correlations are often difficult because of the unfavorable distribution of trait expressions in populations of different racial origin. For instance, the cusp number of lower M_1 often shows too little reduction in Mongoloid populations, and the deflecting wrinkle c6 and c7 are usually too rare for an effective analysis in Caucasoid populations. The Icelanders show more variation than most other Caucasoid populations particularly in regard to the deflecting wrinkle, c6, and c7 (Axelsson and Kirveskari 1977, 1978).

Because the variation of dental traits is nearly always continuous or quasi-continuous, presenting the frequencies as dichotomies necessarily reduces the information obtained. However, the size of the hypoconulid is difficult to categorize, let alone quantify, and the same pertains to the deflecting wrinkle. The existence of a true + pattern is largely hypothetical (Erdbrink 1965), thus making the fissure pattern essentially a Y–X dichotomy (the length of the basal cusp contact would represent a quantitative measure of the pattern). Classification of c6 and c7 sizes is possible (Turner et al. 1969, 1970), but as the cusp expressions in Icelanders were for the most part at the lower end of the scale (Axelsson and Kirveskari 1978), the present/absent classification corresponds well with the true situation. Thus, all the variables studied can be presented as dichotomies without much loss of information.

The variation of cusp number and fissure pattern was within expectations. If the Icelandic frequencies are plotted on a diagram showing the distribution of these two frequencies in several other populations of different races (see Jørgensen 1955; Suzuki and Sakai 1973), they fall in the

Caucasoid area. In fact, only the cusp number seems to have the racial discriminatory power. The fissure patterns are not significantly different in different races.

Although there was no correlation between cusp number and fissure pattern in Icelanders studied, the analysis suffers from the rarity of 4-cusped first molars and Y patterns on the second molars (table 17.2). The different way of categorizing data perhaps explains why Sofaer et al. (1972) found a significant correlation between these variables. Another phenomenon that may have a bearing in the question is the variation of stability of the different cusp number and groove pattern combinations (Erdbrink 1965). For the same reason combining the data on M_1 and M_2 would perhaps lead to erroneous conclusions.

The finding that the c6 was significantly more often present on 4-cusped M_1 than on 5-cusped M_1 is intriguing and contrary to expectations. We re-examined all casts that were registered as 4-cusped M_1 with a c6, but that did not change the picture. Both the size and location of the hypoconulid vary considerably, and the same is true of the c6. There are cases in which the c6 is large, and the hypoconulid very small. It is logical to assume that the hypoconulid could also be absent when the c6 is present. When a single cusp is located at the midline on the posterior border, it seems impossible in individual cases to identify the cusp. Sometimes dm_2 can give a clue; if it shows a c6, M_1 is also likely to show one. The same can be suspected if M_2 or M_3 show a c6. We made the decision in favor of the hypoconulid if nothing clearly suggested that the cusp in question was a c6, because the hypoconulid is phylogenetically one of the main cusps and because the hypoconulid most often is the larger of the two when both are present.

The deflecting wrinkle was significantly associated with the Y pattern on M_1 (table 17.4). While the finding clearly lends support to Morris' (1970) argument, it should be noticed that the coefficient of correlation is low, and that the deflecting wrinkle does also occur with the + and X patterns. We think the error probably present in reports of lower molar patterns without regard to the deflecting wrinkle is not of major importance, but at the same time agree with Morris (1970) in urging a simultaneous reporting of the two traits.

A surprisingly high correlation was present between the fissure pattern of M_2 and the occurrence of the c7 on M_1. However, the number of second molars with the Y pattern was as low as 20 in this sample, so chance may have played a part in spite of the relatively large total sample. The lack of correlation between the fissure pattern and the c7 on M_1 suggests the same.

According to the field concept (Dahlberg 1945) a positive correlation could be expected between the frequencies of a given trait on the different teeth of the molar series. The deflecting wrinkle showed a relatively high correlation on dm_2 and M_1, but the c6 and c7 failed to do so.

The few significant but low correlations between cusp number, fissure pattern, accessory cusps, and deflecting wrinkle obviously do not make unjustified the continued use of these traits in population studies. Part of the correlations may still reflect past selective pressures on tooth morphology, part are probably due to chance.

References

Axelsson, G. and P. Kirveskari. 1977. The deflecting wrinkle on the teeth of Icelanders and the Mongoloid dental complex. *Am. J. Phys. Anthrop.* 47:321–324.

——1979. Sixth and seventh cusp on lower molar teeth of Icelanders. *Am. J. Phys. Anthrop.* 51:79–82.

Berry, A. C. 1976. The anthropological value of minor variants of the dental crown. *Am. J. Phys. Anthrop.* 45:257–268.

Dahlberg, A. A. 1945. The changing dentition of man. *J. Am. Dent. Assoc.*, 32:676–690.

Erdbrink, D. P. 1965. A quantification of the *Dryopithecus*- and other lower molar patterns in man and some of the apes. *Z. Morph. Anthrop.* 57:70–108.

Garn, S. M., A. A. Dahlberg, A. B. Lewis, and R. S. Kerewsky. 1966a. Groove pattern, cusp number, and tooth size. *J. Dent. Res.* 45:970.

——1966b. Cusp number, occlusal groove pattern and human taxonomy. *Nature* 210:224–225.

Hanihara, K., T. Kuwashima, and N. Sakao. 1964. The deflecting wrinkle on the lower molars in recent man. *Zinruigaku Zassi* 72:1–8.

Jørgensen, K. D. 1955. The Dryopithecus pattern in recent Danes and Dutchmen. *J. Dent. Res.* 34:195–208.

Morris, D. H. 1970. On deflecting wrinkles and the *Dryopithecus* pattern in human mandibular molars. *Am. J. Phys. Anthrop.* 32:97–104.

Scott, G. R. 1975. Association between nonmetrical tooth crown characteristics. *Am. J. Phys. Anthrop.* 42:328.

——1977. Classification, sex dimorphism, association, and population variation of the canine distal accessory ridge. *Hum. Biol.* 49:453–469.

Sofaer, J. A., C. J. MacLean and H. L. Bailit. 1972. Heredity and morphological variation in early and late developing human teeth of the same morphological class. *Archs. Oral Biol.* 17:811–816.

Suzuki, M. and T. Sakai. 1973. Occlusal surface pattern of the lower molars and the second deciduous molar among the living Polynesians. *Am. J. Phys. Anthrop.* 39:305–316.

Turner II, C. G., G. R. Scott, and T. A. Rose. 1969. Mandibular molar cusp 6 plaque and definitions of variation. Dept. Anthropology, Arizona State Univ., Tempe.

Turner II, C. G., G. R. Scott, and M. Larsen. 1970. Mandibular molar cusp 7 plaque and definitions of variation. Department of Anthropology, Arizona State Univ., Tempe.

18. The Dental Morphology of North American Whites A Reappraisal

JOHN T. MAYHALL, SHELLEY R. SAUNDERS
AND PATRICIA L. BELIER

Descriptive studies of the morphology of the dentitions of various racial groups of North America have been published during this century. These have included Eskimos in Alaska (Turner 1967) and Canada (Mayhall 1979a), Pima Indians (Dahlberg 1963), Indians of the American Southwest (Scott 1973) and Aleuts (Moorrees 1957), to name only a few. In many of these studies the results have been compared with North Americans of European origin. However, the description of a North American "white" population has been almost secondary to the target population description. When the comparisons have been made they have been for only a few selected traits and, generally, there appears to have been little control over the genetic background of the "white" population. G.V. Black (1902), Hellman (1928), Dahlberg (1951, 1963) and more recently Scott (1973), have provided us with the best descriptions of trait frequencies for this group that we have today. Of these, Dahlberg's are the most comprehensive and useful. His "Chicago whites" are primarily of northern European ethnic origin (Dahlberg 1979, pers. comm.). Dahlberg's pioneering work in trait frequency description has primarily been concerned with aboriginal groups in North America, but in his classic 1951 paper on the dental morphology of the American Indian he provides us with one of the best descriptions of the dental morphology of "whites." Scott's (1973) description of western United States whites has added another valuable dimension to morphology studies of regional groups of North Americans.

As with many disciplines of anthropology in North American we became increasingly interested in the population with whom we live—thus the present study of one of the less "exotic" populations—North American whites. Our aim is to describe a well-documented population from a representative area of North America and compare these results with the preceding studies.

Materials and Methods

We have chosen the residents of Burlington, Ontario, in Canada as our focus of study. In the 1950s the University of Toronto began a longitudinal growth study of the children of this community, and dental stone casts were available for both the children in the study and their parents. The present paper deals with the parents; subsequent papers will deal with the children involved in this 20-year study. In a paper describing the demographics of Burlington it was noted that Burlington was comparable with other communities in Ontario, the province itself and the country as a whole (Popovich 1959). We might add that the demographics of Canada and the United States are similar, so Burlington seems ideal for our purposes.

When the study began, Burlington was a town of 6,000. The predominant racial group in Burlington was Caucasian, and while there were some North American Indians, Orientals, and Negroes living in Burlington they were so few that through chance they weren't represented in the sample. In terms of family size, educational levels, income, and age, the Burlington sample was representative of the general population of other larger political units. While in this paper we are dealing with the parents of the growth sample, it may be of interest to note that 85 to 90 percent of the children residing in Burlington took part in the study.

The present sample consists of 280 males and females who were included in the data collection, with the sole criterion being that they were the parents of the children in the growth study. None of the parents are related to each other by any closer ties than that of cousins. Practically all the parents traced their ancestry to the British Isles.

We decided to study morphological traits which would encompass as many different teeth as possible and yield as wide a range of observations as was practical. The majority of the traits are ones which had been reported previously for other groups of whites in North America. In order to reduce inter-investigator error to the minimum, most traits were scored using the standard plaques prepared by Dahlberg (n.d.), Turner (n.d.), or Hanihara (1960). All observations were done by a single experienced investigator, and she constantly referred to the standards for consistency in observation. Any observations that were dubious because of wear or cast inaccuracies were not included in the results. Thus, we are confident that the observations reported here are as consistent as possible with those of the originator of the standards.

Table 18.1. Axial rotation of central incisors in Burlington whites.

	Males	Females	Chicago Whites[a]
Straight	87(68%)	87(66.4%)	(46%)
Bilateral winging	5(3.9)	6(4.6)	(3)
Unilateral winging			(3)
right only	0(0.0)	2(1.5)	
left only	1(0.8)	1(0.8)	
Bilateral counter-winging	18(14.1)	20(15.3)	(43)
Unilateral counter-winging			(5)
right only	13(10.2)	5(3.8)	
left only	4(3.1)	9(6.9)	
N	128	131	

[a]Percentages are approximate and based on those presented in Dahlberg (1963, figure 8, page 157).

Observations were made of both left and right sides of each arch but the results here are from the left side only. The observations were statistically evaluated to identify any significant differences between the antimeres. None were found.

Incisors

In the incisor series, two traits were selected for study, the axial rotation of the incisors (Dahlberg, Enoki, and Nakamura 1958) and the expression of shovel shaping. Our results for the amount of axial rotation of the incisors can be seen in the first table (table 18.1). The comparison of our results with those of Dahlberg (1963) indicates general agreement in the frequency of bilateral winging but a wide divergence of results in the frequency of bilateral counterwinging. Dahlberg reported that 43 percent of his Caucasoid sample displayed bilateral counterwinging while we noted only 15 percent. The difference in the results is also noted in our 20 percent higher frequency of a straight alignment of the incisors. While these

Table 18.2. Shovel-shaped incisors in Burlington whites (left side only).

	Males		Females	
	Lateral	Central	Lateral	Central
Absent	50(63.3%)	57(69.5%)	57(67.9%)	56(64.4%)
Trace	17(21.5)	20(24.4)	21(25.0)	29(33.3)
Semi-shovel	10(12.7)	5(6.1)	6(7.1)	2(2.3)
Shovel	1(1.3)	0(0.0)	0(0.0)	0(0.0)
Marked	1(1.3)	0(0.0)	0(0.0)	0(0.0)
N	79	82	84	87

differences are difficult to explain, it is possible that differing arch form may be one of the influences. This requires further study, however.

For the study of the frequencies of shovel-shaped incisors (table 18.2) we used the standard categories proposed by Hrdlička in 1920 with the addition of a marked category to record the extreme manifestations of the trait. Our results were compared with Hrdlička's. While some studies, including Hrdlička's, have indicated the shoveling was more frequently found in females, the present results do not support such a contention. There was no statistically significant difference in the degrees of expression of the trait between the sexes.

It is obvious from the table that North American whites can be characterized as displaying only minimal manifestations of shoveling. None of the central incisors showed any shovel-shaped or marked expressions while only two individuals out of almost 160 showed these larger expressions. The comparisons with Hrdlička's results revealed no differences that were highly statistically significant, although the female central incisor and male lateral incisor comparisons were just significant at the 95 percent level. Therefore, it seems reasonable to state that these results agree generally with earlier studies confirming the general absence of shoveling in permanent incisors in North American whites.

Premolars

Three traits were chosen for observation on the premolars. Maxillary and mandibular premolars were studied for the presence of occlusal tubercles and the number and size of cusps, while the mandibular premolar groove patterns were observed.

We thought it of interest to record the frequency of occurrence of premolar occlusal tubercles in whites (table 18.3), since this trait has been observed almost exclusively in Mongoloid populations. This trait has been identified by numerous titles such as evaginated odontome, tuberculated

Table 18.3. Premolar occlusal tubercles in Burlington whites (left side only).

	Males		Females	
Tooth	Present	Absent	Present	Absent
Maxillary first	0(0.0%)	40(100.0%)	0(0.0%)	32(100.0%)
Maxillary second	0(0.0)	27(100.0)	0(0.0)	20(100.0)
Mandibular first	0(0.0)	108(100.0)	0(0.0)	104(100.0)
Mandibular second	0(0.0)	47(100.0)	0(0.0)	43(100.0)
N	0	223	0	199

Table 18.4. Burlington whites cusp number—mandibular premolars (left side only).

	First Premolar		Second Premolar	
Number	Male	Female	Male	Female
1	6(4.9%)	7(5.7%)	0(0.0%)	0(0.0%)
2	108(88.5)	108(88.5)	55(73.3)	57(74.0)
3	8(6.6)	6(4.9)		
DM = ML[a]			9(12.0)	9(11.7)
DM½ML[b]			9(12.0)	7(9.1)
DL < ½ML[c]			2(2.7)	4(5.2)
4	0(0.0)	1(0.8)		
N	122	122	75	77

[a]Distolingual cusp is equal in size to mesiolingual cusp.
[b]Distolingual cusp is one-half the size of the mesiolingual cusp.
[c]Distolingual cusp is less than half the size of the mesiolingual cusp.

premolar, occlusal enamel pearl and axial core odontome. Our results clearly confirm the lack of expression of this trait of the Caucasoid population studied here. Not one tubercle was observed in over 400 premolars examined.

The maxillary premolars invariably exhibited two cusps, a buccal one and a lingual one. In the mandibular premolars (table 18.4) the first premolar exhibited two cusps in the vast majority of instances while the second premolar showed somewhat more variation. From the table it can be seen that in instances where three cusps are present the mesiolingual cusp is always as large as or larger than the distolingual cusp. This result is in agreement with the work of Kraus and his co-workers (1969).

Kraus et al. (1969) identified three primary groove patterns in the mandibular second premolar; the Y, H, and U configurations (table 18.5). The majority of the groove patterns were the H type as would be expected, since this pattern is generally associated with the two-cusped second premolar. As in the incisors, none of these traits displayed any statistically significant differences in expression between the sexes.

Table 18.5. Burlington whites groove patterns-mandibular second premolars (left side only).

Pattern		Males	Females
Y		12(26.1%)	16(40.0%)
H		24(52.2)	20(50.0)
U		10(21.7)	4(10.0)
	N	46	40

Maxillary Molars

For the maxillary molars, two observations were recorded, cusp number and the expressions of Carabelli's trait. The number of cusps on the maxillary second molar alone was recorded (table 18.6) because the first molars invariably had four cusps, and the sample of third molars was too small to yield meaningful results. As with the previously discussed traits, there was no sexual dimorphism in the development of the hypocone. Fifty percent of the males and one third of the females had four well developed cusps. Dahlberg's description (1963) of the Chicago white sample indicated that 26 percent had a fully developed hypocone. Statistical testing of the male results for Dahlberg's and our sample revealed a highly significant difference. However, the differences are not as dramatic as might initially appear. Dahlberg's study indicated that the majority of the expressions were in the 4− category. If we combine the 4 and 4− categories of the two studies we find that both are in the 60 percent range. Thus, the difference may possibly be attributed to slight inter-investigator differences.

The frequency of occurrence of Carabelli's trait for our Burlington sample is presented in table 18.7. There were no statistically significant differences in the incidence of the various expressions between the sexes. Our sample can be characterized as generally showing some expression of the trait on first molars as might be expected in Caucasoids. Generally, this expression took the form of a small cusp or a bulge of tooth material in this area. The only factor preventing us from defining the bulge as a cusp was the lack of clearly defined grooves surrounding the bulge.

The second and third molars usually did not display an expression of the Carabelli's trait. Our results were statistically significantly different from those reported by Dahlberg (1963). His sample for first and second molars displayed a much higher percentage of cusp expressions than we observed. We are not able to explain these differences except through the differing samples studied. However, both studies indicate that the Caucasoids can be characterized as having a high percentage of the larger expressions of the traits when compared with Mongoloid populations.

Table 18.6. Burlington whites cusp number—maxillary second molars (left side only).

Cusps	Male	Female
4	45(50.0%)	32(33.7%)
4−	9(10.0)	8(8.4)
3+	13(14.4)	21(22.1)
3	23(25.6)	34(35.8)
N	90	95

Table 18.7. Carabelli's trait in Burlington whites (left side only).

	Males			Females		
	M^1	M^2	M^3	M^1	M^2	M^3
Absent	22(24.4%)	75(75.0%)	12(75.0%)	34(36.6%)	93(86.9%)	12(85.7%)
Pit	1(1.1)	0(0.0)	0(0.0)	1(1.1)	0(0.0)	0(0.0)
Single furrow	20(22.2)	10(10.0)	2(12.5)	21(22.6)	9(8.4)	2(14.3)
Double furrow	3(3.3)	9(9.0)	0(0.0)	6(6.5)	5(4.7)	0(0.0)
Y	3(3.3)	0(0.0)	0(0.0)	7(7.5)	0(0.0)	0(0.0)
Bulge	35(38.9)	5(5.0)	2(12.5)	16(17.2)	0(0.0)	0(0.0)
Small cusp	5(5.6)	1(1.0)	0(0.0)	5(5.4)	0(0.0)	0(0.0)
Large cusp	1(1.1)	0(0.0)	0(0.0)	3(3.2)	0(0.0)	0(0.0)
N	90	100	16	93	107	14

Table 18.8. Sixth cusp in Burlington whites (left side only).

	Males			Females		
	M_1	M_2	M_3	M_1	M_2	M_3
Absent	33(94.3%)	66(98.5%)	13(86.7%)	19(95.0%)	44(97.8%)	8(100.0%)
Small	2(5.7)	1(1.5)	0(0.0)	0(0.0)	1(2.2)	0(0.0)
Medium	0(0.0)	0(0.0)	0(0.0)	0(0.0)	0(0.0)	0(0.0)
Large	0(0.0)	0(0.0)	1(6.7)	0(0.0)	0(0.0)	0(0.0)
Present (size undeterminable)	0(0.0)	0(0.0)	1(6.7)	1(5.0)	0(0.0)	0(0.0)
N	35	67	15	20	45	8

Mandibular Molars

The expressions of the sixth cusp and seventh cusp on mandibular molars is only occasionally reported. The sixth cusp, also known as the tuberculum sextum, and c_7, known as tuberculum intermedium, were observed along with the occurrence of the protostylid trait. Cusp number and groove patterns for the mandibular molars were not included because of the amount of occlusal wear, which would make these observations unreliable as a result of the small sample sizes.

It seems safe to say that the sixth cusp is virtually nonexistent in the Burlington sample (table 18.8). There were only nine manifestations of anything resembling a cusp out of 190 observations for males and females. These results are in concert with those of Dahlberg (1963) and Hellman (1928) and in contrast to the relatively high number of the cuspal expressions in Mongoloids.

Essentially, the same results accrue from the figures for the seventh cusp (table 18.9). Almost no expressions of the cusp were noted in the Burlington study. The absence of both of these traits are further evidence of the simplification of the Caucasoid mandibular molar pattern.

Finally, the last trait to be described is the protostylid (table 18.10). This trait, located on the mesiobuccal cusp of the mandibular molars (Dahlberg 1950), is only rarely evident in this North American white sample. Practically no manifestations were evident on the first molars while there was a tendency toward some larger expressions as one moved distally. However, even third molars showed no evidence of large cusps. These results are in general agreement with those of Dahlberg (1963), except that he noted a large number of buccal pits which we were unable to observe in our sample. Some questions have been raised previously about the inclusion of the pit category in the range of the trait expressions (Kirveskari 1974; Mayhall, 1976, 1979b). If we regard the pits as part of the absent category, the two studies are remarkable in their agreement.

Conclusions

We titled this paper a reappraisal of the dental morphology of North American whites because we felt it was appropriate to briefly compare our results with some of the previous studies of similar populations. The characterization of the North American whites as a single rather homogeneous population is obviously impossible. Because of the diversity of ethnic backgrounds found in North America, we didn't expect our results to totally agree the studies of Hellman, Dahlberg or Scott, among others. As noted earlier, there are differences in the expressions of the traits between the studies. Many of these differences are attributable to the fine

Table 18.9. Seventh cusp in Burlington whites (left side only).

	Males			Females		
	M_1	M_2	M_3	M_1	M_2	M_3
Absent	65(97.0%)	80(100.0)	22(100.0)	46(93.9%)	68(100.0)	15(100.0)
Small	1(1.5)	0(0.0)	0(0.0)	0(0.0)	0(0.0)	0(0.0)
Medium	1(1.5)	0(0.0)	0(0.0)	0(0.0)	0(0.0)	0(0.0)
Large	0(0.0)	0(0.0)	0(0.0)	1(2.0)	0(0.0)	0(0.0)
Present size undeterminable	0(0.0)	0(0.0)	0(0.0)	2(4.1)	0(0.0)	0(0.0)
N	67	80	22	49	68	15

Table 18.10. Protostylid in Burlington whites (left side only).

	Males			Females		
	M_1	M_2	M_3	M_1	M_2	M_3
Absent	74(100.0%)	96(94.1)	26(86.7)	66(95.7%)	92(94.8)	15(71.4)
Pit	0(0.0)	0(0.0)	0(0.0)	0(0.0)	0(0.0)	0(0.0)
Distal deviation of groove	0(0.0)	0(0.0)	0(0.0)	1(1.4)	0(0.0)	0(0.0)
Furrow	0(0.0)	5(4.9)	2(6.7)	1(1.4)	3(3.1)	5(23.8)
Prominence without groove	0(0.0)	0(0.0)	0(0.0)	0(0.0)	0(0.0)	0(0.0)
Prominence with groove	0(0.0)	0(0.0)	1(3.3)	1(1.4)	1(1.0)	0(0.0)
Small cusp	0(0.0)	1(1.0)	1(3.3)	0(0.0)	1(1.0)	1(4.8)
Large cusp	0(0.0)	0(0.0)	0(0.0)	0(0.0)	0(0.0)	0(0.0)
N	74	102	30	69	97	21

distinctions that have been made in observing the various expressions of the traits. If we had resorted to presence-absence categorization or reduced the number of categories to only two or three, we would not have encountered the statistical differences between studies. It is fair to state that our results basically confirm the observations of the previous studies.

One of the advantages that this study has over most of the previous ones is the known ethnic composition, the use of readily available observational standards, and the known genetic composition of the study group.

Utilizing the results of this study, we propose a Caucasoid Dental Complex for the permanent dentition. We aren't presumptous enough to suggest that this is the first attempt to do this, nor is the idea ours. Hanihara (1967, 1970), in his excellent studies of the deciduous dentition illustrated a Mongoloid Dental Complex, and this has been useful in the characterization of the frequencies for numerous morphological traits. Like Hanihara's, this complex includes traits which are spread throughout the dentition and appear to occur in frequencies differing as widely as possible from Mongoloid and Negroid frequencies where these are known. Taken as a group, they may be useful in suggesting the "racial" background of, if not an individual, a group of individuals.

The Caucasoid Dental Complex (table 18.11) should include the following: (a) the absence of, or at most a trace expression of shovel-shaped incisors; (b) a high frequency of bilateral counter-winging or straight axial alignment of the central incisors; (c) the absence of premolar occlusal tubercles; (d) the high frequency of Carabelli's cusp (including the "bulge" category) on the first molar; (e) the absence of expressions of the protostylid; (f) the absence of expressions of the sixth cusp, and (g) the absence of expressions of the seventh cusp.

The traits included in the complex should be viewed as a basis for further additions as more traits are studied. We intend to critically evaluate the inclusion of further traits in future papers. The first of these will deal with

Table 18.11. Caucasoid dental complex permanent dentition—North American whites.

Trait	Tooth	Expression
Shovel-shaping	Incisors	Absent or trace
Axial rotation	Incisors	Bilateral counter winging or straight
Occlusal tubercle	Premolars	Absent
Carabelli's trait	Max. first molar	Cusp or bulge expressions
Protostylid	Mand. molars	Absent
Sixth cusp	Mand. molars	Absent
Seventh cusp	Mand. molars	Absent

the permanent and deciduous dentition of the children included in the Burlington study.

This initial, descriptive paper serves as a basis for the future analysis of trait frequencies not only from a descriptive viewpoint but also from a genetic standpoint. It is the beginning of a long-term study which may finally provide a few answers to long standing questions about the expressions of traits, in both the permanent and deciduous dentition, and about the variation of frequencies within a tooth class.

References

Black, G. V. 1902. *Descriptive Anatomy of the Human Teeth*, 4th ed. Philadelphia: S.S. White.

Dahlberg, A.A. 1963. Analysis of the American Indian dentition. In D.R. Brothwell, ed., *Dental Anthropology*. Oxford: Pergamon Press.

—— 1951. The dentition of the American Indian. In W.S. Laughlin and S.L. Washburn, eds., *The Physical Anthropology of the American Indian*, pp. 138–176. New York: Viking Fund.

—— 1950. The evolutionary significance of the protostylid. *Am. J. Phys. Anthrop.* 18:15–25.

—— n.d. Materials for the establishment of standards for classifications of tooth characters, attributes, and techniques in morphological studies of the dentition. University of Chicago. Mimeo.

Dahlberg, A.A., K. Enoki, and E. Nakamura. 1958. Rotated maxillary central incisors. *J. Jap. Orthodon. Soc.* 17:157–169.

Hanihara, K. 1970. Mongoloid dental complex in the deciduous dentition with special reference to the dentition of the Ainu. *Zinruigaku Zassi* 78:3–17.

—— 1966. Mongoloid dental complex in the deciduous dentition. *J. Anthrop. Soc. Nippon* 74:61–72.

—— 1960. Standard models for classification of crown characters in the human deciduous dentition. Mimeograph, Univ. of Chicago.

Hellman, M. 1928. Racial characters in the human dentition. *Proc. Am. Phil. Soc.* 67:157–174.

Hrdlička, A. 1920. Shovel-shaped teeth. *Am. J. Phys. Anthrop.* 3:429–465.

Kirveskari, P. 1974. Morphological traits in the permanent dentition of living Skolt Lapps. *Proc. Finn. Dent. Soc.* 70:Supp II.

Kraus, B.S., R.E. Jordan and L. Abrams. 1969. *Dental Anatomy and Occlusion.* Baltimore: Williams and Wilkins.

Mayhall, J.T. 1979a. The Dental Morphology of Canadian Inuit. Ossa. 6:199–218.

—— 1979b. The biological relationships of Thule culture and Inuit populations: an odontological investigation. In A.P. McCartney, ed., *Thule Eskimo Culture: An Anthropological Retrospective*, Archaeological Survey of Canada Paper No. 88, pp. 448–473. Ottawa: National Museums of Canada.

—— 1976. The Morphology of the Permanent Dentition of Prehistoric and Modern Central Arctic Eskimoid Peoples: A Study of Their Biological Relationships. Ph.D. dissertation, University of Chicago.

Moorrees, C.F.A. 1957. *The Aleut Dentition.* Cambridge: Harvard University Press.

Popovich, F. 1959. Demography of Burlington population and the status of the Burlington sample. In Burlington Orthodontic Research Centre Progress Report Series, pp. 12–22, No. 4, 1958–1959.

Turner, II, C.G. 1967. The Dentition of Arctic Peoples. Ph.D. dissertation, University of Wisconsin.

—— n.d. Dental stone models for classification of crown characters in the permanent dentition.

19. Microdifferentiation in Tooth Crown Morphology Among Indians of the American Southwest

G. RICHARD SCOTT
AND
ALBERT A. DAHLBERG

From the time of J. F. Blumenbach's *On the Natural Variety of Mankind* (1775) to W. C. Boyd's *Genetics and the Races of Man* (1950), physical anthropologists directed much of their energy toward describing human biological variation and classifying populations into broad racial categories. Emphasis was placed on comparing groups which had been isolated for long periods of time and which showed obvious and pronounced biological differences in skin color, hair form, stature, etc. Concern thus focused on the macro-differentiation of human populations at the level of subspecies or geographical race. Because of varying assumptions and methodologies, the number of classificatory schemes proposed was almost

We would like to express our appreciation to several individuals who contributed significantly to this study. First, many of the dental casts examined were obtained through the tireless efforts of Thelma Dahlberg. The energy and enthusiasm she displayed throughout many years of fieldwork made her an indispensable part of the long-term Southwest Indian dental project carried out by one of us (Dahlberg). Christy Turner and Donald Morris were also kind enough to allow the senior author to examine large collections of casts they obtained in the Southwest (and elsewhere). The advice and encouragement of Turner and Morris received by the senior author while a graduate student at Arizona State University is also much appreciated. The basic research for this paper was conducted when one of us (Scott) was receiving the support of a Public Health Service genetics training grant (5 TO1 GM01433-03). The director of this grant, Charles M. Woolf, is kindly acknowledged for both this financial support and intellectual stimulation.

coincident with the number of different workers setting up the classifications (Count 1950).

While broad patterns of human variation and differentiation continue to be of interest to human biologists, especially when concern is with long term human evolution (Cavalli-Sforza and Edwards 1964; Cavalli-Sforza and Bodmer 1971), classifications of geographical races such as the one proposed by Garn (1971) serve primarily a communicative rather than an analytical function. To better understand how evolutionary mechanisms have acted and interacted to generate human diversity, the movement in human population biology has been toward more finely grained studies of local races and the variation among their constituent subpopulations. Much of the impetus for this movement can be attributed to the increasing interest in theoretical models of population structure. These models, developed by Wahlund (1928), Wright (1931, 1943), and Malecot (1948) among others, were not practically applied to human groups until recently (for reviews, see Schull and MacCluer 1968; Morton 1969; Cannings and Cavalli-Sforza 1973). Application in part has been stimulated and enhanced by the discovery of more genetic markers and the increased use of computers in processing large quantities of data, both genetic and demographic.

Ethnographic observations indicate that tribal populations are rarely, if ever, distributed continuously throughout their geographical range. Rather, a tribal population will be divided into a series of subpopulations of varying sizes, each inhabiting and utilizing a given part of the tribal range. The rate of gene flow between any two subpopulations can be influenced by a variety of sociocultural factors and also by geographical distance. The interplay of marriage rules and distance essentially precludes random mating within a tribal population. Because of this, genetic drift will generate diversity among subpopulations; this differentiation can be impeded or enhanced for specific traits or alleles by the action of selection (Cannings and Cavilli-Sforza 1973).

Until recently, physical anthropologists placed most of their emphasis on making comparisons between tribal groups, and in the process, failed to take into account the extent of within-group (or between subpopulation) differences. When geneticists began collecting data on several subpopulations of a single population system, a rather striking discovery showed that a significant proportion of the genetic variation found between groups was already evident within a group (Neel and Salzano 1966; Neel and Ward 1970). To illustrate the ramifications of this observation, consider that many of the serological surveys undertaken during the 1950s and 1960s were interested in between group variation. In these surveys, a population or tribal unit was often characterized as a whole on the basis of data obtained from a single sample of one subpopulation. Now, considering the extent of within-group variation, an ideal sampling design is not measured solely by the number of individuals (n) studied but must also involve sampling from an adequate number of subpopulations (k).

The approach of dental anthropologists interested in tooth crown morphology has paralleled that of their colleagues who deal with other biological systems. That is, in historical sequence, comparative studies first centered on geographical races, then on local races, and finally, on the variation within populations. The classic papers of Hrdlička (1920) on shovel shaped incisors and Hellman (1928) on lower molar morphology illustrate the early interest in finding differences in crown trait frequencies between geographical races. Beginning in the 1940s, interest shifted in part to describing the extent of dental variation among local races. These efforts are exemplified by the studies carried out among East Greenland Eskimos (Pedersen 1949), American Indians (Dahlberg 1951, 1963), Aleuts (Moorrees, 1957; Turner 1967), and Pacific peoples (Riesenfeld 1956). Although these studies were concerned with more detailed problems than, say, finding the difference in shoveling frequencies between Europeans and Asiatics, they still centered on dental comparisons between groups.

Only recently has interest been shown by dental researchers in determining the degree to which crown trait frequencies vary within a population. Not surprisingly, the two best examples of work on dental microdifferentiation were both allied with larger team projects that involved the collection of extensive genealogical, genetic, anthropometric, and other biological data which was analyzed from the standpoint of population structure and distances. These studies, by Brewer-Carias et al. (1976) on the Yanomama and by Harris (1977) on Solomon Islanders, both revealed a considerable degree of variation in crown trait frequencies between subpopulations.

Recently, Palomino et al. (1977) addressed the potential of tooth crown morphology in discriminating among human populations at varying levels of differentiation. Utilizing data taken from the literature, they considered morphological variation at the following four levels of differentiation: village, village cluster, tribal, major racial. The village and village cluster samples were represented by data obtained among the Yanomama (Brewer-Carias et al. 1976). The information on tribal and major racial variation was derived from the reports of several workers.

Employing a method of apportioning genetic diversity within and between populations developed by Nei (1973), Palomino et al. (1977) found similar coefficients of diversity for the four levels of differentiation. Although they qualify their conclusions by noting the limitations imposed by the qualitative treatment of the traits and the inclusion of small samples, they still suggest that nonmetric tooth crown traits are not sensitive indicators of biological affinity at low levels of group differentiation. The authors state that "there are some general features in tooth morphology which make them useful indicators of biological distance when one is concerned only with differentiation at a broad level" (Palomino et al. 1977:67).

Because of the limitations of their analysis, especially regarding the uncontrolled effects of interobserver error, the authors feel that the conclusion reached by Palomino et al. (1977), regarding the utility of tooth crown morphology in discriminating between populations, may be premature. As there are many advantages in studying dental crown features (observability, preservability, heritability, etc.), this trait system could potentially provide an important adjunct to genetic, anthropometric, and other sources of data on group differentiation. For that reason, dental

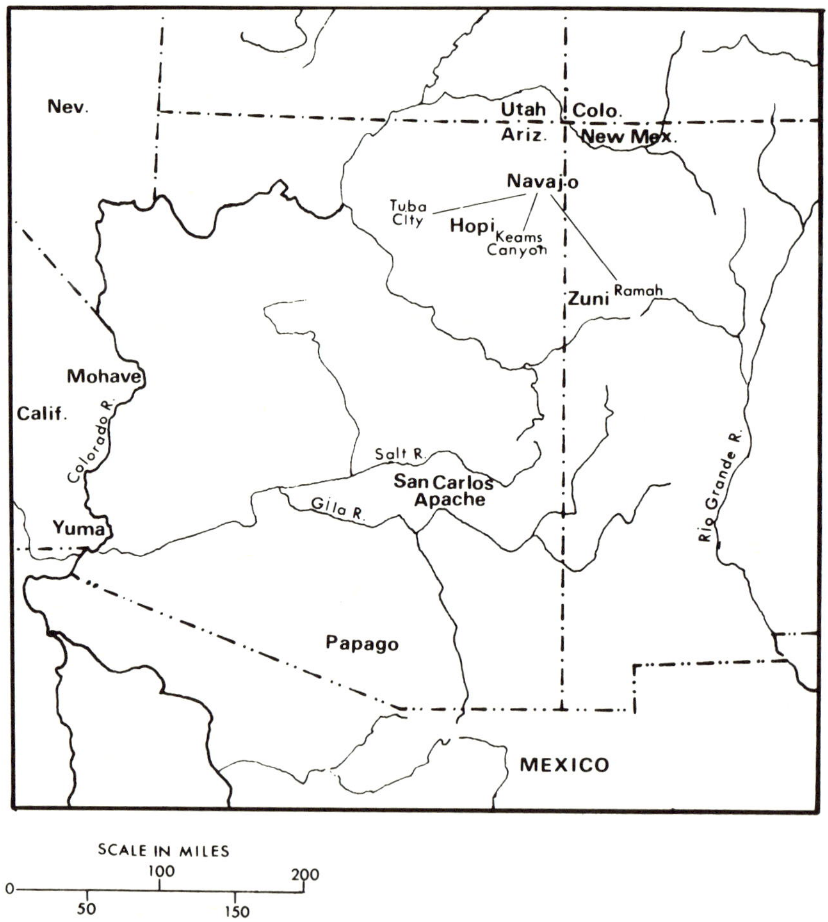

Figure 19.1. Geographic distribution of groups in the American Southwest.

Table 19.1. Linguistic families and sample sizes of Southwest Indian groups (code names are used in subsequent tables).

Group	Code Name	Linguistic Family	Sample Size Male	Sample Size Female	Sample Size Total
Papago	PAP	Uto-Aztecan	94	90	184
Hopi	HOP	Uto-Aztecan	124	204	328
Zuni	ZUN	Zuni	28	37	65
Yuma	YUM	Hokan	31	42	73
Mohave	MOH	Hokan	28	24	52
Apache	APA	Athapaskan	[a]	[a]	53
Navajo	NAV	Athapaskan	207	289	496
		Totals	512	686	1251

[a] Information on sex unavailable.

anthropologists should examine more closely morphological variation at the microgeographical level.

To determine the efficacy of tooth crown traits in reflecting varying levels of population differentiation, the longstanding interest and work of A. A. Dahlberg in the American Southwest provided an areal focus for this study. Our basic purposes are to: 1) provide an overall characterization of dental morphological variation for groups from a circumscribed geographical region; 2) analyze the pattern of crown trait variation at two levels of population divergence, i.e. within single tribes and among several distinct tribal groups; and 3) compute measures of distance to determine the extent to which dental morphology can discriminate between populations and reflect known and inferred historical relationships.

Materials and Methods

Samples

Nonmetric tooth crown characteristics were observed on dental casts (full-mouth impressions) from 1251 Indians from the Southwest United States. This total number is broken down into samples from seven tribes representing four linguistic families. The geographical distribution of the samples is shown in figure 19.1. The linguistic family of each group and sample sizes are presented in table 19.1.

Total sample size is unevenly distributed across the seven samples. There are basically four small and three large samples ranging in size from 53 (Apache) to 496 (Navajo). In terms of total population size, the Navajo, with over 130,000 individuals (Morgan 1973) outnumbered all other groups combined, so their greater representation fairly reflects this situation.

Intragroup analysis was made possible by subdividing the Navajo and

Table 19.2. Subsample divisions for Navajo and Hopi.

Group	Area	Code Name	Sample Size M	Sample Size F	Sample Size Total
Navajo	Tuba City	NAT	97	101	198
	Keams Canyon	NAK	45	125	170
	Ramah	NAR	65	63	128
Hopi[a]	First mesa	1st			38
	Second mesa	2d			49
	Third mesa	3d			53
	Tewa	TEW			24

[a] Not broken down by sex for analysis.

Hopi samples (table 19.2). The Navajo sample was broken down into three subsamples, but these do not represent village populations per se. Instead, they reflect groupings of individuals from numerous small communities within limited geographic areas of the Navajo reservation. The Ramah Navajo group, in being removed from the major part of the Navajo reservation, most closely approximates a breeding population, albeit not a closed one (Morgan 1973). The Tuba City and Keams Canyon samples are more artificial in the sense that government agency services (medical, educational, etc.) and/or employment attracted individuals from outlying communities to these more concentrated settlements. Keams Canyon, for example, is on the Hopi reservation, but Navajos on the eastern fringe of this reservation utilize the available medical/dental services of the Public Health Service hospital located there. Dental impressions were obtained during such visitations. In the parlance of population structure models, the isolation by distance model with its concept of "genetic neighborhoods" would seemingly be more applicable than the island model (Wright 1969) to the Navajo population.

In contrast to the Navajo, who are widely distributed in northeast Arizona and northwest New Mexico, the Hopi population, totaling about 6,000 individuals, is found within a tightly circumscribed area. The majority of the population resides in several villages on or at the base of three mesas, which are separated by only 10 to 25 miles. While the three mesas do not represent isolated breeding populations, they do provide a means of breaking the group down in the absence of extensive genealogical data. It should be pointed out that the total Hopi sample is actually comprised of two samples collected by different workers, and for only one of these samples could individuals be assigned to specific mesas. This is why the summed mesa samples do not equal the total sample. Moreover, the Hopi subsamples were too small to divide into males and females, so only totals are considered.

In addition to the three mesa samples, it was also possible to include a small sample of Hopi-Tewa. Originally a Rio Grande Pueblo population, the Tewa migrated west and settled among the Hopi of First mesa in 1696. Pride in their own cultural heritage in conjunction with matrilocal residence rules have allowed the Tewa to maintain their sociocultural identity for almost 300 years (Dozier 1966). Regarding Hopi-Tewa admixture, Dozier (1966:26) notes that "While intermarriages are now common, they were forbidden in the past. Marriage restrictions were lifted toward the closing decades of the last century coincident with the establishment of the United States Government Agency at Keams Canyon, the construction of schools, and the advent of white traders." Because of the recency of gene flow between Hopi and Tewa the sample from the latter group is treated separately from the Hopi samples.

It is beyond the purview of this study to consider the many controversies surrounding the interpretation of Southwest prehistory, particularly regarding all the suggested relationships between prehistoric cultural traditions and recent human populations. However, as our primary concern is with the biological divergence of groups within the Southwest, it is important to address briefly the question of time depth in the area for the samples under consideration.

Of the seven groups, the Athapaskans (Navajo and Apache) represent the most recent immigrants into the Southwest. While there is debate as to the specific time of their entry (Hester 1962), the consensus is that they did not arrive in the area until A.D. 1000 or later. The ancestral population from which they initially diverged was probably located somewhere along the broad range currently inhabited by Athapaskans in western Canada and central Alaska.

The point of divergence between Athapaskans and other Southwest Indians can probably be placed at a very early date (e.g., 10,000 years ago or earlier). In all likelihood, their common ancestor would have been an eastern Asiatic group with differentiation into Athapaskan and non-Athapaskan stocks preceding their respective migrations into the New World.

On linguistic and archaeological grounds, it is almost a certainty that the five non-Athapaskan groups diverged following entry into the New World. To go one step further, the common ancestry for all five groups can probably be traced to the general area of the American Southwest, especially if the southern section of the Great Basin, southern California, and northern Mexico are included in this designation. While this suggestion lacks specificity, at least it is reasonable to infer that the Papago, Hopi, Zuni, Yuma, and Mohave populations have a time depth in the area ranging somewhere between 2,000 and 12,000 years ago.

Establishing relationships between living groups and earlier populations in the American Southwest is not without its problems. All the living tribes considered here have been linked with cultural traditions going back approximately 2,000 years; the Papago have purported ties with the Hohokam (Ezell 1973) or Ootam (DiPeso 1956) of southern Arizona; the Yuma and Mohave tribes may have connections with the Hakataya (Schroeder, 1957) or Patayan (Colton 1945) tradition of the lower Colorado River; and the western Pueblo groups, Hopi and Zuni, although distinct linguistically, may represent amalgamations of peoples from the Anasazi and Mogollon traditions (Reed 1950).

All the above prehistoric archeological cultures may have their roots in the Desert tradition (Willey 1966), and this would potentially extend the ancestry of recent groups back 9,000 years in the Southwest. The most likely exceptions to this extended time depth would be for the Yuma and Mohave. Some authors suggest that these groups moved into the area of the lower Colorado River from southern California (Rogers 1945; Hayden 1970), probably within the past 2,000 years.

The archeological record amply confirms many interrelationships between the Hohokam, Mogollon, Anasazi, and Hakataya traditions (McGregor 1965; Willey 1966). Within the area there was an extensive movement of ideas and in all probability, people. In other words, there is no pretense that we are dealing with individual tribes that evolved in isolation for several thousand years. While gene flow cannot easily be documented, it certainly must have existed at varying levels between the ancestors of recent tribal populations.

One point that is important to reiterate, even if the details of their population histories are incomplete, is that five of the groups have resided within or around the Southwest for an extended period of time whereas Athapaskans clearly represent a recent migration into the area.

Crown Traits

Thirteen morphological dental traits were scored to assess the pattern of dental variability among the seven Southwest Indian groups. With the exception of upper central incisor winging which is related to variations in the conformation of the alveolar sockets, the traits are surficial features of the tooth crowns.

The morphological traits examined are all presence/absence variables which show varying degrees of expression for the presence phenotype. Initially, observations were made using ranked scales developed by Dahlberg (1956), Turner (1970), and Scott (1973), but for the purposes of analysis and presentation, the data were dichotomized into affected and nonaffected phenotype frequencies. Generally, the total frequencies of presence expressions, representing the affected phenotypes, were used to measure within and between group variation and compute biological distance estimates. The only exceptions to this breakdown were made for shoveling and the hypocone which both attained frequencies close to 100 percent on I^1 and M^1, respectively. Because total frequencies were effectively invariant for these two traits, they were dichotomized by moderate (nonaffected) and pronounced (affected) expressions.

The justification for utilizing total trait frequencies in the analysis is based on the proposition that morphological crown traits are quasi-continuous variables with polygenic modes of inheritance (Sofaer 1970; Scott 1973; Bailit et al. 1974; Harris 1977). If dental traits are threshold variables with underlying (absence) and visible (presence) scales corresponding to positions along a continuous genotypic distribution, then total trait frequency is the best single measure to characterize this distribution (Falconer 1960). Because there is generally a correlation between total trait incidence and degree of expression, the use of graded expressions adds only a moderate amount of additional information to an analysis based on total trait frequencies. Despite the approach taken here, we strongly recommend that observations be made on the basis of ranked standard plaques. The use of graded scales not only allows more exacting analyses of sex dimorphism, inter-trait association, and inheritance patterns, but also increases the probability of making more refined observations in the critical range of expression between trait absence and presence.

Although it is traditional to present data on all members of a tooth district for a specific crown trait (e.g., Carabelli's trait on M^1, M^2, and M^3), the approach here was to consider trait expression on the "key" tooth of a given morphogenetic field (e.g., shoveling on I^1 but not I^2). Considering one tooth per trait obviates the problem of bias introduced by the high within-field correlations for single traits. Another justification for concentrating on trait expressions of the key teeth is that they seemingly have lower environmental components of variance (Sofaer et al. 1972a) and should thus serve as the best indicators of genetic differentiation. The two traits not observed on key teeth were lingual cusp number of P_2 and the hypoconulid on M_2. Multiple lingual cusps were scored on P_2 rather than P_1 because they exhibit more pronounced expressions and are in higher frequency on P_2. These two conditions generally characterize the situation found for trait development on the key teeth. The reason for using hypoconulid frequencies on M_2 rather than M_1 can be found under the description of that trait.

All morphological features considered excepting Carabelli's trait are expressed independently of the other twelve traits (Scott 1973). Carabelli's trait shows significant associations with both the hypocone of the upper first molar (Keene 1968; Scott 1979) and the protostylid of the lower first molar (Scott 1978).

All observations were made by G. R. Scott, so the data are internally consistent. This obviates problems relating to interobserver error.

A brief description of each trait and the criteria used for ascertaining affected and non-affected phenotypes are provided below. Abbreviations of trait names that are used in the tables and the tooth on which each trait was observed are shown in parentheses.

Shoveling (SHOV, I^1). One of the most intensively studied traits of the human dentition is shoveling, or lingual marginal ridging, of the incisors and canines (although emphasis is generally on the upper incisors). Hrdlička, who reviewed thoroughly the nature of this trait in 1920, considered not only the presence or absence of shoveling but also dealt with the variable development of this enamel rim. An eight-grade classification ranging from absence to pronounced marginal ridges (Scott 1973) was employed to collect data on the degree of shoveling expression in Southwest Indians. However, as this trait was effectively invariant (100 percent) on I^1 in these Indians, the data were dichotomized for the sake of consistency between slight to moderate shoveling (grades 1–3; unaffected) and pronounced shoveling (grades 4–7; affected).

Winging (WING, I^1). The sockets of the upper central incisors can develop in such a way that the crowns of these teeth rotate unilaterally or bilaterally with the distal margins everted labially (winging) or lingually (counter-winging) (Dahlberg 1963). The phenotype classified as present was bilateral winging, with unilateral winging, either type of counterwinging, and no winging considered absent. The rationale for this procedure is based on the observation that bilateral winging expression shows the most consistent patterning and also appears to be the least influenced by anterior tooth crowding; thus, it is probably the best reflection of an underlying genotype.

Canine Tubercle (CT, C^1). This lingual tubercle emanates from the basal cingulum of the upper canines and is developmentally related to phenotypically similar ridges and tubercles of the incisors (Scott 1977a) which have been traditionally termed *tuberculum dentale* (Pedersen 1949; Moorrees 1957). When the trait is expressed, it ranges from a slight groove and ridge to a large tubercle with a free apex; all such expressions were grouped in the present category. Only when there was a complete lack of ridge-tubercle manifestation on the cingulum was the trait considered absent.

Hypocone (HYP, M^1). The distolingual cusp or cusp 4 of the upper molars can be absent but, when present, ranges from a faint cusp outline on the distolingual border of the protocone to a large free-standing cusp. As with shoveling, the hypocone was essentially invariant (ca. 100 percent) on Southwest Indian upper first molars. Thus, to discern the overall variability of hypocone expression among the samples, the dichotomy used was between moderate (unaffected) and pronounced (affected) expressions. As a point of reference, moderate expressions include grades 3, 3+, and 4− of the Dahlberg scale (1951).

Carabelli's Trait (CARA, M^1). This intensively studied trait arises from the cingulum of the mesiolingual cusp of the upper molars with trait presence ranging from slight grooves and ridges to large cusps with free apexes (Dahlberg 1963). Only when there was a total lack of patterned cingular development was an individual classified in the absence category.

Cusp 5 (c5, M^1). This supernumerary cusp is located between the distal aspect of the hypocone and metacone of the upper molars. The cusp shows a limited degree of size variation, and any manifestation was classified as present.

Distal Accessory Ridge (DAR, C_1). This ridge is located between the mediolingual ridge and distal marginal ridge of either the upper or lower canine. There is a positive association between this trait on both canines, but as the ridge exhibits greater intergroup variation on the lower canine (Scott 1977b), observations on this trait were limited to this tooth. Again, the presence of a ridge of any size warranted inclusion in the affected category.

Lingual Cusp Number (LCN, P_2). Kraus and Furr (1953) and Ludwig (1957) defined multiple lingual cusps for the lower first and second premolars, respectively, in terms of the presence of two or more cusps or even cusplets without regard to the position or size of the cusps. Following their definition, a single lingual cusp with two small cusplets at any position on the occlusal ridge would be scored as multiple lingual cusps. This unrestricted definition obscures the patterning that is evident in the expression of two (and more rarely three) lingual cusps on the lower premolars. When there is more than a single lingual cusp, the additional cusp(s) generally lies distal and adjacent to a primary lingual cusp. Of the two cusps, the one in the distal position is generally smaller in size and does not reach the occlusal height of its mesially placed counterpart. This is especially true for the lower first premolar as it is much more common on the second premolar to find two lingual cusps of equal size and height. A classification with absence and eight grades of trait presence which took the patterning of multiple lingual cusps into account was utilized to score the expression of this trait on the lower second premolar (Scott 1973).

Deflecting Wrinkle (DW, M_1). The deflecting wrinkle is found on the occlusal surface of the metaconid or cusp 2 of the lower first molar. The absence phenotype was characterized by a central ridge which coursed in a direct line from the tip of the metaconid to the central occlusal pit. When this ridge was bent or deflected about halfway along its length, a deflecting wrinkle was considered present (Morris 1970).

Hypoconulid (CNUM, M_2). The hypoconulid or cusp 5 was found in a frequency approximating 100 percent on the lower first molars of Southwest Indians. As only the presence or absence of this cusp was recorded, it was not possible to dichotomize expression into moderate and pronounced categories for the lower first molar, the procedure for the hypocone of the upper first molar. Although violating the procedure of using observations on key teeth (M_1 in this case), the exclusion of this important trait was avoided by considering the presence or absence of the hypoconulid of the lower second molar.

Protostylid (PROT, M_1). This cingular trait is expressed on the buccal surface of the mesiobuccal cusp of the lower molars. It ranges in expression from a slight oblique groove and fissure from the buccal groove to a fairly well developed tubercle (Dahlberg 1950). Buccal pits are sometimes considered as part of the protostylid complex, but due to the high frequency of fillings in the buccal groove (and pit?) of Southwest Indian lower molars, only positive manifestations of the cingulum were classified as affected or present.

Cusp 6 (c6, M_1). Other terms for this supernumerary cusp of the lower molars are *tuberculum sextum* (Hellman 1928) and the entoconulid (Turner 1970). When present, cusp 6 is situated between cusps 4 and 5 and can range in expression from a very slight cuspal elevation on the lingual aspect of cusp 5 to the point where it actually exceeds cusp 5 in size. Following the Turner classificatory scheme, cusp 5 had to be observable before cusp 6 was scored as present.

Cusp 7 (c7, M_1). Synonyms for cusp 7 are *tuberculum intermedium* (Hellman 1928) and the metaconulid (Turner 1970). This supernumerary cusp is exhibited on the lower molars between cusps 2 and 4. Its presence expressions vary from a slight vertical groove on the distolingual aspect of cusp 2 to a wedge-shaped cusp with an independent apex which extends to the central occlusal pit. All such expressions were classified as present.

Counting Procedure

Because teeth are bilaterally occurring structures, there are several methods that can be used to score the expression of morphological features on antimeres. The problems introduced by asymmetry and missing antimeres can be avoided by using a tooth count method which involves scoring trait expression on all observable teeth; sample size is determined by either the total number of teeth observed or preferably, the number of individuals with one or both antimeres scorable. Asymmetry can also be circumvented by consistently scoring trait expression on only the right *or* left antimere. Often, if the designated antimere is not present, authors using this unilateral count method do include the expression of the other antimere to maximize sample size. A third approach is the individual count method in which the expression of both antimeres is recorded, but for the computation of total trait frequencies, an individual is classified relative to the antimere showing the greatest degree of trait expression.

The individual count method was used in this study, and since emphasis is on total trait frequencies, the only cases of asymmetry of concern are those where one antimere exhibits the trait but the other does not (presence/absence asymmetry). If, for example, there were two individuals exhibiting left (+), right (−) and left (−), right (+), a tooth count or unilateral count would give trait frequencies of 50 percent. An individual count, on the other hand, would give a frequency of 100 percent. On developmental grounds, it seems most plausible that individuals exhibiting presence/absence asymmetry can be assumed to have the underlying genotype for trait development. Although there is some controversy over the most appropriate counting method for tooth crown traits, it has been shown that the three methods yield closely comparable results (Scott 1980).

Analysis

As it has yet to be demonstrated that methods of analysis developed for use with gene frequency data are strictly applicable to total trait frequencies for quasicontinuous variables, more conventional procedures were followed to measure the extent and/or significance of crown trait variation within and between groups.

For the within-group comparisons, contingency analysis was used to determine if the frequencies for any single trait differed significantly among Navajo males, Navajo females, total Navajo, and total Hopi. These analyses involved 2 × 3 contingency tables, so the use of chi-square was deemed satisfactory in this instance. Because of the larger number of samples, contingency analysis was not used for the between-group analysis.

For both within- and between-group comparisons, trait frequency means and variances were computed. Variance was derived using the standard working formula for this measure of dispersion (Woolf 1968) and was not weighted by sample size.

Although many distance statistics or measures of similarity are now available for use with qualitative data, the statistic employed here was the mean measure of divergence (MMD) (Berry and Berry 1967). Incorporating the suggestions of Green and Suchey (1976), the following formula was developed:

$$\mathrm{MMD} = \frac{\sum_{i=1}^{r} (\theta_{1i} - \theta_{2i})^2 - [(1/\mathrm{n}_{1i} + 1/2) + (1/\mathrm{n}_{2i} + 1/2)]}{\mathrm{r}}$$

r = number of traits considered
θ_{1i} *and* θ_{2i} = transformed frequency of trait i in populations 1 and 2, respectively
n_{1i} *and* n_{2i} = number of dentitions examined for trait i in populations 1 and 2, respectively

As indicated by the formula, raw percentages are not used in the calculations. Instead, trait frequencies are transformed to the angle θ (measured in radians) through an inverse sine transformation. With k equal to the number of individuals expressing the trait (affected phenotypes) out of n observable individuals in a sample (so that k/n = observed trait frequency), the formula for θ is: $1/2 \sin^{-1} (1 - 2K/(n + 1)) + 1/2 \sin^{-1} (1 - 2(k + 1)/(n + 1))$.

Although the distance statistics used with qualitative data often give similar results, the mean measure of divergence has two attributes which make it appropriate for this study. First, considering the range in sample sizes noted earlier, this statistic does include a correction factor for sample size. Sanghvi's G, another distance measure (Vyas et al. 1962), which has been utilized in conjunction with nonmetric dental traits (Sofaer et al. 1972b; Scott 1973), does not take into account sample size and hence ignores the variance in trait frequencies introduced by sampling error. Second, assuming that sample size remains the same, the inverse sine transformation will give equal measures of divergence for equivalent percentage differences between two samples anywhere along the scale from 0 to 100 percent. That is, if samples 1 and 2 have hypocone frequencies of 80 and 90 percent and cusp 5 frequencies of 10 and 20 percent, either comparison will yield the same MMD. This is not true for many other distance measures, including Sanghvi's G.

Within-Group Variation

To assess the extent of intragroup crown trait variation for the Navajo, the total sample was subdivided according to sex and geographic locale (i.e., Keams Canyon, Tuba City, Ramah). The frequency data for these Navajo subsamples are presented in table 19.3. The Hopi subsamples were too small for partitioning males and females so the crown trait frequencies for the three Hopi mesas and the Tewa shown in table 19.4 are for combined sexes.

Table 19.3. Crown trait frequency variation among Navajo male, female, and total subsamples.

		Males			Females			Total		
Trait	Tooth	NAK	NAT	NAR	NAK	NAT	NAR	NAK	NAT	NAR
SHOV	I^1	.531	.590	.479	.429	.663	.415	.459	.629	.449
		(32)	(83)	(48)	(77)	(95)	(41)	(109)	(178)	(89)
WING	I^1	.294	.183	.175	.271	.253	.286	.277	.218	.230
		(34)	(93)	(57)	(96)	(95)	(56)	(130)	(188)	(113)
CT	C^1	.793	.763	.467	.697	.581	.486	.717	.686	.477
		(29)	(59)	(30)	(109)	(43)	(35)	(138)	(102)	(65)
HYP	M^1	.814	.783	.656	.792	.656	.710	.799	.719	.683
		(43)	(92)	(64)	(106)	(93)	(62)	(149)	(185)	(126)
CARA	M^1	.684	.570	.632	.724	.570	.473	.713	.570	.554
		(38)	(86)	(57)	(105)	(93)	(55)	(143)	(179)	(112)
C5	M^1	.111	.225	.102	.208	.287	.250	.182	.258	.169
		(27)	(71)	(49)	(72)	(80)	(40)	(99)	(151)	(89)
DAR	C^1	.516	.596	.344	.423	.367	.333	.445	.485	.339
		(31)	(52)	(32)	(97)	(49)	(30)	(128)	(101)	(62)
LCN	P_2	.238	.300	.122	.209	.373	.139	.217	.333	.130
		(42)	(60)	(41)	(110)	(51)	(36)	(152)	(111)	(77)
DW	M_1	.562	.396	.451	.500	.347	.273	.525	.373	.368
		(16)	(53)	(51)	(24)	(49)	(44)	(40)	(102)	(95)
CNUM	M_2	.972	.617	.630	.750	.667	.600	.821	.639	.614
		(36)	(47)	(27)	(76)	(36)	(30)	(112)	(83)	(57)
PROT	M_1	.571	.309	.228	.375	.435	.260	.435	.379	.243
		(35)	(68)	(57)	(80)	(85)	(50)	(115)	(153)	(107)
C6	M_1	.760	.442	.400	.536	.413	.244	.596	.429	.333
		(25)	(77)	(55)	(69)	(91)	(45)	(94)	(168)	(100)
C7	M_1	.262	.136	.212	.191	.161	.192	.213	.149	.202
		(42)	(81)	(52)	(94)	(93)	(52)	(136)	(174)	(104)

This within-group analysis of dental morphological variation centers primarily on two questions. First, is there a difference in the pattern and degree of microdifferentiation shown by Navajo males and females? Second, is there a difference in the degree of divergence among the subsamples of the Navajo and Hopi? In the following discussion, these questions are addressed through the use of three different but complementary approaches: contingency analysis, unweighted mean variances, and mean measures of divergence.

As shown in table 19.3 and 19.4, moderate to pronounced frequency variation exists among Navajo males and females and among the total Navajo and Hopi subpopulations. As the nonmetric crown traits are treated as binary variates, contingency analysis was employed to determine whether or not the variation exhibited by the subsamples was statistically significant.

A summary of the contingency analysis is provided in table 19.5. First, considering the variation among Navajo subsamples broken down by sex,

Table 19.4. Crown trait frequency variation among Hopi subsamples and Tewa.

Trait	Tooth	1st mesa	2d mesa	3d mesa	Tewa
SHOV	I^1	.296	.395	.364	.476
		(27)	(38)	(33)	(21)
WING	I^1	.346	.486	.310	.136
		(26)	(37)	(42)	(22)
CT	C^1	.583	.733	.738	.957
		(24)	(45)	(42)	(23)
HYP	M^1	.879	.907	.756	1.000
		(33)	(43)	(41)	(22)
CARA	M^1	.710	.810	.791	.870
		(31)	(42)	(43)	(23)
C5	M^1	.118	.158	.200	.231
		(17)	(19)	(25)	(13)
DAR	C^1	.560	.649	.583	.444
		(25)	(37)	(36)	(18)
LCN	P_2	.147	.250	.122	.217
		(34)	(40)	(41)	(23)
DW	M_1	.857	.286	.500	—
		(7)	(7)	(10)	
CNUM	M_2	.739	.828	.758	.600
		(23)	(29)	(33)	(20)
PROT	M_1	.448	.395	.353	.444
		(29)	(38)	(34)	(18)
C6	M_1	.500	.643	.423	.611
		(24)	(28)	(26)	(18)
C7	M_1	.286	.286	.227	.500
		(35)	(42)	(44)	(20)

Table 19.5. Summary of contingency analysis of crown trait variation in Navajo and Hopi subsamples.

Group	Sex	Number of Significant Differences	Traits Showing Significance
Navajo	Male	4/13 (31%)	CT, CNUM, PROT, c6
	Female	4/13 (31%)	SHOV, CARA, LCN, c6
Navajo	M + F	7/13 (54%)	SHOV, CT, CARA, LCN, CNUM, PROT, c6
Hopi	M + F	0/12[a] (0%)	—

[a]Deflecting wrinkle not included because of small samples sizes

it was found that both males and females showed significant differences for four of the thirteen traits. Cusp 6 showed significant intragroup differences for both males and females, but the remaining three traits which exhibited significant differences were not the same for the sexes. However, when the data on sexes were combined to form total Navajo subsamples, all the traits that differed significantly for either males or females continued to show significant differences. Thus, for the total Navajo subsamples, seven of thirteen traits, or 54 percent, differed significantly among the three Navajo groups. In marked contrast to the Navajo situation, contingency analysis did not indicate any of the trait frequency variation among the three Hopi mesas (Tewa excluded here) to be statistically significant.

If degree of microdifferentiation is measured by the number of traits differing significantly among subpopulations, it would follow that Navajo males and females show equal divergence, and that the Navajo exhibit much more within-group divergence than the Hopi. Regarding the latter conclusion, however, the Navajo subsamples are much larger than their Hopi counterparts, and sample size does have an effect on chi-square values. For this reason, unweighted variances were computed for all traits for Navajo males and females, and for the total Navajo and Hopi samples. Mean variance was then computed for the entire array of traits.

In contrast to the contingency analysis which suggested similar degrees of divergence among the Navajo male and female subsamples, the mean variances for the two sexes are distinctly dissimilar. Several authors have made the point that differences, even when nonsignificant, may still yield important biological information. That case may pertain here because even though males and females show the same number of traits differing significantly among subsamples, male variances exceed those of females for nine of the thirteen traits. This overall pattern is reflected in a larger mean variance for males (.0151) than for females (.0084).

When the data on sexes are combined, the mean variance among the Navajo subsamples is .0086, which is more in line with the mean female variance. Still, this mean variance value, which is free of the effects of sample size, is considerably larger than the mean variance of .0048 for the Hopi subsamples. Thus, it would appear on the basis of both contingency analysis and unweighted variance values that there is more crown trait differentiation within the Navajo than within the Hopi.

As a final check on intragroup variation, the mean measure of divergence described previously, was employed to estimate the relative degrees of similarity for the Navajo and Hopi subsamples. Table 19.6 shows the MMD values for the three pair-wise comparisons for Navajo males and females and for the total subsamples.

For the Navajo, the MMD values are consistently larger for the male subsamples. The mean of the MMD values is more than twice as great for males as for females, which corresponds to the results indicated by the means for unweighted variances. That is, for these Navajo subsamples, male crown trait frequencies evidence more within-group variation or divergence than corresponding female frequencies.

The matrix of MMD values for the Hopi is presented in table 19.7. Two points are apparent. First, the MMD values among the three mesas are very small, thus corresponding to the lack of significant differences among the mesas and the low unweighted variances. The mean of the MMD values among the three mesas is .031, which is not quite half the value of .063 found among the total Navajo subsamples. Second, there are markedly divergent MMD values for all three mesas when compared with the Tewa sample. This finding was unexpected considering Dozier's statement (1966:23) that "Tewa and Hopi marry freely. Biologically the two populations are now completely mixed." As the Tewa sample is fairly small (n ranges from 13 to 25), this relatively greater divergence may result from sampling error. A more optimistic conclusion would be that nonmetric crown traits reflect the recency of Hopi-Tewa admixture and accurately discriminate among populations at even this low level of differentiation.

In summary, two noteworthy points emerged from this within-group analysis of crown trait frequencies. First, there is more frequency variation among Navajo male subsamples than among female subsamples. This

Table 19.6 Mean measures of divergence among Navajo male, female, and total subsamples.

Sex	NAK-NAT	NAK-NAR	NAT-NAR	$\bar{x}$ MMD
Male	.122	.175	.062	.120
Female	.042	.073	.050	.055
M + F	.050	.088	.051	.063

Table 19.7. Mean measures of divergence among Hopi subsamples and Tewa.

	1st	2d	3d	TEW
1st	—			
2d	.020	—		
3d	.027	.047	—	
TEW	.107	.103	.082	—

$\bar{x}$ MMD = .031 (minus TEW)
.064 (with TEW)

finding is in agreement with the work of Harris (1977), who studied crown trait variation among Bougainville Island village populations. His distance estimates showed male subsamples to be more divergent (or less similar) than female subsamples. Cluster analysis, based on these distance estimates, revealed refined clustering for the male subsamples but broad clusters for females. The implications of such a sex dimorphism in rates of differentiation are significant considering the fact that most workers combine male and female data in their analyses. It is hoped that this question will be followed up over a greater range of populations and subpopulations.

The difference in rates of within-group divergence for the Navajo and Hopi samples is the second point to stress. As discussed earlier, the Hopi mesa samples do not actually correspond to isolated breeding populations as there is mate exchange throughout the Hopi reservation. However, even if viewed as three random samples, the data on the Hopi are revealing when compared to those on the Navajo.

The Hopi samples are invariably smaller than the Navajo samples, and this would allow for more sampling error among the Hopi. However, despite smaller sample sizes, every measure of variation indicates less differentiation within Hopi than within Navajo. Even though the Navajo samples also do not represent well defined breeding populations, there is more potential among these groups for isolation by distance. While geography does not preclude mate exchange between the three areas of the reservation, distance can contribute to slowing the rate of gene flow. Because of this, one would predict more variation among the Navajo groups, and that finding was consistently indicated. Thus, it does appear that tooth crown traits exhibit predictable patterns of variation for groups at low levels of differentiation.

Between-Group Variation

Intragroup analysis of dental morphological variation focused on frequency diversity within the Navajo and Hopi samples. Intergroup analysis centered on the variation in trait frequencies among the seven Southwest Indian tribes. Contingency analysis was not employed for this purpose as a 2 × 7 table yields chi-square values that are difficult to interpret and can be biased by a single marked contrast between just two of the seven groups. For this reason, discussion is restricted to unweighted variances and mean measures of divergence.

Total trait frequencies for the seven Indian groups are shown in table 19.8. The only trait to exhibit a significant degree of sex dimorphism was the distal accessory ridge of the lower canine (Scott 1977b), so the data for males and females were combined in this table. However, as male trait frequencies tended to be more variable among the Navajo subsamples, the male and female frequencies are summarized by unweighted means and variances in table 19.9 to determine if this same pattern applies at a higher level of differentiation.

In table 19.9, the means and variances of nine traits are shown for male, female, and total samples. As the summary statistics are unweighted, four

Table 19.8. Crown trait frequency variation among Southwest Indian tribal groups.

		Tribal Group						
Trait	Tooth	PAP	HOP	NAV	APA	ZUN	YUM	MOH
SHOV	I^1	.509	.448	.537	.613	.474	.642	.646
		(171)	(250)	(376)	(31)	(57)	(67)	(48)
WING	I^1	.353	.314	.239	.172	.200	.182	.327
		(170)	(277)	(431)	(29)	(60)	(66)	(49)
CT	C^1	.529	.737	.656	.737	.909	.622	.684
		(153)	(236)	(305)	(19)	(11)	(37)	(19)
HYP	M^1	.736	.843	.735	.837	.705	.889	.896
		(178)	(286)	(460)	(43)	(44)	(63)	(48)
CARA	M^1	.719	.803	.613	.583	.745	.762	.723
		(160)	(279)	(434)	(48)	(47)	(63)	(47)
C5	M^1	.222	.189	.212	.154	.029	.125	.068
		(126)	(180)	(339)	(39)	(34)	(56)	(44)
DAR	C_1	.746	.627	.446	.500	.790	.434	.650
		(158)	(208)	(291)	(18)	(15)	(57)	(25)
LCN	P_2	.252	.154	.235	.174	.304	.286	.308
		(179)	(260)	(340)	(23)	(23)	(49)	(26)
DW	M_1	.613	.378	.397	.667	.260	.571	.486
		(106)	(119)	(237)	(39)	(50)	(63)	(35)
CNUM	M_2	.700	.763	.714	.632	.571	.473	.533
		(141)	(194)	(252)	(19)	(7)	(36)	(15)
PROT	M_1	.351	.344	.357	.292	.575	.324	.250
		(137)	(241)	(375)	(48)	(40)	(68)	(40)
C6	M_1	.526	.498	.445	.562	.452	.279	.098
		(135)	(215)	(362)	(48)	(42)	(68)	(41)
C7	M_1	.335	.246	.184	.082	.222	.182	.268
		(158)	(277)	(414)	(49)	(45)	(66)	(41)

Table 19.9. Means and variances of crown trait frequencies for Southwest Indian tribal groups.

		Male Trait Frequency		Female Trait Frequency		Total Trait Frequency	
Trait	Tooth	Mean	Variance	Mean	Variance	Mean	Variance
SHOV	I^1	.543	.0035	.552	.0164	.553	.0066
WING	I^1	.209	.0035	.325	.0133	.255	.0056
HYP	M^1	.771	.0098	.830	.0077	.806	.0062
CARA	M^1	.739	.0120	.712	.0118	.707	.0064
C5	M^1	.105	.0076	.176	.0056	.143	.0054
DW	M_1	.496	.0101	.417	.0282	.482	.0211
PROT	M_1	.392	.0074	.344	.0193	.356	.0108
C6	M_1	.402	.0357	.376	.0233	.409	.0270
C7	M_1	.300	.0065	.185	.0020	.217	.0063
Mean variance:			.0107		.0142		.0106
Mean variance for Navajo considering same traits:			.0116		.0095		.0069

Note: PAP, HOP, NAV, ZUN, YUM, MOH; total also includes APA

traits (CT, DAR, LCN, CNUM) were omitted because one or more samples numbered less than 25 individuals. As sex provenience was not available for the Apache sample, it was considered only for the computation of total means and variances.

While the means are instructive for characterizing trait variation in Southwest Indians, our interest here is in the dispersion of frequencies. First, it can be noted that variance values themselves vary widely between traits. This is particularly striking for the male and female frequency variances, which show ranges of .0035–.0357 and .0020–.0282, respectively. Moreover, the variances for either sex do not tend to cluster around any particular set of values. For the total samples, however, the effect of combining males and females is to homogenize the variances. Six of the nine traits now fall in the limited range between .0054 and .0066. The protostylid exhibits a variance somewhat above this range at .0108, while the deflecting wrinkle and cusp 6 both maintain the exceptionally large variances (.0211 and .0270) found for the sexes separately. Although it is difficult to state precisely why different traits exhibit differing degrees of between group variation, there may be some relationship to mean trait frequency. That is, variance values tend to be higher for those traits approximating a mean of 50 percent.

When mean variance values were computed for the nine traits, it was found that the male value was smaller than the corresponding female mean. This finding for the between-group variances is not in agreement with the greater degree of variation found among males within the Navajo. Interestingly, when the Navajo are considered for the same nine traits, the intragroup mean variance of the Navajo exceeds the intergroup variance of all samples for the males. This situation is reversed for females as the intergroup variance is notably larger than the intragroup variance for Navajo females. This lack of a consistent pattern may be primarily due to the relatively small number of samples and subsamples studied.

As a final note, one prediction which does appear to be validated is that between-group divergence, as measured by mean variance for the total samples, is greater than within-group divergence. The mean variance over the seven samples is .0106, while the same value for the Navajo is .0069. This means that approximately 65 percent of the divergence between groups can be found within a single population. Using distance measures derived from blood group gene frequencies rather than variances, Neel and Ward (1970) found that within-group mean distances were 68 percent of mean distances among groups. Even though the traits and measures of dispersion are not strictly comparable, these studies do correspond in showing the relatively large magnitidue of differentiation found within populations.

Although the full distance matrix will be presented and discussed in the following section on population affinities, MMD values are also instructive in measuring the degree of between-group variation. The mean MMD values for each group based on the six possible pair-wise comparisons are: Navajo (.062), Hopi (.064), Yuma (.071), Papago (.081), Apache (.084), Mohave (.101), and Zuni (.108).

The mean MMD values indicate that the Navajo and Zuni are the least and most divergent, respectively, from all other groups. However, the point to stress here is that the overall mean measure of divergence among the seven Southwest Indian groups is .082. In contrast to the comparable measures within Hopi (.031) and within Navajo (.063), this value is larger, as expected. Of course, there is the problem of determining just how much more differentiation one would expect to find between groups than within groups. Based on MMD values, one could state that there is 77 percent as much divergence within a single group as there is between all Southwest groups. Unfortunately, blood group gene frequencies are of uneven quality for these same populations (Merbs 1965), so there is no good biological referent to compare to this value. Further work in this area may help clarify the situation and provide some answer as to the expected degree of microdifferentition in the Southwest.

Phenetic Distance and Historical Inference

As Livingstone (1973) has noted, the use of measures of similarity or divergence among several groups yields a large table of pair-wise values which cannot all be adequately explained by our limited knowledge of historical relationships and earlier evolutionary processes. Distance statistics, when used for inferring phylogeny, are based on the assumption that evolution is a branching process with the diverging populations showing constant rates of change through time. Genetic drift is assumed to be the primary mechanism generating diversity between groups because gene flow would result in the convergence of even distantly related populations, and depending on its effects, natural selection would either impede or enhance the process of differentiation.

Because the evolutionary history of most populations is poorly understood, it is exceedingly difficult to estimate how gene flow and natural selection have contributed to the patterns of biological variation evident among living populations. Because of this, a distance matrix invariably shows values which are not readily explained by the random differentiation model. The interpretations of such matrices tend to center on those pairwise values which do seem to correspond to what is known about the population histories of two or more groups based on linguistic, archaeological, and other biological data (Livingstone 1973). With this limitation in mind, we will now review some of the MMD values based on crown trait frequencies shown in table 19.10.

Apache. The prediction for the Apache sample, based on linguistic and archeological evidence, would be that they should be most similar to the Navajo. Both groups represent relatively recent Athapaskan migrations into the American Southwest, and although the precise time of their divergence from a common ancestor is still unknown, it is likely that it will be numbered in hundreds of years rather than thousands of years (Vogt 1961), which would be the case for the Athapaskan/non-Athapaskan divergence. The Navajo-Apache MMD value of .034 does reflect the close affinity of these two groups, as this is the second smallest value on the table.

Zuni. The Zuni are so distinct linguistically, and apparently archeologically, that they are generally thought of as being quite divergent from other living Southwest Indian groups. This distinctiveness is reflected in the

Table 19.10. Mean measures of divergence among Southwest Indian tribal groups.

	PAP	HOP	NAV	APA	ZUN	YUM	MOH
PAP	—						
HOP	.040	—					
NAV	.070	.042	—				
APA	.069	.072	.034	—			
ZUN	.097	.064	.087	.138	—		
YUM	.095	.065	.042	.065	.128	—	
MOH	.115	.103	.099	.124	.134	.031	—

consistently high MMD values between Zuni and the other six groups. The mean distance of Zuni from the other groups was .108, which represents the highest measure of differentiation for any sample. Interestingly, the distinctiveness of the Zuni is better reflected in tooth crown morphology than in blood group frequencies, where one of their few points of distinction is in having a relatively high frequency of the B allele (Workman et al. 1974). Also, as one would predict, the closest similarity Zuni does show is with the Hopi, another Western Pueblo group of the Colorado plateau.

Mohave. The River Yuman groups, Yuma and Mohave, are closely allied linguistically, culturally, historically, and geographically (Forbes 1965). This affinity also extends to crown trait frequencies as their pair-wise MMD value of .031 is the lowest on the entire table. An interesting but enigmatic point is that the Mohave are almost as consistently divergent from all other groups as Zuni (mean MMD = .101), but the Yuma pair-wise MMD values with the remaining five groups range from low (.042) to high (.128). This array of values cannot be easily explained, and their close similarity to the Navajo is an unexpected finding.

Uto-Aztecans. The two Uto-Aztecan speaking groups, Hopi and Papago, are quite similar in tooth crown morphology as indicated by the low MMD value of .040. Despite the broad linguistic tie, these two groups belong to different branches of the widespread Uto-Aztecan family (Newman 1954), and it would seem that the historical relationships between these groups would not have been close for 2,000 years or more. There is no ready explanation for their dental similarity.

Navajo-Hopi. While the Navajo–Apache similarity was predictable on the basis of other data, the close similarity between the Hopi and Navajo (MMD = .042) was surprising, considering the long-term population histories of these groups. However, over the past several hundred years the Navajo have been strongly influenced culturally by Pueblo populations (Vogt 1961; Hester 1962), and it is possible that genetic exchange has also been significant. If, for example, gene flow occurred when the total Navajo population was small (2,000–3,000 around 1700; Hester 1962), this genetic influence could have been carried into the recent Navajo population, which now numbers over 130,000. Pueblo gene flow into the Navajo population today would have little impact because of the large population size, but it could have had a significant effect in the past. In essence, dental data support the notion of anthropologists who hold that the Navajo are a composite population now and not merely a continuously isolated immigrant population (Vogt 1961). If overall differentiation is measured by the mean MMD values from all other groups, the Navajo are on the average, more similar to the remaining groups than any other sample (mean MMD = .062).

Table 19.11. Comparison of Southwest Indian and American white crown trait frequencies.

Trait	Tooth	Southwest Indians	American Whites
SHOV	I^1	.521	.000
WING	I^1	.273	.041
CT	C^1	.660	.720
HYP	M^1	.781	.908
CARA	M^1	.696	.795
C5	M^1	.185	.104
DAR	C^1	.571	.219
LCN	P_2	.220	.509
DW	M_1	.456	.018
CNUM	M_2	.704	.131
PROT	M_1	.352	.048
C6	M_1	.448	.061
C7	M_1	.223	.245

Population Divergence at a High Level

The primary thrust of this paper has been to consider the subtle degrees of crown trait variation within populations and the somewhat greater but still moderate degrees of difference between seven Indian tribes. To provide a broader perspective on this fine-grained differentiation, data were collected by G. R. Scott on crown trait variation in a sample of American whites. The overall frequencies for Southwest Indians and the sample frequencies for American whites are contrasted in table 19.11. Sample size is not noted for whites, but for any given trait, the sample always exceeded 100 individuals.

For the 13 traits under comparison, Southwest Indians and American whites show similarities in frequency and form for only the canine tubercle, cusp 5, and cusp 7. Moderate differences in frequency exist for the hypocone and Carabelli's trait but, for the latter variable, there is a major difference in terms of degree of expression (Kraus 1959; Scott 1980). While Southwest Indians have a relatively high frequency of Carabelli's trait, expression is almost entirely limited to intermediate expressions; tubercles have an incidence of only about 2 percent. This clearly contrasts American whites who show tubercles in a frequency of about 25 percent.

The remaining eight traits show pronounced differences between whites and Indians. Shoveling, winging, the distal accessory ridge, the deflecting wrinkle, M_2 hypoconulid, protostylid, and cusp 6 are all far more common in Indians than in whites. The higher frequencies of all of these surficial crown features is the reason that Indian dentitions appear to show more overall crown sculpting than one observes in the dentitions of European and European-derived populations. The only instance where whites show a significantly higher trait frequency is for multiple lingual cusps of the lower second premolar.

The mean measure of divergence between Southwest Indians and American whites is .639, a value almost eight times greater than the mean MMD of .082 among the seven Indian groups. This corresponds to the longstanding observation of dental anthropologists that major geographical races differ markedly in their crown trait profiles.

Discussion

There have been several recent attempts to determine the extent to which tooth crown morphology can be utilized to determine predictable patterns of biological relationships among human populations (Barksdale 1972; Sofaer et al. 1972b; Brewer-Carias et al. 1976; Berry 1976; Harris 1977; Palomino et al. 1977). The methods and samples of these studies vary widely, but the researchers are in general agreement that the utility of such traits can be viewed with "cautious optimism." This conclusion is tempered with comments on the difficulties in consistently scoring the expression of morphological traits and the lack of more detailed knowledge on their inheritance mechanisms. What is not considered is that the major impediment to the use of crown morphology in population studies may not relate to the nature of the crown traits but to the widely contrasting methodologies employed in their study.

Workers who deal with blood groups, serum proteins, and other simple genetic markers generally share a similar format for the presentation of phenotype and allele frequencies. Dental morphological characterizations by different workers show precisely the opposite in an almost total lack of standardization. This applies to the specific traits studied, the definitions of the traits, the teeth on which each trait is examined, and the method of scoring trait expression.

Most studies of dental morphology include a consideration of shoveling, Carabelli's trait, upper molar cusp number, lower molar cusp number, and lower molar groove pattern. This is as close as dental researchers come to using a standard battery of traits. The inclusion of additional variables seemingly reflects the experience and inclination of the individual researcher.

Regarding trait definitions, one problem arises when crown variants which are listed as single traits are actually composites of two or more traits. For example, if one characterizes the affected phenotype for lower molar cusp number as "6 cusps or more" (Sofaer et al. 1972b:360), the presence of cusp 6 and/or cusp 7 would warrant classification in the affected category. Cusp 6 and cusp 7 are not correlated variables (Scott 1973), so combining them into one category obscures the presence of two genetically independent features. The reverse of this situation occurs when the variable expressions of a single trait are treated as two separate traits. An illustration of this is in Berry's (1976:260) listing of "31" dental variants where she treats "Carabelli's pit present" and "Carabelli's cusp present" as two variants. The same situation applies to her viewing Y and X groove patterns on the lower molars as two different traits. Neither combining traits to form a single trait nor dividing single traits into two or more traits can be justified on genetic-developmental grounds. It is critical to delineate and define precisely individual traits which are expressed independently of other such traits.

One of the more consistent aspects of data collection, presentation, and analysis pertains to the teeth on which specific traits are scored. Workers almost invariably include in their analyses specific traits on two (or three) teeth within a single field. That is, shoveling is considered on I^1 and I^2, lower molar groove pattern on M_1 and M_2, etc. The problem is not only that there are high within-field correlations, but using trait frequencies for the variable teeth as opposed to the key teeth potentially introduces a significant amount of environmentally induced variation into an analysis. If interest is in genetic differentiation and population affinities, the inclusion of traits on the more variable teeth could obscure predicted relationships and lead to the conclusion that dental traits are of limited utility in population studies, especially at the microgeographical level.

The difficulty in scoring trait expression is often noted as a problem in dental morphological studies. This does not relate simply to problems of ranking trait expression, which some authors imply, but also to making consistent observations on trait presence and absence. Low level expressions which subtly reflect the presence of a trait are the most difficult observations to make, and yet are by far the most critical if analysis is based on total trait frequencies.

There is always some degree of intra-observer error (c. 5–10 percent) in scoring crown traits, but this can be minimized with observational experience and the use of standardized classification schemes. Inter-observer differences pose a much larger and as yet unresolved problem, which resides primarily in the varying definitions of different workers on the point of distinction at which a trait is classified as present. This has been a problem with even intensively studied characteristics such as Carabelli's trait (Scott 1980). It is only necessary to compare the frequencies reported in this study with those of Sofaer et al. (1972b) on Zuni, Papago, and Pima samples to obtain an idea as to the potential magnitude of inter-observer error. For some traits, this error is several hundred percent, e.g. Zuni protostylid frequencies of 0.1 percent (Sofaer et al. 1972b) and 57.1 percent (this study).

One possible way to reduce inter-observer error would be through a conference or workshop where dental morphologists could interact directly and attempt to standardize their trait definitions and observations. Until standardization is achieved through some forum, inter-observer error will continue to hinder the full utilization of dental morphological data in human population studies.

In this study, inter-observer error did not pose a problem as all observations were made by a single worker. Intra-observer error was minimized through the use of standard plaques and the gaining of observational experience on dentitions from many different geographic groups. This error was measured at between 3 and 10 percent, depending on the trait.

Granted that their precise modes of inheritance are unknown, we have attempted to maximize the genetic information that can be derived from crown traits by considering clearly defined traits on the most developmentally stable teeth. This approach, in conjunction with the minimization of intra-observer error, was directed at reducing the sources of error variance in the estimates of crown trait frequencies as much as possible.

Palomino et al. (1977), using crown trait frequencies collected by several different workers, found similar degrees of diversity at four levels of population differentiation. Inter-observer differences in scoring trait expression probably had a significant effect on their analysis. We feel that the data presented here do suggest that nonmetric crown traits exhibit the relative, but not necessarily absolute, degrees of variation expected at different levels of group differentiation with MMD values ranging from within Hopi (.031), within Navajo (.063), among all Southwest groups (.082), to Southwest Indians and American whites (.639). We do not dispute the proposition that polygenic traits, including tooth crown characteristics, diverge at a slower rate and thus show less intra- and inter-group variation than monogenic traits, because they are less susceptible to the influences of genetic drift. It is possible to agree with that suggestion but still contend that crown traits are useful in analyzing variation at the microgeographical level and not just at broad levels of differentiation.

Regarding the utility of crown traits in measuring population affinity, some of the distance estimates could not be easily explained. This is not surprising, however, in light of the extensive movements of peoples throughout the Southwest over a broad temporal span. The relationships most firmly established on other lines of evidence were indicated by the mean measures of divergence for the comparisons between Yuma and

Mohave, Navajo and Apache, and Hopi and Zuni. Additionally, crown traits showed the general distinctiveness of the Zuni, which corresponds to the observations of many Southwest archaeologists and ethnographers. The composite nature of the Navajo population is suggested by their low level of divergence from other groups in general and their close similarity to Hopi in particular. Blood group distributions do indicate that the Navajo are generally distinct from indigenous Southwest Indians (Merbs 1965), so hopefully future work will clarify this discrepancy. Despite such problems, we are more optimistic than Palomino et al. (1977) on the potential of crown traits as indicators of biological affinity between groups at more than high levels of differentiation.

When crown trait frequencies were contrasted between Southwest Indians and American whites, major differences were indicated by a divergence measure eight times larger than the mean divergence among Southwest Indians. With such a large difference in distance estimates between local groups and major geographical races, an interesting question is what groups should be compared to bridge the gap between these extremes. If crown traits are selectively neutral, as many claim, and if they evolve at slower rates than monogenic traits, they might be useful in estimating rates and even times of divergence for groups compared at many different levels of differentiation. As there is some reluctance to accept the random differentiation model for blood group alleles and anthropometric traits, the type of phylogenetic analyses proposed by Cavalli-Sforza and Bodmer (1971) might be especially appropriate for tooth crown traits. The added advantage of dental characteristics that has not been fully exploited to date is the opportunity to evaluate their rates of change through dental remains in the fossil record.

In sum, while the authors are fully aware of their limitations, we believe dental morphological traits can contribute significantly to studies of human population structure and history. Rather than stating that their potential can be viewed with cautious optimism, we would prefer to think that if the methodology employed in their study is cautious, their potential in human population biology can be viewed with unquestioned optimism.

References

Bailit, H. L., S. Anderson, and D. Kolakowski. 1974. Quasi-continuous variation: the genetics of tooth morphology. *Am. J. Phys. Anthrop.* 41:468.

Barksdale, J.T. 1972. A descriptive and comparative investigation of dental morphology. In R. A. Littlewood, ed., *Physical Anthropology of the Eastern Highlands of New Guinea.* Seattle: University of Washington Press.

Berry, A. C. 1976. The anthropological value of minor variants of the dental crown. *Am. J. Phys. Anthrop.* 45:257–267.

Berry, A. C. and R. J. Berry. 1967. Epigenetic variation in the human cranium. *J. Anat.* 101:361–379.

Boyd, W. C. 1950. *Genetics and the Races of Man.* Boston: Little, Brown.

Brewer-Carias, C. A., S. LeBlanc, and J. V. Neel. 1976. Genetic structure of a tribal population, the Yanomama Indians. XIII. Dental microdifferentiation. *Am. J. Phys. Anthrop.* 44:5–14.

Cannings, C. and L. L. Cavalli-Sforza. 1973. Human population structure. *Advances in Hum. Genet.* 4:105–171.

Cavalli-Sforza, L. L. and A. W. F. Edwards. 1964. Analysis of human evolution. *Proc. 11th Int. Cong. Genet.* 3:923–933.

Cavalli-Sforza, L. L. and W. Bodmer. 1971. *The Genetics of Human Populations.* San Francisco: Freeman.

Colton, H. S. 1945. The Patayan problem in the Colorado River valley. *Southwest J. Anthrop.* 1:114–121.

Count, E. W. 1950. *This is Race.* New York: Henry Schuman.

Dahlberg, A. A. 1950. The evolutionary significance of the protostylid. *Am. J. Phys. Anthrop.* 8:15–25.

——1951. The dentition of the American Indian. In W. S. Laughlin, ed., *The Physical Anthropology of the American Indian.* New York: Viking Fund.

——1956. Materials for the establishment of standards of classification of tooth characters, attributes, and techniques in morphological studies of the dentition. Zoller Laboratory of Dental Anthropology, University of Chicago.

——1963. Analysis of the American Indian dentition. In D. R. Brothwell, ed., *Dental Anthropology.* New York: Pergamon Press.

DiPeso, C. C. 1956. *The Upper Pima of San Cayetano del Tumacacori.* Publication No. 7. Dragoon, Arizona: Amerind Foundation.

Dozier, E. P. 1966. *Hano: A Tewa Indian Community in Arizona.* New York: Holt, Rinehart, and Winston.

Ezell, P. H. 1963. Is there a Hohokam-Pima culture continuum? *Am. Antiq.* 29:61–66.

Falconer, D. S. 1960. *Introduction to Quantitative Genetics.* New York: Ronald Press.

Forbes, J. D. 1965. *Warriors of the Colorado.* Norman: University of Oklahoma Press.

Garn, S. M. 1971. *Human Races,* 3d ed. Springfield, Ill.: C. C. Thomas.

Green, R. F. and J. M. Suchey. 1976. The use of inverse sine transformations in the analysis of non-metric cranial data. *Am. J. Phys. Anthrop.* 45:61–68.

Harris, E. F. 1977. Anthropologic and Genetic Aspects of the Dental Morphology of Solomon Islanders, Melanesia. Ph.D. dissertation, Arizona State University, Tempe.

Hayden, J. D. 1970. Of Hohokam origins and other matters. *Am. Antiq.* 35:87–93.

Hellman, M. 1928. Racial characters in human dentition. *Proc. Am. Phil. Soc.* 67:157–174.

Hester, J.J. 1962. Early Navajo migrations and acculturation in the Southwest. *Papers in Anthrop.*, No. 6. Santa Fe: Museum of New Mexico.

Hrdlička, A. 1920. Shovel-shaped teeth. *Am. J. Phys. Anthrop.* 3:429–465.

Keene, H.J. 1968. The relationship between Carabelli's trait and the size, number and morphology of the maxillary molars. *Archs. Oral Biol.* 13:1023–1025.

Kraus, B.S. and M.L. Furr. 1953. Lower first premolars. Part I. A definition and classification of discrete morphologic traits. *J. Dent. Res.* 33:554–564.

Kraus, B.S. 1959. Occurrence of the Carabelli trait in Southwest ethnic groups. *Am. J. Phys. Anthrop.* 17:117–124.

Livingstone, F. 1973. Gene frequency differences in human populations: some problems of analysis and interpretation. In M.H. Crawford and P.L. Workman, eds., *Methods and Theories of Anthropological Genetics*. Albuquerque: University of New Mexico Press.

Ludwig, F.J. 1957. The mandibular second premolars: morphologic variation and inheritance. *J. Dent. Res.* 36:263–273.

Malecot, G. 1948. *Les Mathématiques de l'Hérédité.* Paris: Masson.

McGregor, J.C. 1965. *Southwestern Archaeology*, 2d ed. Urbana: University of Illinois Press.

Merbs, C.F. 1965. Blood group distributions in the American Southwest. Paper presented at 34th annual meeting of the American Association of Physical Anthropologists, University Park, Pennsylvania.

Moorrees, C.F.A. 1957. *The Aleut Dentition.* Cambridge: Harvard University Press.

Morgan, K. 1973. Historical demography of a Navajo community. In M.H. Crawford and P.L. Workman, eds., *Methods and Theories of Anthropological Genetics.* Albuquerque: University of New Mexico Press.

Morris, D.H. 1970. On deflecting wrinkles and the *Dryopithecus* pattern in human mandibular molars. *Am. J. Phys. Anthrop.* 32:97–104.

Morton, N.E. 1969. Human population structure. *Annual Rev. Genet.* 3:53–74.

Neel, J.V. and F.M. Salzano. 1966. Further studies on the Xavante Indians. X. Some hypotheses-generalizations resulting from these studies. *Am. J. Hum. Genet.* 19:554–574.

Neel, J.V. and R.H. Ward. 1970. Village and tribal genetic distances among American Indians, and the possible implications for human evolution. *Proc. Nat. Acad. Sci.* 65:323–330.

Nei, M. 1973. Analysis of gene diversity in subdivided populations. *Proc. Nat. Acad. Sci.* 70:3321–3323.

Newman, S. 1954. American Indian linguistics in the Southwest. *Am. Anthrop.* 56:626–644.

Palomino, H. R. Chakroborty, and F. Rothhammer. 1977. Dental morphology and population diversity. Hum. Biol. 49:61–70.

Pedersen, P.O. 1949. The East Greenland Eskimo dentition. *Meddel. om Grφnland* 142:1–256.

Reed, E. K. 1950. Eastern-central Arizona archaeology in relation to the Western Pueblos. *Southwest J. Anthrop.* 6:120–138.

Riesenfeld, A. 1956. Shovel-shaped incisors and a few other dental features among the native peoples of the Pacific. *Am. J. Phys. Anthrop.* 14:505–521.

Rogers, M. J. 1945. An outline of Yuman prehistory. *Southwest J. Anthrop.* 1:167–198.

Schroeder, A. H. 1957. The Hakataya cultural tradition. *Am. Antiq.* 23:176–178.

Schull, W. J. and J. W. MacCluer. 1968. Human genetrics: structure of population. *Annual Rev. Genet.* 2:279–304.

Scott, G. R. 1973. Dental Morphology: A Genetic Study of American White Families and Variation in Living Southwest Indians. Ph.D. dissertation, Arizona State Univ., Tempe.

——. 1977a. Lingual tubercles and the maxillary incisor-canine field. *J. Dent. Res.* 56:1192.

——. 1977b. Classification, sex dimorphism, association, and population variation of the canine distal accessory ridge. *Hum. Biol.* 49:453–469.

——. 1978. The relationship between Carabelli's trait and the protostylid. *J. Dent. Res.* 57:570.

——. 1979. Association between the hypocone and Carabelli's trait of the maxillary molars. *J. Dent. Res.* 58:1403–1404.

——. 1980. Population variation of Carabelli's trait. *Hum. Biol.* 52:63–78.

Sofaer, J. A. 1970. Dental morphologic variation and the Hardy-Weinberg law. *J. Dent. Res.* 49:1505–1508.

Sofaer, J. A., C. J. MacLean, and H. L. Bailit. 1972a. Heredity and morphological variation in early and late developing human teeth of the same morphological class. *Archs. Oral Biol.* 17:811–816.

Sofaer, J. A., J. D. Niswander, and C. J. MacLean. 1972b. Population studies on Southwestern Indian tribes. V. Tooth morphology as an indicator of biological distance. *Am. J. Phys. Anthrop.* 37:357–366.

Turner, C. G., II. 1967. The Dentition of Arctic Peoples. Ph.D. dissertation, Univ. of Wisconsin, Madison.

—— 1970. New classifications of non-metrical dental variation: Cusps 6 and 7. Paper presented at 39th annual meeting of the American Association of Physical Anthropologists, Washington, D.C.

Vogt, E. Z. 1961. Navaho. In E. H. Spicer, ed., *Perspectives in American Indian Culture Change.* Chicago: University of Chicago Press.

Vyas, G. N., H. M. Bhatia, P. K. Sukumaran, V. Balakrishnan, and L. D. Sanghvi. 1962. Study of blood groups, abnormal haemoglobins and other genetical characters in some tribes of Gujarat. *Am. J. Phys. Anthrop.* 20:255–266.

Wahlund, S. 1928. Zusammensetzung von populationen und korrelation serscheinungen vom standpunkt der vererbungslehre aus betrachtet. *Hereditas* 2:65–106.

Willey, G. R. 1966. *An Introduction to American Archaeology.* North and Middle America, Vol. 1. Englewood Cliffs: Prentice-Hall.

Woolf, C. M. 1968. *Principles of Biometry.* Princeton: Van Nostrand.

Workman, P. L., J. D. Niswander, K. S. Brown, and W. C. Leyshon. 1974.

Population studies on Southwestern Indian tribes. IV. The Zuni. *Am. J. Phys. Anthrop.* 41:119–132.

Wright, S. 1931. Evolution in Mendelian populations. *Genetics* 16:97–159.

——1943. Isolation by distance. *Genetics* 28:114–138.

——1969. *Evolution and the Genetics of Populations*, vol. 2., *The Theory of Gene Frequences.* Chicago: University of Chicago Press.

20. The Pima Indian Studies of the Inheritance of Dental Morphological Traits

ALBERT A. DAHLBERG, PENTTI KIRVESKARI, AND THELMA DAHLBERG

Simple pedigree observation was the extent of genetic studies of tooth morphology and size in the earlier part of this century. These are still highly important, but the techniques and methodology of statistics and other branches of science have carried these studies forward at an encouraging rate. Analysis of intra-family data for sex linkage, the Hardy-Weinberg procedures, multivariate analyses, intercorrelations, twin studies, and above all the gene environment interaction studies in the developmental process are but a few.

However, much is to be learned by the interplay of these with direct observations of pedigrees of critically classed and determined traits of the teeth.

The substance of this presentation is confined to such an analysis of fourteen traits in the individuals of seven of the pedigrees drawn from the Pima Indian study collections of the University of Chicago. Many of the pedigrees consisted of particularly complete successions of parent-sibling units with sizable numbers of offspring that seemed to have good observation-recording potential. Other pedigrees are being studied and described in subsequent reports.

Most of the individuals in the pedigrees were represented by casts of the teeth that had been made yearly over a period of time covering the changing of the dentition from deciduous to permanent. Some, but not all, also had casts which included the third molars.

In the seven pedigrees selected, several traits were found to be pan-human and common to all or almost all of the members. For example, the fully developed distolingual cusp of lower molars (entoconid) is reduced in size in many individuals of other populations. In the Indian group observed, the cusp was nearly always of a robust character. Fourteen other traits were studied, and in each instance the class of the trait was discussed and agreed upon by the authors of this presentation. These included the finger-like projections of upper permanent incisor teeth, the nature of the incisor cingula and shape of lateral incisors, the classification of hypocones, Carabelli's cusps, metacones, shovel shape, protostylid, mandibular occlusal surface patterns, the distal trigonid crest, the labial surfaces of incisors, the deflecting wrinkle, and the sixth and seventh cusps of lower molars. Nine of these are summarized in the following paragraphs.

Carabelli's Cusps

Carabelli's cusp is an auxiliary cusp which has its origin and roots on the anterior aspect of the lingual surfaces of upper molars. Although seven classes of expression of this cusp were identified, for the purposes of this report it was thought best to bulk them into three major groups. These groupings differentiated mainly between the weak categories of furrows and "Y" modifications (2, 3, 4, and 5) and the stronger characteristics of bulge and cusp formation (6 and 7). The third group (1) included those having no visible expression of the cusp. In pedigree 5, having parents with weak Carabelli's cusps, the progeny classes were all of the weak grouping. In pedigree 4, with one weak and one strong parent, of 12 progeny 8 females and 3 males were strong, and 1 male was weak. In other pedigrees in which at least one parent was strong, the preponderance of 6 of 6, 6 of 10, and 5 of 6 siblings were strong. In pedigree 1, unfortunately, the parents' Carabelli's trait could not be evaluated; only 3 of the 9 siblings were strong and 6 were weak. In pedigree 7, both parents were weak; of 4 siblings, 2 females and 1 male were strong on the deciduous molars but all except the male were weak on the first permanent molars.

Hypocones

The hypocones in the pedigree subjects were observed in individuals who had casts showing both deciduous and permanent upper molars. In only one, pedigree 7, did a hypocone reduction to the 3+ category occur in the first molars. The female parent had three-cusped first and second molars. Two of her 4 children, one of each sex, had three-cusped (3+) first molars. All the other individuals had either the 4 category or slightly reduced 4− type of hypocone in the first molars. All had reductions on the more distally located molars in accordance with the field concept expectations.

Metacones

The distobuccal cusp (metacone) of upper molars has a tendency to be less accented and somewhat reduced in size in most populations, particularly in the second upper molar. This is a strong trait in some groups, especially in eastern Mediterraneans and their derivatives. Most American Indians have well-developed metacones, with reduction in size mainly in the upper second, and sometimes with less reduction, in the upper third molars. Only 5 of the 56 individuals in these pedigrees had any degree of metacone reduction in the first molars. The fact that in the three upper permanent molars in modern man the reduction is exhibited most frequently and in greatest degree in the second molar requires that particular attention to be paid to the second molar. In pedigree 3, 5 of 10 progeny have slight reduction in the second molars, as does the female parent. The other 5 have no reduction; the male parent has no reduction. In pedigree 4, however, only 2 of 12 (one of each sex) have a slight reduction on first and second molars. Two others, also one of each sex, had slight reductions on only the upper second molars. In pedigree 7 this same degree was found in the second molars of the male parent and 2 progeny, one of each sex. In 2 other progeny, one of each sex, there was some reduction in the first molars.

Shovel Shape

The class of shovel shape in the progeny generally followed an intermediate degree between the parents and a range of plus or minus one class. Classification by sight and understanding of definition is at best a problem. Here, absent and full shovel were observed as extremes and a trace was a very positive shovel but not as emphasized as the very substantial semi category. They were compared to the plaques as none, trace, semi, and full shovel.

In pedigree 1, with parents having the semi-shoveled central incisors, 1 of 9 of the progeny completely lacked the trait. This individual showed the semi category in the lateral incisor. Three of the 9 had very substantial full shovels and 5 followed the semi pattern of the parents. In pedigree 3, one parent having semi and the other full shovel central incisors, 2 of the 10 progeny showed traces and 4 of the 10 had full shovel, the remainder being semi. Pedigree 4 had the parents semi and full, respectively, and the progeny had 4 of 12 full shovel individuals and 8 semi, 2 of whom had different categories of semi and full for the laterals or centrals.

Protostylids

The protostylids on the anterior part of the buccal surfaces of lower molars have a range in form carrying through sequences of seven classes, similarly to the Carabelli's cusps of the upper molars. Here again, though the individual classes were carefully determined, the overview for the purposes of this report seemed best served by bulking some of the categories into weak and strong expressions with detailed analysis withheld for the future.

In pedigree 1 the male parent had total absence of any discernible protostylid, and the female parent exhibited an eminence and a small cusp in the second and third molar. In the 9 progeny, 2 showed at the most a pit, 2 showed distal deviations of the buccal grooves, 4 had the strong eminence-cusp categories, and 1 could not be determined.

In pedigree 2 the male parent again lacked trait manifestation, and the female parent showed strong traits in the three permanent molars. Of the 6 progeny, 1 male exibited no structure in the permanent teeth, 1 male showed a strong cusp in the lone deciduous second molar, and the 4 others showed strong protostylid characterizations on the deciduous and permanent molars available. The other pedigrees showed a range close in classes to that of the parents and siblings where available.

In pedigree 7 all 4 siblings had higher rates of categories of expression than did the parents.

Lower Molar Occlusal Surface Patterns

Arbitrary pattern designations are made combining the groove pattern formed (Y, + or X) with the number of primary cusps present that are recognized in the phylogenetic sequence. The number of cusps are somewhat independent of the groove patterns, but the latter can be altered by size variation of the individual cusps. The cusp number, however, has been shown to vary with overall tooth size.

The progeny of the pedigrees generally followed the occlusal patterns of the parents, although there were also some wide variants. Pedigree 1 had a Y pattern in the first molars for 7 of 9 progeny, the other two having a + pattern. Four of the 9 had X patterned second molars. Four had + patterned second molars. Pedigree 2 showed inconsistency in the parent-sibling display of groove patterns, 5 children having the Y and only one the + patterned first molars, with both parents having the + pattern. With one parent having a +, pedigree 3 had three X-patterned molars and six

Y-patterned among the 9 siblings. Similar variations occured in the other pedigrees.

Distal Trigonid Crest

The distal trigonid crest is the distal boundary of the ancestral trigonid of the lower molars in phylogeny. The crest can be present as a single continuum or it can have a break or interruption at a point midway across the occlusal surface. All the pedigrees demonstrated a majority of interrupted crests, a few being uninterrupted without any specific pattern. A continuous crest occurred in one instance where both parents had interrupted crests.

Barrel Shape

The barrel shape in incisor teeth is generally considered to be a greater manifestation of the prominent and somewhat frequent display of a thickened or elevated cingulum found on the gingival aspect of lingual surfaces of some upper lateral incisors. It is found in Mongoloid populations in about the same frequency as the peg shape is found in other groups. The barrel shape occurred in some siblings of the Pima pedigrees without being present in the parents. However, when the cingulum was strongly elevated in one or both parents, there was a greater likelihood of having a barrel shape in the progeny (one male of 12 siblings in pedigree 3, 2 males of pedigree 6). On the other hand, elevated cingula in both parents was sometimes accompanied by smooth cingula in the children (pedigree 7). No teeth in the Pima were peg shaped. In pedigree 4 only one (a male) of the 12 progeny was barrel shaped, and only the father had a barrel-shaped lateral incisor. Other pedigrees not included in this report show other variant combinations.

Discussion

The observations presented in this report on the traits of the dentition bring several points into focus. First, the traits ordinarily observed and included here are continuous variables but, by their individual nature, offer information that can be used in the analyses. The polygenic background generally makes for complications in the phenotype and can also be complicated by developmental environmental factors, along with the aspects of what we know about variability in tissues and structures. The most important of these is what Bateson, Butler, Dahlberg, Garn, and others have been discussing and exploring for many years.

Variability in structure at the beginnings and ends of biological meristic series modifies the phenotype, not only at these points in the series but also in many instances of the entire series. Consequently, when trait studies such as these are made, it is likely that the results greatly reflect the variability factors of the structures. This does not mean that the traits are valueless; it means rather that our sights must be adjusted accordingly. This is exactly what we see in the sequence patterns of the protostylid, Carabelli's cusp, upper and lower molar patterns, and most of the other traits.

At times it is possible to penetrate the maze and recognize the primary genetic factors, such as in the hypocone reduction of the female parent and the two siblings in pedigree 7.

There are a few recorded instances in which mutations have caused changes in tooth morphology of special value for genetic interpretations. One of these was the pedigree and dental descriptions of the tooth trait morphology resulting from a single gene mutation described by Robbins and Keene in *Oral Surgery* magazine. In that family, major, easily recognized changes occurred involving phylogenetically important concepts of cusps and genetics.

These concepts of primary and secondary dental characters as discussed in W. B. Laughlin, ed., *The Physical Anthropology of the American Indian*, (New York: The Viking Fund), pp. 139–40; and the other concepts relating to modifications from growth limitations of tooth form, such as found in G. Kurth, *Evolution and Hominisation* (Stuttgart, 1968), pp. 274–279, are additional important basic ideas necessary to these problems.

Statistical and population genetic studies are being continued on the Pima cast series described in this report.

Conclusion

The observations made on the traits in the Pima pedigrees so far do not give an exact clue or picture of the patterns of inheritance. The evidence is that the traits are inherited, but on different bases, with factors of penetrance and expressivity yet to be explored. The discrete details of classes may prove to be useful, as the data is put to differing techniques by genetric and statistical consultants.

21. Dental Morphology in Turner's Syndrome (45,X Females)

PENTTI KIRVESKARI AND LASSI ALVESALO

ABSTRACT

Dental morphology in 54 45,X females, their 48 female relatives and their 14 male relatives, was examined. The expressions of four crown traits—shovel shape, hypocone reduction, Carabelli's trait, and lower molar cusp number—were scored visually, and the depth of lingual fossa of maxillary incisors was measured. A comparison of the trait expressions suggested morphological dental reduction in the 45,X females.

A size factor has been recognized in some studies of dental crown traits (Dahlberg 1961; Sofaer et al. 1972). There seems to be a tendency of small teeth showing a more simplified crown morphology than large teeth. Turner's syndrome patients (45,X females) suffer from growth retardation from a very early age (Brook et al. 1974). There is a general reduction in size except for the head measurements (Park 1977), but the heritability of stature seems to be unaffected (Brook et al. 1974). The permanent teeth have a tendency to an early eruption (Filipsson et al. 1965), and tooth size is clearly reduced both in the permanent (Filipsson et al. 1965) and in the deciduous (Kari et al. 1980) dentition. These findings suggest that the crown morphology of the teeth might also be affected.

Material and Method

We examined the dental hard stone casts of 54 45,X females, 48 of their female relatives, and 14 male relatives. The observer was unaware of the caryotype while scoring the casts. Because of fillings, wear, and missing teeth, the number of scorable teeth was often considerably lower than that

Table 21.1. Shovel shape of maxillary incisors.

		No Shovel		Trace Shovel		Semi-Shovel		Full Shovel		P
		n	%	n	%	n	%	n	%	
I^1	45,X females	23	(54.8)	14	(33.3)	5	(11.9)	0	(00.0)	
	Male relatives	3	(33.3)	5	(55.6)	1	(11.1)	0	(00.0)	no sig.
	Female relatives	15	(55.6)	8	(29.6)	3	(11.1)	1	(3.7)	no. sig.
	45,X females	31	(70.5)	13	(29.5)	0	(00.0)	0	(00.0)	
I^2	Male relatives	2	(25.0)	4	(50.0)	1	(12.5)	1	(12.5)	$0.05 > p > 0.01$
	Female relatives	11	(47.8)	9	(39.1)	2	(8.7)	1	(4.3)	$0.1 > p > 0.05$

of the persons originally examined. We scored the traits on both sides, but used only the right side for statistics. The left side scores were used only to fill in missing right side data.

Wherever applicable in the scoring we used the standard plaques issued by the Zoller Laboratory of Dental Anthropology (Dahlberg 1956). The maximum depth of the lingual fossa of upper incisors was measured with a dial gauge at right angles to the lingual surface.

The student's t-test was applied to the results of metric measurements, and the chi-square test to the visual scorings.

Results

Shovel Shape of Maxillary Incisors

The shoveling of the central incisors did not differ much between the 45,X females and their relatives, but there was a statistically significant difference in the lateral incisor, the 45,X females showing less shovel shape than the relatives (table 21.1).

Depth of Lingual Fossa of Maxillary Incisors

In comparison with the relatives, the 45,X females have a shallower fossa on the central incisor. The difference between them and their female relatives was statistically almost significant. On the lateral incisor the difference between the 45,X females and their male and female relatives is statistically significant, and highly ($p < 0.001$) significant if the data on the relatives are combined (table 21.2).

Reduction of Hypocone

Reduction of the hypocone (distolingual cusp) of the first molar was more common in the 45,X females than in their relatives. Computed with

Table 21.2 Depth of lingual fossa of maxillary incisors.

		n	$\bar{x}$ (mm/100)	S.D.	P
I^1	45,X females	33	54.70	27.21	
	Male relatives	7	58.57	16.76	no sig.
	Female relatives	16	70.31	26.68	$0.1 > p > 0.05$
	All relatives	23	66.74	24.34	$0.1 > p > 0.05$
I^2	45,X females	34	27.50	14.63	
	Male relatives	8	55.63	30.52	$0.05 > p > 0.01$
	Female relatives	17	42.06	19.29	$0.05 > p > 0.01$
	All relatives	25	46.40	23.70	$0.001 > p$

Table 21.3. Hypocone reduction of first and second molars.

		4		4−		3+		3		P
		n	%	n	%	n	%	n	%	
	45,X females	25	(65.8)	12	(31.6)	1	(2.6)	0	(00.0)	
M^1	Male relatives	10	(100.0)	0	(00.0)	0	(00.0)	0	(00.0)	0.05 > p > 0.01
	Female relatives	25	(92.6)	2	(7.4)	0	(00.0)	0	(00.0)	0.05 > p > 0.01
	45,X females	1	(3.7)	7	(25.9)	4	(14.8)	15	(55.6)	
M^2	Male relatives	2	(25.0)	4	(50.0)	1	(12.5)	1	(12.5)	0.01 > p > 0.01
	Female relatives	1	(7.1)	6	(42.9)	3	(21.4)	4	(28.6)	no sig.

4 full-sized hypocone
4− reduced hypocone
3+ rudimentary hypocone
3 absence of hypocone

Table 21.4. Carabelli's trait on maxillary first molar.

	Absent		Pits and Grooves		Protuberance or Cusp		P
	n	%	n	%	n	%	
45, X females	11	(29.7)	9	(24.3)	17	(45.9)	
Male relatives	1	(8.3)	3	(25.0)	8	(66.7)	no. sig.
Female relatives	3	(11.5)	7	(26.9)	16	(61.5)	no. sig.

one degree of freedom (unreduced vs. all reduced forms), the differences are statistically significant, and highly significant if the data on relatives are combined. On the second molar the situation is almost alike. Only the difference between 45,X females and the female relatives did not reach statistical significance, and correspondingly, the relatives combined yields a significance only at a 5 percent risk level (table 21.3).

Carabelli's Trait

The trait was originally scored according to an eight-degree scale, but the expressions were regrouped into three for presentation: total absence, pits or grooves, and protuberances or cusps. The occurrence of pits and grooves was uniform in the different groups. The absence of the trait was more common in the 45,X females than in their relatives, and the opposite was true of the occurrence of protuberances and cusps. The difference was not statistically significant (table 21.4).

Cusp Number of Lower Molars

The presence and absence of the hypoconulid on the first and second permanent molars showed the same trend as the previous traits. The absence of the cusp is more common in the 45,X females than in their relatives, although the difference was not statistically significant (table 21.5).

Table 21.5. Cusp number of mandibular first and second molar.

		5 cusps		4 cusps		P
		n	%	n	%	
M_1	45,X females	26	(76.5)	8	(23.5)	
	Male relatives	5	(83.3)	1	(16.7)	no sig.
	Female relatives	19	(90.5)	2	(9.5)	no sig.
M_2	45,X females	3	(15.8)	16	(84.2)	
	Male relatives	1	(25.0)	3	(75.0)	no sig.
	Female relatives	3	(21.4)	11	(78.6)	no sig.

Discussion

It was expected that the 45,X females would show a less complicated crown morphology than their relatives. However, in order to evaluate the degree of size-independent morphological reduction in the 45,X females, the size factor should be eliminated. This can be done by comparing the depth measurements of incisor lingual fossae after an adjustment of the data according to tooth size. We have earlier shown that the relationship between the mesiodistal tooth diameter and the depth of the lingual fossa is positively allometric. Tooth size explains 15–20 percent of the fossa depth variation (Kirveskari and Alvesalo 1980). In view of the rather large differences in the fossa depths it seems safe to assume that the morphological reduction cannot be explained by the size difference alone.

In conclusion, the consistency with which 45,X females showed reduced expressions of non-metric traits, and the magnitude of the fossa depth differences, suggest that a size-independent morphological reduction of tooth crowns is associated with the sex chromosome aberration of 45,X females.

References

Brook, C. G. D., G. Mürset, M. Zachmann, and A. Prader. 1974. Growth in children with 45,X0 Turner's syndrome. *Archs. Disease Childh.* 49:789–795.

Dahlberg, A. A. 1956. Materials for the establishment of standards for classifications of tooth characters, attributes and techniques in morphological studies of the dentition. Zoller Laboratory of Dental Anthropology, Univ. of Chicago.

—— 1961. Relationship of tooth size to cusp number and groove conformation of occlusal surface patterns of lower molar teeth. *J. Dent. Res.* 40:34–38.

Filipsson, R., J. Lindsten and S. Almqvist. 1965. Time of eruption of the permanent teeth, cephalometric and tooth measurement and sulphation factor activity in 45 patients with Turner's syndrome with different types of X chromosome aberrations. *Acta Endocrinologica*, 48:91–113.

Kari, M., L. Alvesalo and K. Manninen. 1980. Sizes of deciduous teeth in 45,X females, *J. Dent. Res.* 59:1382–1385.

Kirveskari, P. and L. Alvesalo. 1979. Quantification of shovel shape of incisor teeth. *Ossa* 6:151–156.

Park, E. 1977. Body shape in Turner's syndrome. *Hum. Biol.* 49:215–223. 1977.

Sofaer, J. A., C. J. MacLean, and H. L. Bailit. 1972. Heredity and morphological variation of early and late developing human teeth of the same morphological class. *Arch. Oral Biol.* 17:811–816.

IV

EVOLUTION

22. Morphogenesis of Molar Cusps and Tubercles in Certain Primates

TAKURO SAKAI

The cusp homology of mammalian molar teeth in phylogeny has been much debated. The cusp nomenclature (Vandebroek 1967; Hershkovitz 1971; Butler 1978) has also been in considerable dispute, and results of studies connected therewith have been reported every time the symposium on dental morphology has been held.

The second subject on the phylogenesis of teeth concerns the system of cusp differentiation itself. That is, in the process of phylogenesis, from which part of a tooth and in which form does differentiation of a new cusp begin? through which transformation does it develop into a complete form? and what factor brings about the development of a new cusp in a certain area?

Our study circle has been pursuing the following two investigations, which have been aimed at the study of the second subject. One is a comparative anatomical investigation of existing mammals, such as Insectivora (Sakai and Hanamura 1969, 1973, 1979), Chiroptera (Sakai and Hanamura 1976, 1978), Primates (Sakai et al. 1979) and the like. As Butler (1937, 1939) has already pointed out, there exists a progressive gradient following teeth of an animal from mesial to distal, especially in cheek teeth. Besides, cheek teeth in many species can be ranged in a gradient from a simple tooth to a complex one. With these clues to follow, a process of a cusp transformation can be pursued.

The other investigation concerns the form of dentin-enamel junction of human teeth (Sakai and Hanamura 1965, 1967, 1969, 1971, 1973). In the process of the ontogenesis, the formation of dentin-enamel junction begins at an earlier phase than that of the enamel surface, and induces the form of enamel surface built in the later phase. Judging from the recapitulation theory that the forms of human ontogenesis are more primitive and simpler, the dentin-enamel junction retains a more primitive form than does the enamel surface the earlier human teeth are formed. Korenhof (1960) points out the possibility of interpretating outer tooth form from the dentin-enamel border on the endocast of maxillary molars. He concludes that "some results illustrate the significance of the dentin-enamel border as an important factor in the understanding of the evolution of the surface pattern." That is, the process of cusp transformation can be studied by comparing the form of enamel surface of a cusp with that of the dentin-enamel junction.

On the basis of these investigations, I reported our findings on the morphogenesis of the mammalian molar cusp (Sakai 1974). In summary, our findings showed that most cusps of mammalian molar teeth are derived from cingula, and revealed each developmental phase, from cingulum to completion of cusp. The purpose of this report is to extend our findings on cusp morphogenesis, primarily on the comparative anatomical investigations on teeth of primitive Primates.

As shown in table 22.1, the materials used in these investigations were plastic teeth models mady by Dr. Berbeley, of a total of 44 species of primitive primates. For observation, I used binocular stereomicroscopes.

Results

Protocone

The differentiation of the protocone begins as the cingulum surrounding the lingual base of the eocone widens and thickens lingually in its center or near its center. Then, the part that protrudes the most lingually also

Table 22.1. Material.

	No. of Genera
Paromomyiforms	
Picrodontidae	1
Paromomyidae	5
Microsyopidae	4
Plesiadapidae	2
Carpolestidae	2
Lemuriformes	
Adapidae	7
Tarsiiforms	
Omomyidae	10
Tarsiidae (Microchoerinae)	3

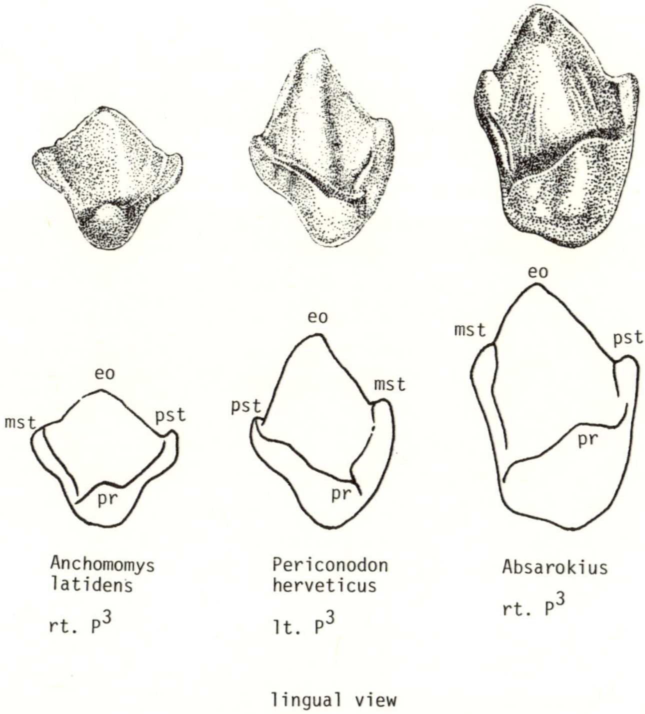

Figure 22.1. Protocone.

protrudes occlusally in some degree, thus forming a triangle, and the cusp tip becomes distinguishable. Such an early phase can be seen in the third premolar of *Periconodon, Anchomomys* and the like. In the phase when parastyle and metastyle are underdeveloped and when the differentiation has just begun, this cingulum connects with these stylar cusps and then with the buccal cingulum. Accordingly, the eocone is surrounded by a cingulum. However, as is shown in the third premolar of *Absarokius*, when the parastyle and metastyle are highly developed, the direct connection between lingual cingulum which has built the protocone and between these two styles is lost gradually. The mesial and distal ends disappear naturally on the sides of these two styles respectively.

Mills (1967) concluded that the protocone arose from a cingulum around the lingual margin of the upper molar, and that the occlusion of protocone with talonid developed into a grinding mechanism. Vandebroek (1967) also pointed out that the main lingual cusp arises at the contact point of the epicrista with the lingual cingulum, and is formed by individualization from the cingulum.

Metacone

The differentiation of the metacone begins as the central part of the ridge that stretches distobuccally protrudes around to the occlusal side. This projecting part protrudes higher gradually and becomes independent of the ridge. The cusp tip can be distinguished. The metastyle which arises at a point of contact between the cingulum and the distobuccal ridge stretching from the eocone is separated from the metacone by a clear groove. When the differentiation is well in progress, and the metacone protrudes highly, the metacone gradually parts from the eocone. Gregory (1927) has already pointed out that the Late Cretaceous *Deltatheridium* seemed to indicate that the metacone had originated on the posterior slope of the paracone.

In premolars, the metacone is situated on the distobuccal side of the eocone, i.e, in the primitive situation. In molars, it moves from the distobuccal to the lingual side and is situated directly on the distal side of the eocone. The first phase of differentiation is observed in the third and fourth premolars of *Plesiolestes*. The movement of metacone location can be judged by comparison between the fourth premolar and the first molar of *Cynodontomys*, for instance. Patterson (1956) pointed out that the metacone shifted inwardly from the pantothere postero-external position to the tribosphenic position posterior to the paracone *pari passu* with the lengthening of the teeth, narrowing of the stylar area and addition of the protocone and basined talonid.

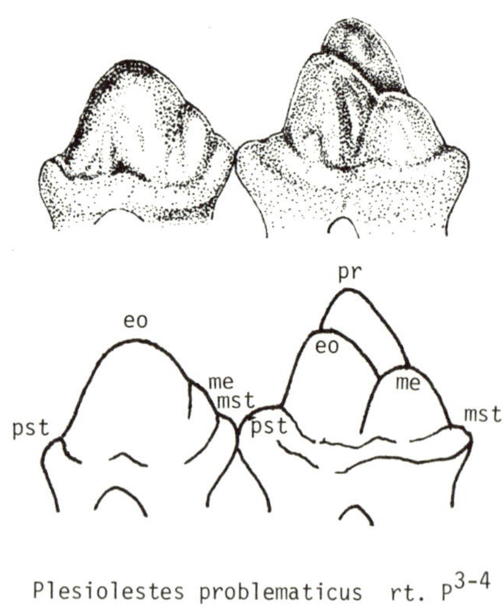

Figure 22.2. Metacone.

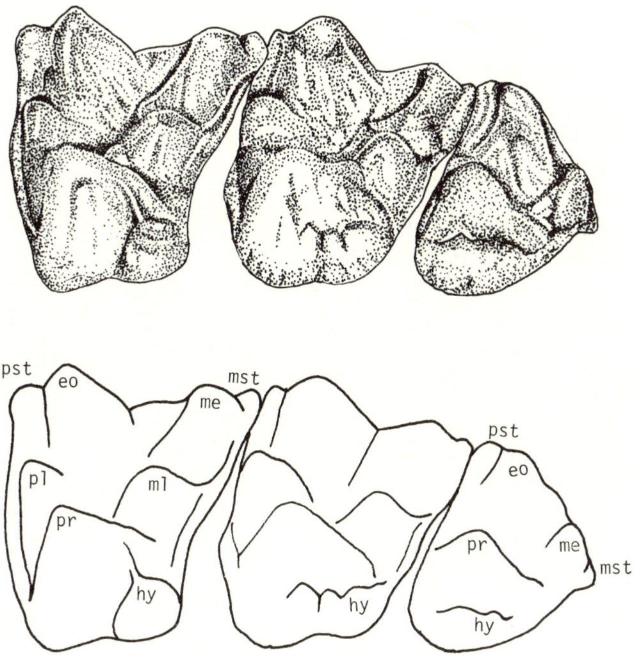

Hemiacodon gracilis lt. M^{1-3}

Figure 22.3a. Hypocone.

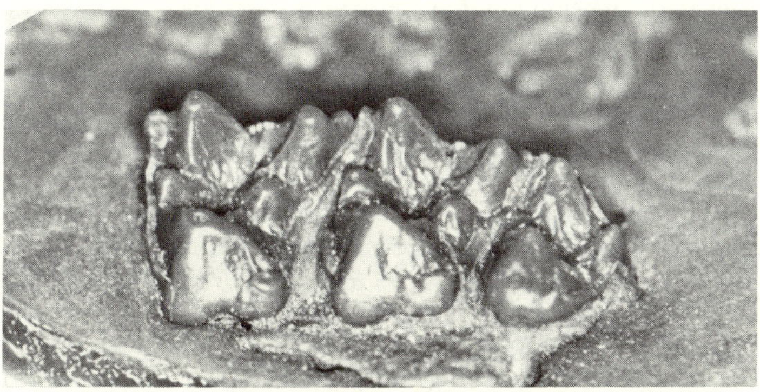

Figure 22.3b. Hypocone.

Hypocone

The differentiation of the hypocone starts from the cingulum surrounding the distolingual base of the protocone. This cingulum arises from the distal surface base of the protocone and reaches the distolingual angle. The terminal part of the cingulum then widens, protrudes like a knob, and becomes the hypocone. Hershkovitz (1971) stated that the hypocone differentiates from the posterior lingual angle of the posterior cingulum.

Mostly, a sharp fine ridge grows from this cusp "tip" and connects with the tip of the protocone. This sharp ridge is designated as the trigonal hypocone crest, and is observed clearly in enamel-dentin border of the upper molars of modern man. It is also observed on enamel surface of molars of *Necrolemur*, *Pliopithecus*, *Dryopithecus* and so on (Korenhof 1960). This crest is analogous with the cingulum-protocen crest.

As observed in *Washakius*, *Hemiacodon* and *Paromomys*, in cases of localized knob-shaped projections of the hypocone, the connection with the cingulum can be clearly distinguished. But, in the case where it has developed into a larger cusp, the relation between cusp and cingulum sometimes becomes indistinct. Thus the original cingulum is transformed into a cusp.

Carabelli's Cusp and Other Lingual Styles

The differentiation of Carabelli's cusp originates from the secondary cingulum surrounding the mesiolingual base of the protocone. This cingulum arises from the area near the mesial surface base of protocone, and reaches the mesiolingual angle, which protrudes into the Carabelli's cusp.

In many cases, a fine sharp ridge stretches from the tip of the Carabelli's cusp toward the lingual surface of the protocone. This ridge, whose nature is the same as that of the trigonal hypocone crest, is called the cingulum protocone crest.

Observing the first molar of *Washakius* from the lingual side, three cones are seen. The central and highest cusp is the protocone, while Carabelli's cusp and hypocone, two cusps much of a size, are observed symmetrically on the mesial and distal sides. In addition, a mesiolingual cingulum which gives rise to Carabelli's cusp is observed in *Purgatorius* and not a few molar teeth. In premolars, however, no Carabelli's cusp exists.

In general, Carabelli's cingulum and the hypocone cingulum disappear after the appearance of their cusps. However, in certain cases, these two cingula surround the lingual side of protocone and connect with each other. From these cingula, spiniform processes may arise and attach themselves to the lingual surface of the protocone.

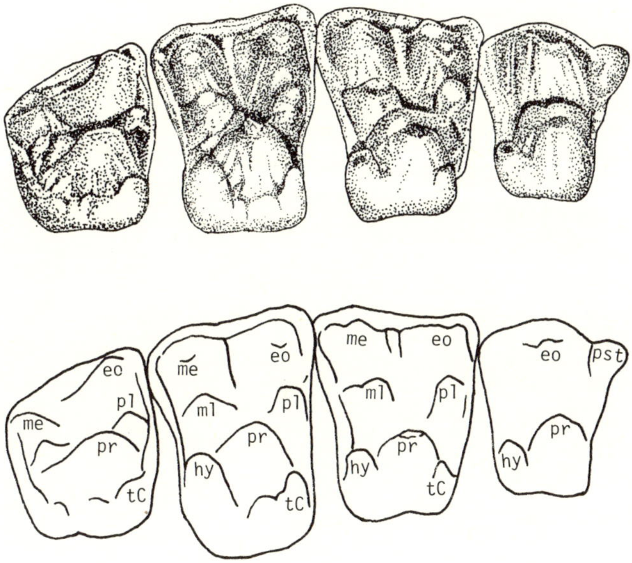

Washakius insignis rt. P^4- M^3

Figure 22.4a. Hypocone and Carabelli's cusp.

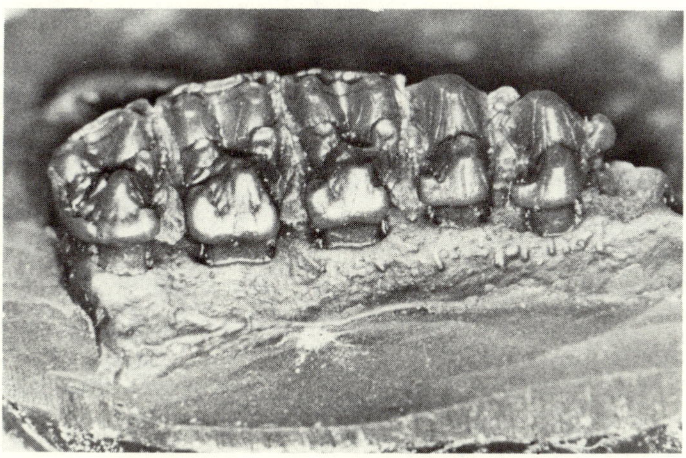

Figure 22.4b. Hypocone and Carabelli's cusp.

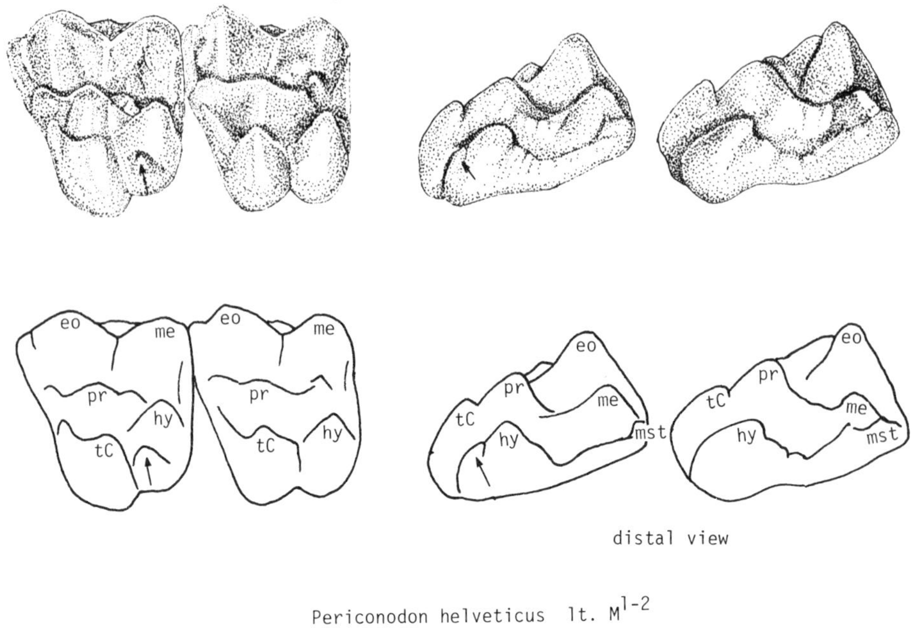

Figure 22.5a. Carabelli's cusp and cingula styles.

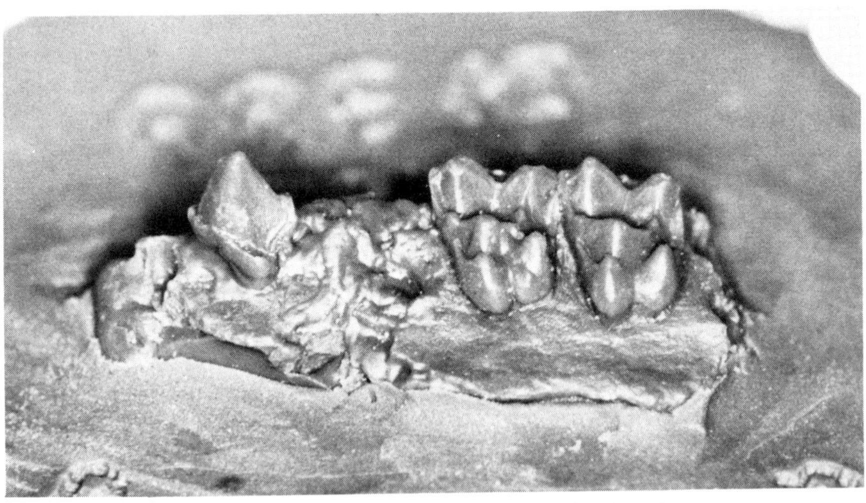

Figure 22.5b. Carabelli's cusp and cingula styles.

In the first molar of *Periconodon*, the third cingulum arises on the lingual side of the hypocone, from which a tubercle that approximates Carabelli's cusp is formed.

These observations suggest that secondary and tertiary cingula have the potential to build tubercles or styles.

Buccal Styles

The differentiation of parastyle, metastyle and mesostyle proceeds from the cingulum surrounding the buccal base of the eocone. The differentiation of parastyle and metastyle begins, as the buccal cingulum thickens in the area where it turns the mesial and distal angle of the teeth. As long as the parastyle and metastyle are underdeveloped, the cingulum continues on the lingual side, surrounds this area, and becomes the protocone cingulum. After that, the thickening part of the cingulum protrudes higher and its tip becomes distinguishable. The tip moves to the buccal side and tends to become hook-like. The early phase of these styles may be observed in the first molar of *Pronothodectes* and *Microsyops.*

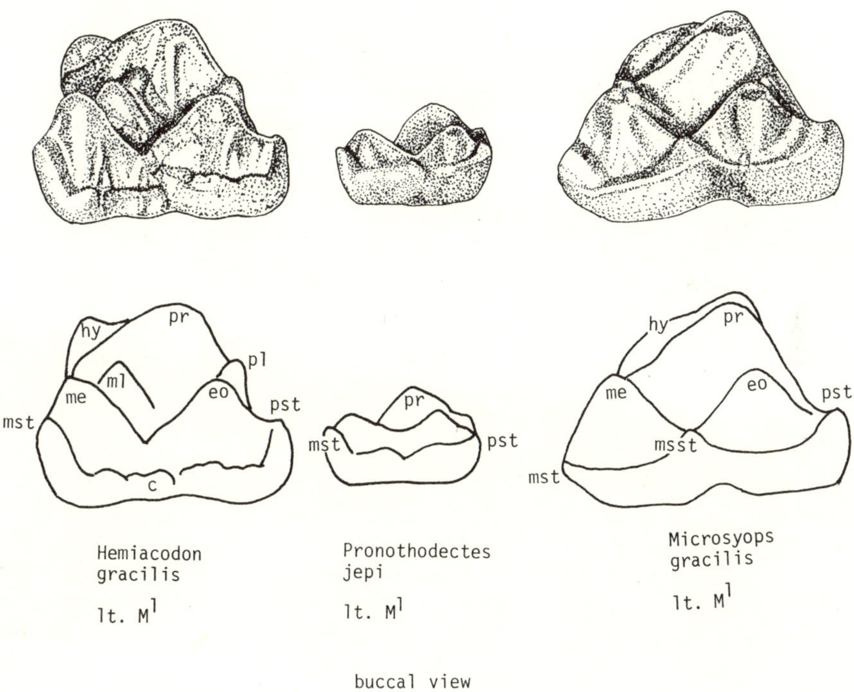

Figure 22.6a. Buccal styles.

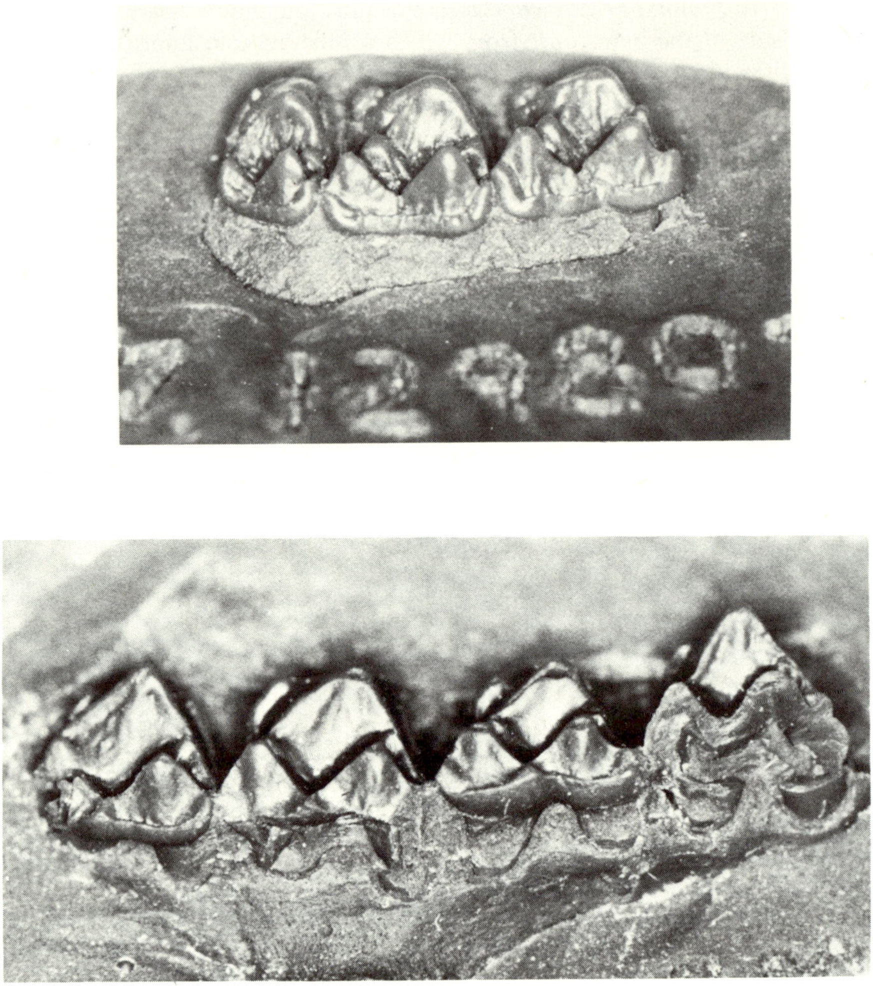

Figure 22.6b. Buccal styles.

The differentiation of the mesostyle begins as this cingulum thickens, and develops a triangle in the center of the buccal cingulum. In this phase, it has no connection with eocone or metacone. When the protruding part of the cingulum has grown higher and its tip is clearly distinguised, the tip then connects with the ridges from the eocone and the metacone, and the ectoloph becomes W-shaped. This form is observed in the first molar of *Microsyops*.

According to Butler's view, the oldest part in the mammalian upper molar is the ectoloph, and is composed of parastyle, paracone (eocone), metacone and metastyle. That is, a haplodont primary tooth has a ridge extending from the eocone to the mesiodistal side, and a cingulum surrounds the base of the tooth. The parastyle and metastyle arise on the contact point of this main crest (ectoloph) with the cingulum. The mesostyle originates from the cingulum independently form the eocone and the metacone. After that, the tip of the mesostyle connects secondarily with the ridges from eocone and metacone.

Paraconid

The differentiation of the paraconid begins as the center of the cingulum surrounding the mesial base of the eoconid makes a high protrusion. In many cases, a fine sharp ridge, called cingulum-cusp crest, extends from the tip of the paraconid toward the mesial surface of the eoconid. This cingulum leads to the buccal and lingual sides in the early phase of paraconid differentiation. As the protuberance from the cingulum becomes higher and gradually moves to the lingual side, it parts from the buccal cingulum.

A gradient of paraconid formation is frequently observed in cheek teeth of *Purgatorius*. On the other hand, no metaconid in premolars of *Purgatorius* can be observed. Each phase of paraconid differentiation can be observed in the fourth premolar of *Cynodontomys*, *Microsyops* or *Teilhardina*.

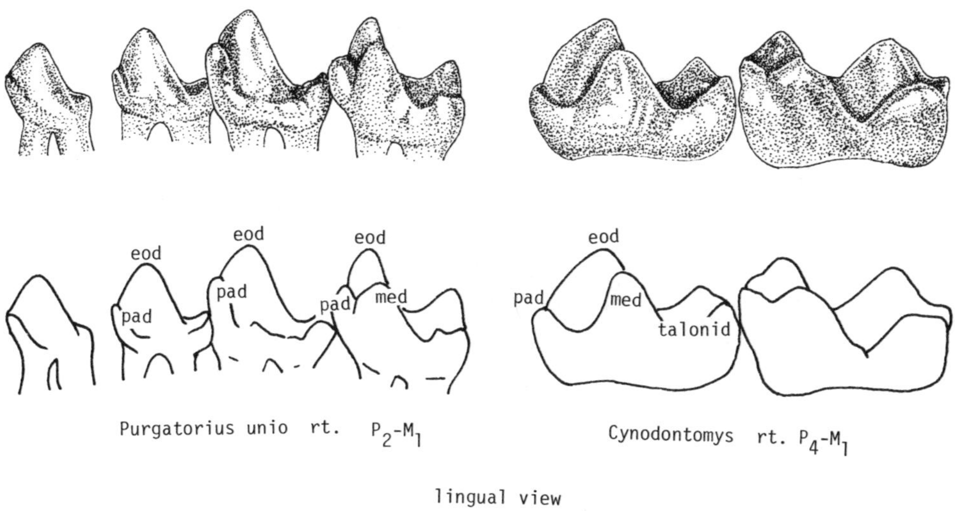

Figure 22.7. Paraconid.

Concerning the differentiation of the paraconid, Hershkovitz (1971) said that the paraconid originates on the eocrista of the anterior eoconid slope in premolars and shifts lingually on molars while enlarging. Vandebroek (1967) concluded that the mesial lingual cusp of the trigonid, here called the mesioconid, arises by individuation from the mesiostyle. Also, the mesioconid is not homologous with the cusp that sometimes develops from a bulge of the eocrista on the mesial slope of the main cusp. He proposed that the noncingular cusp be called plesioconid.

There are two views concerning the origin of the paraconid. One is that the cusp of the paraconid arises on the ridge that stretches from the eocone to the mesial side, and the other is that it originates form the cingulum on the base of the eocone. The paraconid of *Purgatorius* and the other primitive Primates suports the second theory.

Metaconid

The differentiation of the metaconid begins as a small projection on the ridge that stretches from the eoconid to the distolingual side. In this case, the metaconid begins its differentiation independently of the cingulum that surrounds the lingual base of eocone. After that, the projection grows

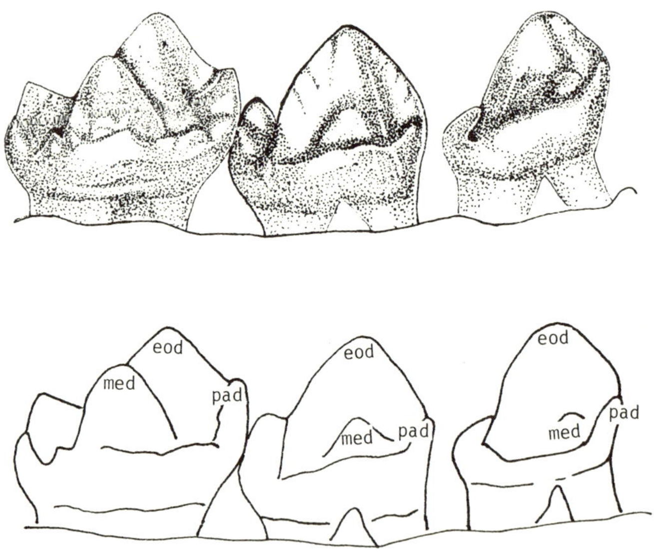

Pelycodus praetutus lt. P_2-P_4

lingual view

Figure 22.8a. Metaconid.

Figure 22.8b. Metaconid.

ridge-like along the distal margin of the lingual surface of the eocone and presents a localized and well-developed marginal ridge. As the metaconid on the ridge increases in height and thickness, it parts from the eoconid and shifts to the lingual side, and the isolated cusp tip can be distinguished.

The gradient of the metaconid formation can be especially well observed in the premolars of *Pelycodus* or *Caenopithecus*. The lingual base of the third and fourth premolars in *Pelycodus* is surrounded by the cingulum connected with the paraconid. The metaconid, which is situated at the inner side of the cingulum, has no direct connection with it. Small spiniform projections from the cingulum toward the metaconid were observed.

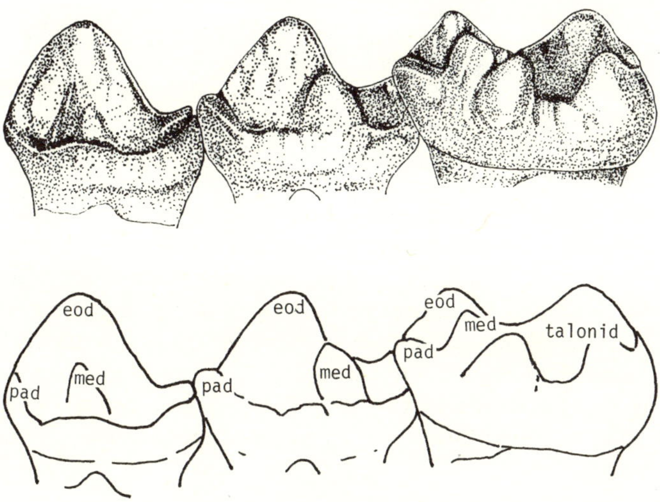

Caenopithecus lemuroides rt. P_3-M_1

lingual view

Figure 22.9a. Metaconid.

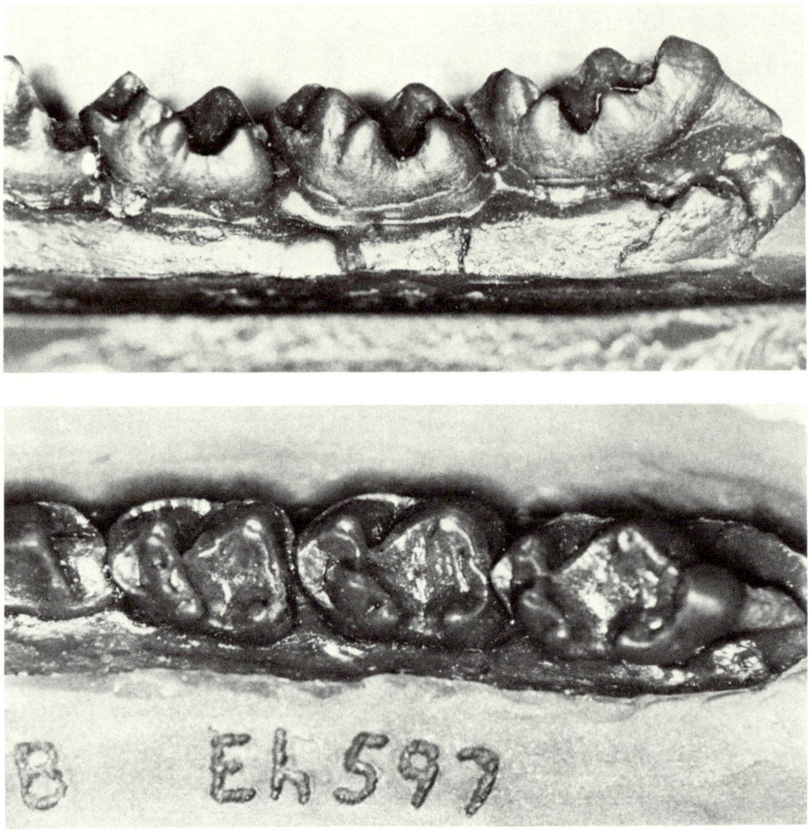

Figure 22.9b. Metaconid.

Vandebroek and Hershkovitz stated that the distal lingual cusp of the trigonid develops on the contact point of the transverse crista or epicrista with the lingual cingulum and is thus an epiconid. However, Butler (1978) stated that the metaconid arises on the ridge extending mesiodistally from the eocone to the longitudinal crista or eocrista, and then shifts lingually.

In molar teeth of *Caenopithecus*, a fairly distinct tubercle is present in the distolingual area of the metaconid. This tubercle has arisen from the projection of the cingulum which has been connected with the paraconid.

Conclusion

There are two ways of new cusp differentiation in molars of primitive Primates. The first is the case where cusps differentiate from a cingulum.

Here, part of the cingulum surrounding the base of the already formed cusp thickens and protrudes, and a small spiniform projection is formed.

In many of these cases, a fine sharp ridge is formed from the tip of the newly differentiating cusp, namely the spiniform projection, and connects with the existing cusp. Frequently observed in particular are the trigonal hypocone crest, the cingulum-protocone crest, and the oblique crest of lower molars.

When the spiniform projection's connection with the cingulum is well-developed, it becomes cusp-like. The cingulum loses its original form and is transformed into a cusp.

In another case of this process of differentiation, the newly differentiated cusp moves from its original situation. Here, the relationship between the cusp and the cingulum becomes indistinct. Further, it is often observed that spiniform or serriform projections or crenellations, arise from the cingulum surrounding a tooth crown. On rare occasions, they develop into a complete style. This seems to indicate a potential in cingula to form cusps.

The second way of cusp differentiation is the case where a new cusp differentiates from a ridge of an existing cusp. Here, one part of a ridge on the already formed cusp slope protrudes or the enamel folds. Then, differentiation of a new cusp begins. In the early phase of differentiation, a small cusp seems to be connate on the existing big cusp. As the newly differentiated cusp develops, its tip becomes distinguishable. In certain cases, it shifts its position.

References

Butler, P. M. 1937. Studies of the mammalian dentition. I. The teeth of *Centetes ecaudatus* and its allies. *Proc. Zool. Soc. London* 107:103–132.

—— 1939. Studies of the mammalian dentition. Differentiation of the post-canine dentition. *Proc. Zool. Soc. London* 109:1–36.

—— 1978. Molar cusp nomenclature and homology. In P. M. Butler and K. A. Joysey, eds., *Development, Function and Evolution of Teeth*, pp. 439–452. London and New York: Academic Press.

Gregory, W. K. 1927. Mongolian mammals of the "Age of Reptiles." *Sc. Month.* 24:225–235.

Hanamura, H. and Sakai, T. 1979. A morphological study on the dentition of *Suncus murinus riukiuanus*. *Growth* 18:28–37.

Hershkovitz, P. 1971. Basic crown patterns and cusp homologies of mammalian teeth. In A. A. Dahlberg, ed., *Dental Morphology and Evolution*, pp. 95–150. Chicago: University of Chicago Press.

Korenhof, C. A. W. 1960. *Morphogenetic Aspects of the Human Upper Molar. A Comparative Study of its Enamel and Dentine Surfaces and Their Relationship to The Crown Pattern of Fossil and Present Primates.* Utrecht: Uitgevermaatschappij Neerlandia.

Mills, J. R. E. 1967. Development of the protocone during the Mesozoic. *J. Dent. Res.* 46:787–791.

Patterson, B. 1956. Early Cretaceous mammals and the evolution of mammalian molar teeth. *Fieldiana Geol.* 13:1–105.

Sakai, T. 1967. Morphologic study of the dentino-enamel junction of the mandibular first premolar. *J. Dent. Res.* 46:927–932.

—— 1974. Morphogenesis of the mammalian tooth, with special reference to the cusp. *Jap. J. Oral Biol.* 16:245–251.

Sakai, T. and Hanamura, H. 1965, 1967, 1969, 1971, 1973. A morphological study of enamel-dentine border on the Japanese dentition. *J. Anthrop. Soc. Nippon* 73:91–109; 75:155–172, 207–223; 77:71–98; 79:297–322; 81:25–45, 87–102.

—— 1969. A morphological study on the dentition of Insectivora. *Aichi-Gakuin J. Dent. Sci.* 7:1–25; 1973. *Jap. J. Oral Biol.* 15:333–346.

—— 1976, 1978. A morphological study on the dentition of Chiroptera. *Jap. J. Oral Biol.* 18:442–445; 20:738–755.

Sakai, T., H. Hanamura, and Y. Toda, 1979. A morphological study on the dentition of *Tupaia glis*. *Jap. J. Oral Biol.* 21:182–192.

Vandebroek, G. 1967. Origin of the cusps and crests of the tribosphenic molar. *J. Dent. Res.* 46:796–804.

23. Patterns of Dental Change in Two Locally Persistent Successions of Fossil Aplodontid Rodents

JOHN M. RENSBERGER

Phyletic successions of mammalian taxa based upon apparent changes in dental structure are not uncommon in the geological record. The phylogeny of the Equidae is a classic example. In most instances, however, the fossils documenting a succession were recovered from deposits separated by appreciable amounts of time or distance. In part, the disjunct record of mammalian fossils is a consequence of the geographically and geologically discontinuous nature of the stream dominated environments of preservation in the terrestrial regime, as well as the likelihood that the mechanisms involved in the transport and deposition of sediments are likely to remove them again. The record is further influenced by the chronological and geographical discontinuity of habitats in terrestrial systems.

As a consequence, the gaps separating apparently related taxa are usually large and of an indeterminant nature, so that one has no way of knowing how much data are missing. The details of the transition must usually be filled in by forms believed to have intermediate structures, but which probably were neither participant species in an actual genetic succession nor competitive species locally replacing one another. Although it may provide a best estimate and therefore be satisfactory in arriving at a useful classification, the resulting phylogeny, does not provide objective

I am grateful to Marguerite Hugueney, who kindly provided casts of a number of specimens of *Allomys ernii*. The Graduate School Research Fund and the College of Arts and Sciences at the University of Washington supported in part the travel costs to the symposium on dental morphology. This research was supported in part by NSF grant 35959.

information about morphological events in the actual successions which gave rise to the changes.

There are exceptions to this paucity of documented transitions, i.e., instances involving assemblages from thick deposits where conditions suitable for preservation persisted for extended intervals. For example, condylarths of the genus *Hyopsodus* have been described in the Big Horn basin in northwestern Wyoming in a 5-million-year-long early Eocene succession, where four or five lineages appeared, within one of which five successive species could be distinguished (Gingerich 1974). The Miocene rodent, *Entoptychus*, is known from an interval of probably less than 2 million years in the Miocene John Day Formation of central Oregon, during which at least eight closely related species succeeded one another (Rensberger 1971). Although each of the above successions as known was based upon dental changes, occlusal configurations of cusps were not involved. In *Hyopsodus* the observed changes involved only tooth size, defined in terms of length and width of M_1. Although the dental differences in *Entoptychus* covered a complex mosaic of structures, including root lengths and enamel heights, the earliest members of the lineage were already moderately hypsodont, and the vestigial cusp structure was apparently no longer a functionally significant part of the masticatory mechanism. This paper analyzes dental changes in two successions of fossil aplodontid rodents from the same region which yielded the samples of *Entoptychus*. In these Miocene aplodontid rodents, unlike *Hyopsodus* and *Entoptychus*, the occlusal surfaces were dominated by evolving complex cusp configurations, and are therefore of special interest from the standpoint of the evolution of occlusal function.

Aplodontids are represented through a long interval of time in the John Day Formation. One of the aplodontid taxa, *Allomys*, persisted for an estimated 4 to 5 million years and another, *Meniscomys*, persisted for probably 2 to 3 million years. The deposits containing *Meniscomys* and *Allomys* are remarkably uniform in physical characteristics, almost entirely lacking sandstones or conglomerates, the rocks which would signify stream activity and erosion. Nor is there appreciable structure to the deposit that would indicate uplift during most of this interval. Throughout all but the final stages, the region was apparently a basin of low relief which received frequent deposition from volcanic sources of fine grained air fall ash (Fisher and Rensberger 1972). Late in this interval, commencing at the time of occurrence of the terminal species of *Allomys*, broad folding resulted in erosion from uplifted areas and deposition in downwarped areas, and the resulting deposits stand in marked contrast to the underlying part of the section.

The aplodontid successions are almost entirely confined to the lower, tectonically inactive section. Absence of river systems, internal drainage, low relief, thickness of the deposit (1,200 m to 760 m, west to east: Peck 1961; Hay 1963; Fisher and Rensberger 1972), and the rather uniform vertical distribution of terrestrial fossils together indicate relative constancy of deposition. If this is correct, the sequences of local populations may represent one of the more sustained records of mammalian succession in the geological record. The successive species of *Meniscomys* and *Allomys* may therefore provide an exceptionally complete documentation of occlusal change at a single geographical position. Knowledge of the multidimensional systems of populations which contributed to these local successions is incomplete without correlative vertical successions from adjoining regions. However, the complete picture may be unattainable, and these local successions provide objective information about the effects of selection operating through extended intervals of time on occlusal structures in a region in which the environment may have been rather stable.

This paper summarizes and contrasts patterns of morphological change which characterize these two successions, and then relates these changes to functional aspects of the respective masticatory systems. The statistical data documenting the evolution of these taxa involve fossil assemblages from numerous localities, some with multiple, superposed stratigraphic levels and others correlated by lithologic or biostratigraphic evidence. The data are being published elsewhere (Rensberger, manuscript).

Morphological Successions

Meniscomys

This genus is represented in the John Day Formation by three successive groups, each of which differs significantly from its predecessor in a number of dental characteristics. The second group, species B (= *Meniscomys hippodus* Cope 1879), contains several populations which differ in one or several characteristics from other local subgroups of this species. At one locality, Picture Gorge 7 (V-6506), the sample of *M. hippodus* from the higher stratigraphic position contained specimens differing from those of the underlying sample but resembling those of samples still lower in the stratigraphic section. This phenomenon was previously observed in the sequence of pocket gophers of the genus *Entoptychus* Rensberger 1971: 45, 55), and indicates that the successions in both genera frequently involve either replacement of pre-existing populations by outside populations or influx of genes from other populations. In spite of such nonlinear aspects of change in *Menicomys*, trends dominate the succession. These trends may be classified as increases in hypsodonty and lophodonty.

Upper dentition. One of the pronounced changes in *Meniscomys* is a

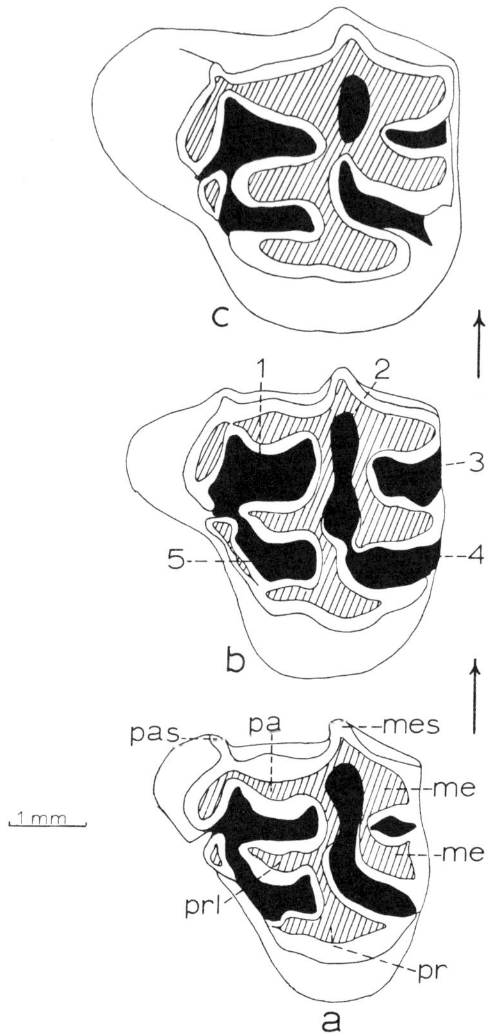

Figure 23.1. Occlusal views of crest and fossette morphology in the left P4 of three local time successive species of *Meniscomys*. *a. M.* sp. A; *b. M.* sp. B; *c. M.* sp. C.

Key: pas = parastyle; pa = paracone; mes = mesostyle; me = metacone; prl = protoconule (paraconule); mel = metaconule; pr = protocone; 1 = anterolabial fossettes; 2 = transverse central fossette (valley); 3 = posterolabial fossette; 4 = posterolingual fossette; 5 = anterolingual fossette. Arrows indicate direction of the succession. Hatched areas = dentine exposed by wear, white = enamel, black = fossettes. Anterior to left, labial toward top.

flattening and lengthening of the occlusal outlines of the maj.. cusps. In P^4, for example, the lingual surface of the paracone becomes less cylindrical (fig. 23.1) and slants in an anterolabial direction, giving it an attitude nearly perpendicular to the direction of occlusal motion (roughly the direction of the hatched lines shading the dentine in the figure). The protocone becomes antero-posteriorly elongated and its lingual surface flattened (fig. 23.1). Similar changes were occurring in other cusps, and the overall antero-posterior length of the tooth was becoming greater relative to the transverse width. All these changes had the effect of increasing the total length of the crest edges measured in directions transverse to the direction of occlusal

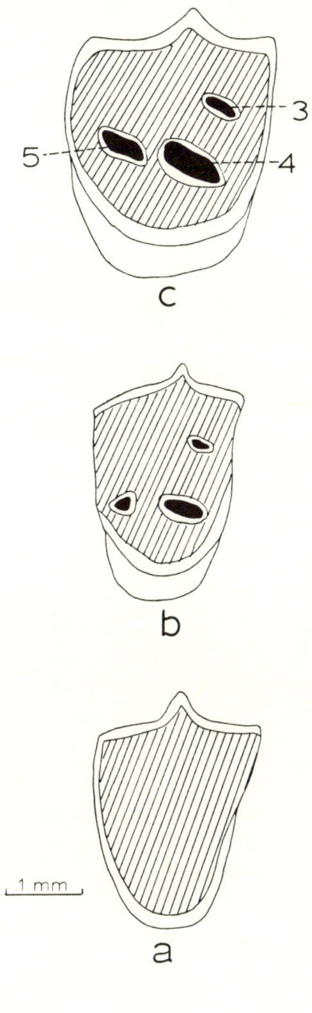

Figure 23.2. Occlusal views of crest and fossette morphology in the left M^1 of the three time successive species of *Meniscomys* shown in figure 23.1. Fossettes numbered as in Figure 23.1.

motion. This is the only effective change that will improve the rate of food subdivision in flat wearing teeth, which depend upon predominantly translatory masticatory motion (Rensberger 1973:520). Similar changes occurred in the upper molars, especially antero-posterior lengthening of the lingual moiety (protocone region), but also relative lengthening of the ectoloph (fig. 23.2). In the upper molar, fossettes, as seen in moderately worn teeth, become increasingly elongate and aligned in a common anterolabial direction, perpendicular to the direction of occlusal motion (fig. 23.2).

Lower dentition. The M_1 in the primitive species tended to lose all traces of the fossettes at an earlier stage of wear than in the posterior molars. In the advanced species the fossettes in M^1 were retained longer, probably almost as long as the fossettes in M^2–M^3 (fig. 23.2). This is interesting because it represents a reversal of a trend in another lineage of meniscomyine aplodontids, the later Miocene family Mylagaulidae, which eventually led to complete loss of M_1^1. Both changes, loss of M_1^1 on the one hand and rejuvenation of M_1^1 on the other, occurred in lineages undergoing increased lophodonty and hypsodonty; in the mylagaulids the P_4^4 acquired lophodonty and antero-posterior length more rapidly than M_1^1, and functionally replaced it.

A conspicuous change in the lower cheek teeth is the emergence of a new lingual cusp on P_4. In the earliest species, the posterior margin of the metaconid descends in the form of a crest (metastylid crest) at about a 45-degree angle and joins the base of the entoconid (fig. 23.3b). However, in some specimens of this species, the metastylid crest bears a small cusp, the mesostylid (fig. 23.3a). Both conditions are present in the lower stratigraphic interval containing *Meniscomys* species A and there is no basis, stratigraphic or geographic, for segregating the two forms as members of distinct populations. In the intermediate species (fig. 23.3c), this cusp is always present, enlarged, and separated from the metaconid by a deep notch. At this stage it clearly represents a positional equivalent of the mesostylid cusp of the lower molars (fig. 23.3d). In most members of the intermediate species the mesostylid is joined to the central structures of the tooth by a crest, as in the molars (fig. 23.4d), but in some it is isolated.

Similar variations in the development of the cusp are present in the early species. In the terminal species the mesostylid is always joined to the center of the tooth by a crest. Functionally, the acquisition of the mesostylid and its connection to the center of the tooth added a loph to the lower premolar in an occlusal position which was previously a rather wide gap.

Increases in hypsodonty, though not as pronounced as those which characterized the gopher *Entoptychus* (Rensberger 1971), are nevertheless apparent in the succession in *Meniscomys*. In the early species, a pair of infoldings or inflections in the lingual enamel of the lower molars on either

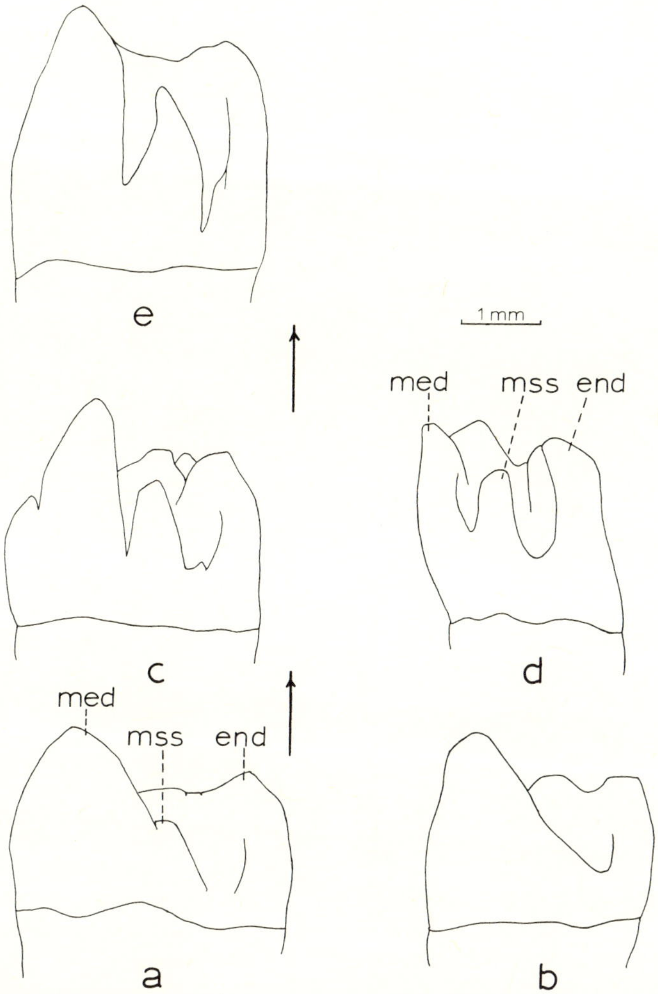

Figure 23.3. *a, c,* and *e* are lingual views of right P_4 crown in successive species of *Meniscomys*: *a. M.* sp. A; *c. M.* sp. B; *e. M.* sp. C. *b.* Right P_4 in another individual of species A, lacking mesostylid; *d.* lingual view of unworn right M_2 in sp. A, showing similarity of mesostylid to that of P_4 in sp. B.
Key: med. = metaconid, mss = mesostylid, end = entoconid.

side of the mesostylid begin a short distance above the base of the crown. The base of these inflections, especially that of the anterior one, becomes progressively higher above the base of the crown through the intermediate and advanced species (fig. 23.4a–c), indicating vertical lengthening of the crown. Similar increases in hypsodonty were occurring in the upper cheek teeth (fig. 23.5).

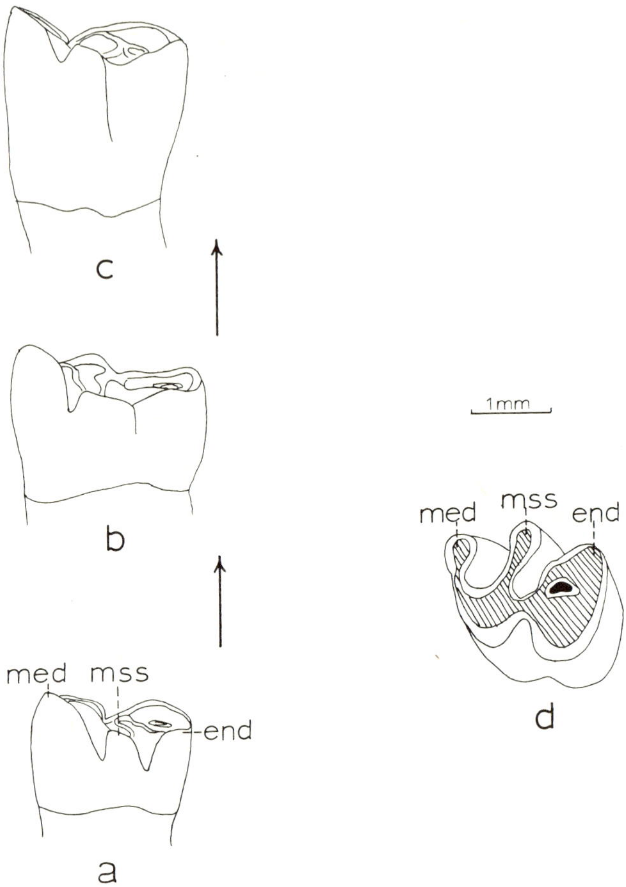

Figure 23.4. *a, b,* and *c* are lingual views of right M_1 in three species of *Meniscomys*, sp. A, sp. B, and sp. C, respectively, showing increasing crown height. *d* is an occlusal view of worn left M_1 in sp. A, showing that differential enamel thickness and lophodonty were already well developed. Abbreviations as in figure 23.3.

Allomys

Upper Dentition. As in the case of *Meniscomys*, after the appearance of the initial species, the presence of *Allomys* seems to be sustained continuously although in the upper part of the range there is an interval in which specimens haven't been recovered (fig. 23.6). Although *Allomys* didn't appear until the intermediate species of *Meniscomys* appeared, it survived much longer than the latter. *Allomys* has been found in deposits occurring after the disappearance of *Meniscomys* but prior to the appearance of the geomyoid *Pleurolicus*, occurs in deposits containing

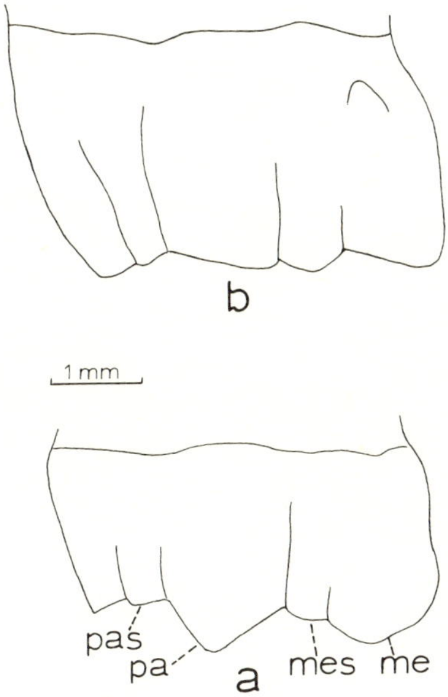

Figure 23.5. Labial view of left P^4 in *a*: *Meniscomys* sp. B; and *b*: *M.* sp. C. Abbreviations as in figure 23.1. Anterior toward left.

Pleurolicus, and is found here and there through the overlying interval containing the eight successive species of the advanced geomyoid, *Entoptychus* (fig. 23.6). From the relative abundances of the species in the overlapping ranges of *Entoptychus, Meniscomys* and *Allomys*, it appears that the teeth of *Entoptychus* were changing most rapidly, whereas those of *Allomys* were changing least rapidly. This inference is probably correct because the systematic decisions in which lines were drawn between successive species were in all cases based upon attainment of a level of significance of difference in dental morphology between adjacent samples (Rensberger 1971, manuscript).

Unlike those in *Meniscomys*, the changes in *Allomys* involved the appearance of numerous thin crests, initially rather radially arranged but later more random in direction as well as shorter and more equidimensional. Also, in later deposits the crests tended to join and form numerous subequal fossettes or fossettids.

In *Allomys* species A, the earliest form, the central transverse valley of the upper cheek teeth is a shallow, elongate, V-bottomed trough, extending from the mesostyle to the protocone (fig. 23.7a). Although short, low crests extend toward the center of the valley from the bordering cusps, these structures hug the cusps and the bottom of the valley, leaving the channel open for the most part. Therefore, the apex of the hypoconid of the lower cheek teeth transits the valley, labial to lingual, almost in contact with the

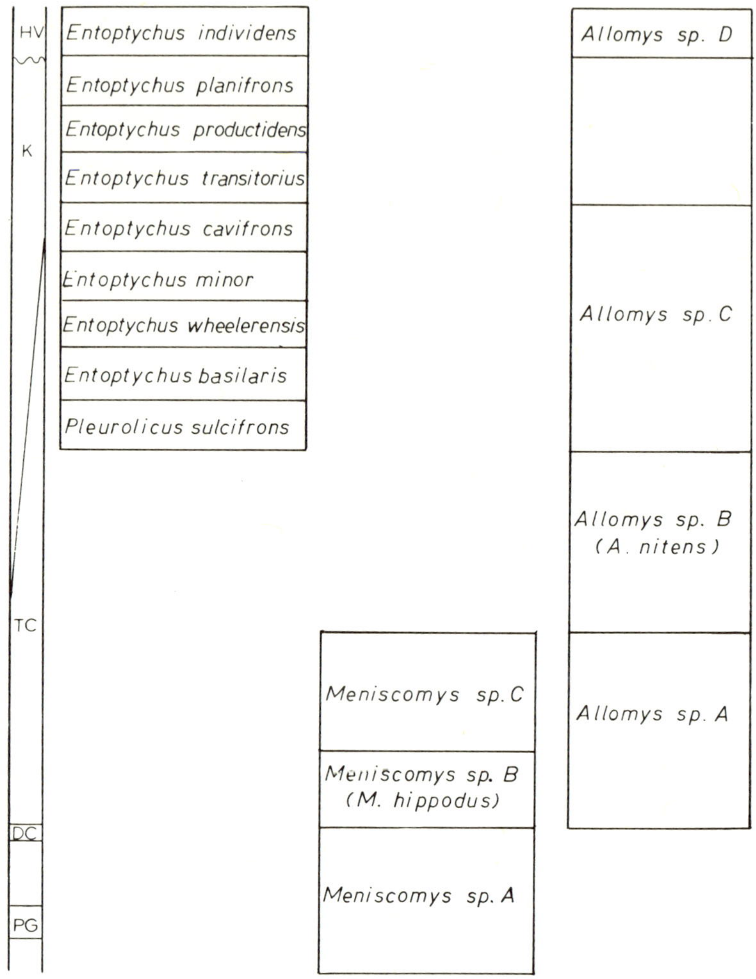

Figure 23.6. Chronological relationships of species of *Meniscomys, Allomys,* and the geomyoids (pocket gophers) *Pleurolicus* and *Entoptychus*. Sequences becoming younger toward the top. Vertical spacing roughly proportionate to thicknesses of strata containing taxa, except upper half of chart disproportionately expanded to accommodate gopher species. *Allomys* sp. A, C, D and *Meniscomys* sp. A, C are new taxa, described systematically elsewhere (Rensberger, manuscript).

Abbreviations of members and key beds of John Day Formation: PG = Picture Gorge ignimbrite, a "glowing avalanche" or welded tuff; DC = Deep Creek tuff, airfall volcanic ash; TC (including deposits between and below DC and PG) = Turtle Cove Member; K = Kimberly Member; HV = Haystack Valley Member, first appearance of extensive stream deposits.

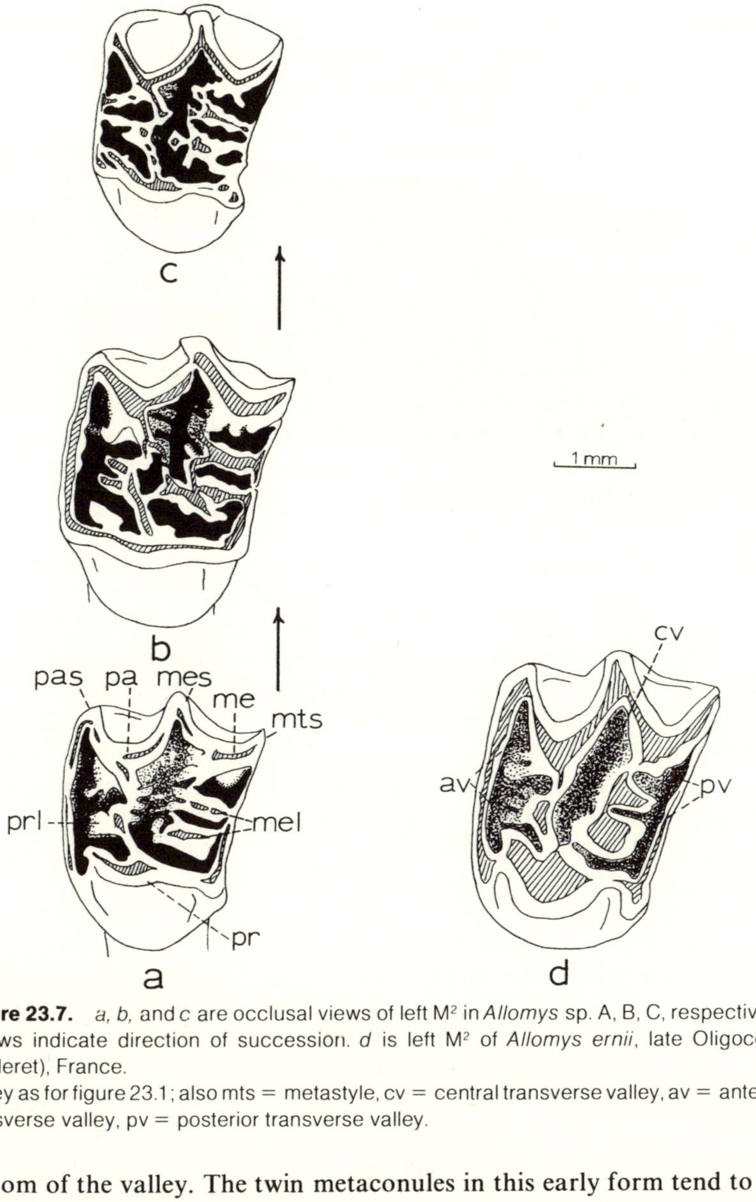

Figure 23.7. *a, b,* and *c* are occlusal views of left M^2 in *Allomys* sp. A, B, C, respectively. Arrows indicate direction of succession. *d* is left M^2 of *Allomys ernii*, late Oligocene (Coderet), France.

Key as for figure 23.1; also mts = metastyle, cv = central transverse valley, av = anterior transverse valley, pv = posterior transverse valley.

bottom of the valley. The twin metaconules in this early form tend to be subrectangular, with relatively straight borders. The protoconule and paracone bear small crests extending only limited distances into the anterior transverse valley, so that this valley is also relatively unobstructed.

The small crests that are characteristic of species B, (= *Allomys nitens* Marsh 1877), are much more elevated above the adjoining valley floors than in species A (fig. 23.7). The hypoconid and protoconid, the opposing cusps which transit these valleys, no longer pass near the valley bottoms in early or intermediate stages of wear, and the worn surfaces slope more gently toward the centers of the valleys. This elevation of structure is confined mainly to the interior crests, and no significant change in the height of the outer cusps has been recognized.

The condition in *Allomys* species C continues the trend for heightened interior accessory crests with no significant increase in overall hypsodonty. The V-shaped wear profiles of the transverse valleys are almost lost in intermediate stages of wear. Another trend now becomes evident, for some crests are reduced and others elongated in such a way as to cause the interior space to be more equally subdivided. For example, the anterior transverse valley is usually divided into three or four fossettes by two or three extended crests which emanate from the protoconule and paracone. The space lingual and labial to the inner metaconule tends to be divided into several fossettes (fig. 23.7c). The rectangular shapes of the twin metaconules are lost, and these cusps are now blending into the interior crest system. The upper dentition of the terminal species is unknown but must have continued these trends, judging from the lower dentition.

Lower Dentition. The lower cheek teeth of *Allomys* are dominated by a broad, elongate basin running from metaconid to hypoconid (fig. 23.8a). Although trigonid and talonid areas of the basin are recognizable, these basins are confluent. Subequally narrow, low crests descend from the apices of cusps toward the centers of the trigonid and talonid basins, and terminate without merging, so that the pattern of the crests is radial. In this form the crest extending posterolingually from the protoconid tends to be long and straight, and to trap an equally elongate space between it and the adjacent crest from the mesoconid (fig. 23.8a).

In *Allomys* species B the interior crests are elevated near the centers of the basins, so that the basins as defined by the occlusal surfaces of the crests are shallower, but the fossettes are deeper (fig. 23.8b) and more vertically walled. More frequently than in species A, there is a tendency for some of the crests to merge terminally with others to form fossettids. Thus the internal protoconid crest, though still straight, tends to unite terminally with the end of the internal mesoconid crest to enclose a triangular fossettid (fig. 23.8b). The more uniformly radial pattern of the crests of the early species is now lost, owing to an increase in small tubercles or short crests. emerging from existing crests, to loss of some of the earlier crests, and to elongation of others. The resulting pattern of fossettids in this species is one of diverse sizes of fossettids, ranging from very small to rather large in a single tooth.

The occlusal pattern in *Allomys* species C is characterized by smaller and more nearly subequal fossettids (fig. 23.8c). For example, the internal

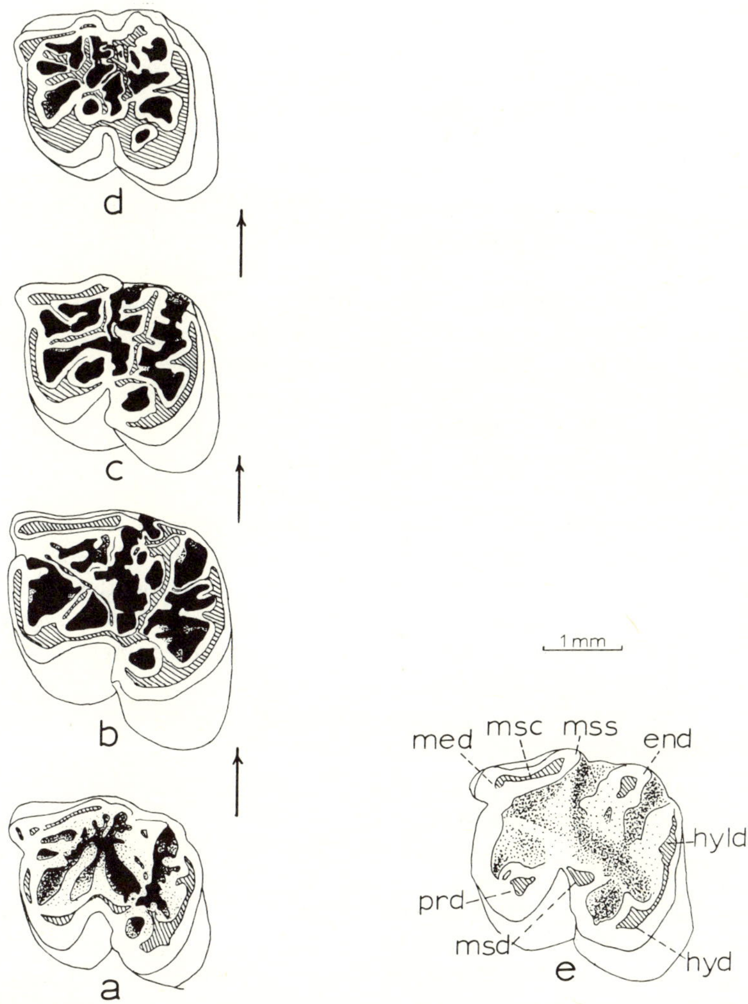

Figure 23.8. *a, b, c,* and *d* are occlusal views of left M_2 in four successive species of *Allomys*, sp. A, B, C, and D, respectively. Arrows indicate direction of succession. *e* is left M_2 of *Allomys ernii*, from late Oligocene (Coderet) of France.

Key: med = metaconid; msc = metastylid crest; mss = mesostylid; end = entoconid; hyld = hypoconulid; hyd = hypoconid; msd = mesoconid; prd = protoconid.

protoconid crest is now shorter, it curves posteriad and encloses a small, oval fossettid with the mesoconid crest (fig. 23.8c). Many of the diminutive tubercles and crests are absent and the remaining crests are of a more uniform size. The crests near the center of the tooth are relatively higher still in this species, and the occlusal surface at a given stage of wear is proportionately less basined.

The terminal species, species D, represents a continuation of these trends with particular decrease in size and increase in number of the fossettids (fig. 23.8d). For example, the internal protoconid crest encloses a still smaller, almost circular fossettid. The tooth illustrated is slightly more worn than the others, but because the walls of the fossettids are more nearly vertical than in the earlier forms, the fossettids of the lingual half of the tooth provide for an accurate comparison. The crests in the center of the tooth are relatively higher in this form than in preceding species, so that the occlusal surface is still flatter.

Although the changes observed continue through a long interval and from this perspective are gradational, the species recognized seem to be distinctive units. For example, the upper molars in species B are anteroposteriorly relatively longer than those of either species A or C, and they have a relatively broader anterior valley. The cheek teeth in species B are larger in absolute measurement than those in species A or C. It therefore appears likely that species C is derived from unknown descendants of species A rather than from its immediate predecessor, species B. Differences in the lingual margins of the molars as well as the small size of the trigonid in species D suggest that it may not be derived from the preceding species or perhaps even from the others. Yet, in spite of evidence suggesting that the vertical sequence of species of *Allomys* was not a single lineage of descent in a strict sense, the species are closely related to one another, and most of the prominent differences form trends. It seems to be a close approximation to a single evolving lineage, perhaps as close as can be found.

Ancestries

Both *Allomys* and *Meniscomys* first appear in this region as forms already rather specialized in occlusal structure. In interpreting the evolution and function of these structures, it is useful to be able to include the antecedent conditions.

In the John Day Formation the genus *Haplomys* occurs in stratigraphically lower strata than *Meniscomys* and *Allomys*, and it has been regarded as an aplodontid of more primitive aspect. Although *Haplomys* was probably not directly ancestral to either *Allomys* or *Meniscomys*, the upper molar bears an ectoloph partially developed in the fashion of that in *Meniscomys* and *Allomys*. Except for its large size, the recently recovered lower dentition of *Haplomys* is very similar to that of the North American prosciurines (Rensberger 1975:8). These relationships indicate that the aplodontids *Meniscomys* and *Allomys* are derived from somewhere within the Oligocene prosciurine complex.

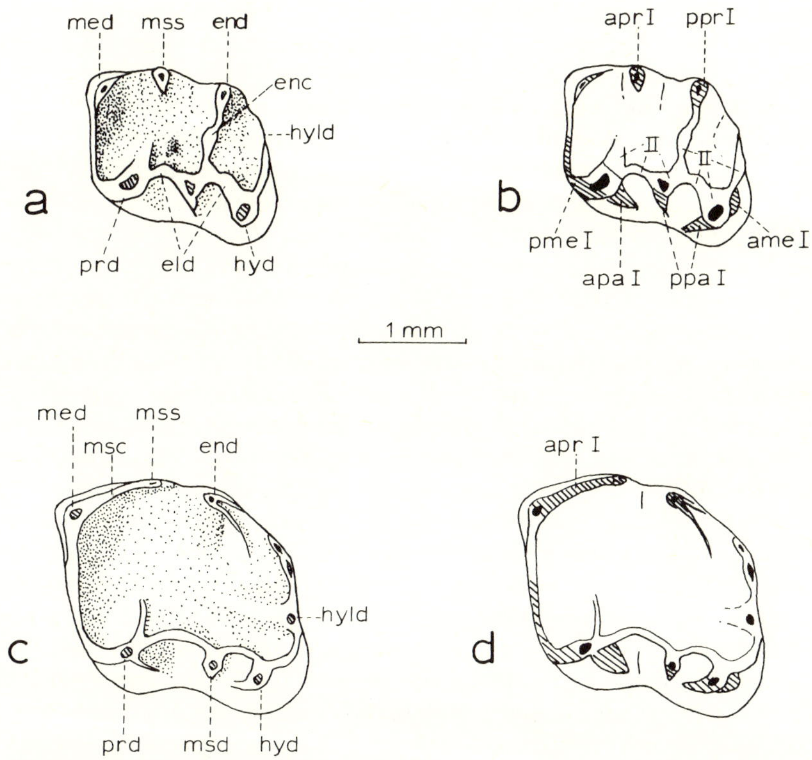

Figure 23.9. *a* and *b* are occlusal views of left M_2 in *Prosciurus relictus*, middle Oligocene, North America. *c, d* are occlusal views of left M_2 in *Plesispermophilus angustidens*, middle or early (?) Oligocene, Phosphorites of Quercy. *b* and *d* show phase I facets (hatched).

Key: apr I = anterior protocone facet; ppr I = posterior protocone facet; pme I = posterior metacone facet; apa I = anterior paracone facet; ppa I = posterior paracone facet; ame I = anterior metacone facet. Other abbreviations as in Figure 23.8 except: eld = ectolophid crests, enc = entoconid crest.

The upper cheek teeth of the North American genus *Prosciurus* are sufficiently like those of *Haplomys* and *Meniscomys* to be regarded as a possible antecedent condition. *Prosciurus* differs in the lesser prominence of the mesostyle and ectoloph and the lower crown height; yet these characteristics were almost certainly present in an ancestor of *Meniscomys*. The lower dentition of *Prosciurus* (fig. 23.9a) lacks prominent development of a mesostylid crest, which in *Meniscomys* (fig. 23.4d) crosses the trigonid and unites with the anterior part of the ectolophid. However, as in *Meniscomys*, the mesostylid of *Prosciurus* is isolated from the metaconid as a distinct cusp, which often has a short internal crest. In addition, a crest—sometimes a strong one—extends from the entoconid to the posterior ectolophid behind the mesoconid (fig. 23.9a). A strong entoconid crest is also characteristic of the North American prosciurine, *Pelycomys* (Galbreath 1953, fig. 17A).

The European species, *Allomys ernii* (Stehlin and Schaub 1951), is the most primitive aplodontid known that is unquestionably related to the North American representatives of *Allomys*. The cheek teeth in *A. ernii* (figs. 23.7d, 23.8e) are more brachyodont than those of the most primitive North American forms, *Allomys* species A and *A. cavatus* (Cope 1881), and bear still lower and fewer interior lophs. The lower molars in *Allomys* are distinguished from those in *Meniscomys* and the North American Oligocene prosciurines by the presence of a continuous crest (metastylid crest) running from the metaconid to the mesostylid, and a small, indistinct, transversely compressed mesostylid. *Allomys* also differs from *Meniscomys* in the absence of a prominent, internal mesostylid crest. The primitive *A. ernii* lacks the crest running from the entoconid to the ectolophid, indicating that the low and variable entoconid crest of the North American *Allomys* was not a primitive condition in this group. These differences suggest that *Allomys* and *Meniscomys* emerged from rather different sets of morphological conditions at a prosciurine grade.

Interestingly, *Allomys ernii* was originally described by Stehlin and Schaub (1951:278) as a species of *Plesispermophilus* on the basis of a partial lower dentition from the Coderet deposit in France. Later, with the discovery of the upper dentition from the same deposit, Viret and Casoli (1961) recognized a strong relationship to *Allomys* and placed the species in the North American genus. Nevertheless, *Allomys ernii* and the primitive North American species of *Allomys* resemble the European prosciurine *Plesispermophilus angustidens* Filhol (1883) in the following characters (fig. 23.9b): mesostylid small, compressed, connected to the metacone by a continuous metastylid crest; metastylid crest with a pronounced posterolingual slant; entoconid with less lingual prominence than the mesostylid; entoconid positioned anteriorly, especially in the posterior molars; broadly basined lower molars, lacking internal crests of any prominence and bordered by moderately compressed major cusps. These structures suggest a closer relationship of *Allomys* to *Plesispermophilus* than to the North American prosciurines.

The European aplodontid *Sciurodon cadurcensis* Schlosser (1884) shares most of the characters which relate *Allomys* to *Plesispermophilus*. Additionally, *Sciurodon cadurcensis* exhibits thin, low crests radiating from the center of the lower molar basin and a biconcave ectoloph in the upper cheek teeth, characters distinctive of *Allomys*. *Sciurodon* also exhibits moderately large labial fossettids on either side of the mesoconid, a condition which may be incipient in *Plesispermophilus* but is not developed in *Allomys*. The referred upper dentition of *Sciurodon cadurcensis* illustrated by Stehlin and Schaub (1951:120, fig. 175) is very much aplodontid-like, and bears crests resembling advanced conditions in *Allomys*, in which thin lophs emanate from the protoconule and metaconule and unite with the anterior and posterior cingula.

However, this form represents a lineage distinct from that leading to *Allomys* because the metaconule is not twinned. Unfortunately, the upper dentition of *Plesispermophilus* is unknown. Wood (1962:236) tentatively referred an upper dentition, *? Prosciurus lohiculus* Matthew and Granger (1923) from the Hsanda Gol Formation of Kazakhstan to *Plesispermophilus*, but did so mainly to avoid erecting a new genus because the specimen is not referable to *Prosciurus*. Although Stehlin and Schaub (1951, fig. 160) described what they believed to represent an upper fourth premolar of *Plesispermophilus angustidens* from Offenbach, that tooth resembles the M^3 of moderately advanced species of *Allomys*. The ectoloph on the paracone is strong but not as deeply invaginated as in *Allomys ernii*, and the crests on the protocone are more prominent than in the latter. It appears to be too advanced to represent *Plesispermophilus*, and is closer in grade to the M^3 of *Allomys* or *Sciurodon* and may represent the latter.

Including *Plesispermophilus angustidens* and *Allomys ernii*, two forms of more primitive aspect than any known North American species with resemblance to *Allomys*, the combinations of characteristics common to the European taxa suggest an origin for *Allomys* from a European form related to *Plesispermophilus*. It should be noted that one species of North American prosciurine, *Prosciurus vetustus* (Matthew 1903) from the early Oligocene, exhibits twin metaconules in the upper cheek teeth, a characteristic of *Allomys*. However, neither of the metaconules in this form are set off on the posterior side of the metaloph (Wood 1962: fig. 85A) as in *Allomys* or the middle to late Oligocene species, *Prosciurus relictus* (Cope 1873). No other known characteristic of *Prosciurus vetustus* as known suggests a special affinity for *Allomys*, and the evidence for a relationship between *Allomys* and *Plesispermophilus* is more convincing.

Functional Relationships

With the implications that *Allomys ernii* represents a primitive condition in the evolution of *Allomys*, that this group is probably derived from a morphology not unlike that of *Plesispermophilus angustidens*, and that *Meniscomys* was probably derived from a distinct group with dental characteristics more like those of *Prosciurus*, it becomes possible to visualize rather long, adaptively divergent pathways.

We see on the one hand in the *Allomys* group, the tendency for fine subdivision of occlusal space by numerous, thin interior lophs, radially arranged, culminating in elevation of the lophs and loss of the radial pattern in favor of a more random one, but no increase in overall hypsodonty. On the other hand we see in *Meniscomys* (and related genera, including *Promylagaulus* and the later Tertiary aplodontines) the tendency for fewer, heavier crests, preferentially aligned with respect to occlusal motion, and prominent trends of overall increase in hypsodonty. The trends in these two adaptive pathways, on closer analysis, do not appear to be as different from a functional standpoint as the resulting structures would suggest, and seem to be a consequence of different adaptive strategies in the ancestral taxa. One of the requisites for interpreting this relationship is understanding the nature of occlusal movements during the chewing stroke.

Occlusal Motion

In all these aplodontids, as in primitive rodents and primitive herbivores in general, occlusal motion during mastication was one in which the lower cheek teeth moved medially with respect to the palatal dentition. During the initial part of this movement, phase I or buccal phase (Mills 1955), motion in *Allomys* was upward and medially in a plane of a moderately high angle (about 45 degrees). This is evident from the attitudes of the phase I facets and directions of contained striae, for example, the facet for the protocone on the metastylid crest (fig. 23.11). During the later part of occlusion (phase II), movement was nearly horizontal but slightly inclined downward, as shown by the attitude of the facets for the apex of the protocone at the center of the talonid valley where the latter forms a sulcus in the posterolophid and small adjacent lophs (figs. 23.10a-b, 23.11). The point of most rapid change in angle of movement in the vertical plane occurs where the internal protoconid crest intercepts the vertical section shown in figure 23.11, and this represents the transition between the two occlusal phases.

When one sights parallel to a long, clearly defined stria on the phase I facet of the lower molar in *Allomys*, the other phase I facets also become almost parallel to this direction of view so that they are seen as edges (fig. 23.10a). Because these facets have opposing attitudes on opposite sides of cusps, their simultaneous alignment parallel to the direction of view defines the direction of motion. As Butler (1972:478) pointed out, this is the most practicable method for seeing the change in direction between phases I and II, especially because striae of sufficient length are often rare. In such a view (fig. 23.10a) it can be seen that the phase II facets on the labial side of the interior basin of a lower molar oppose the phase I motion. If the direction

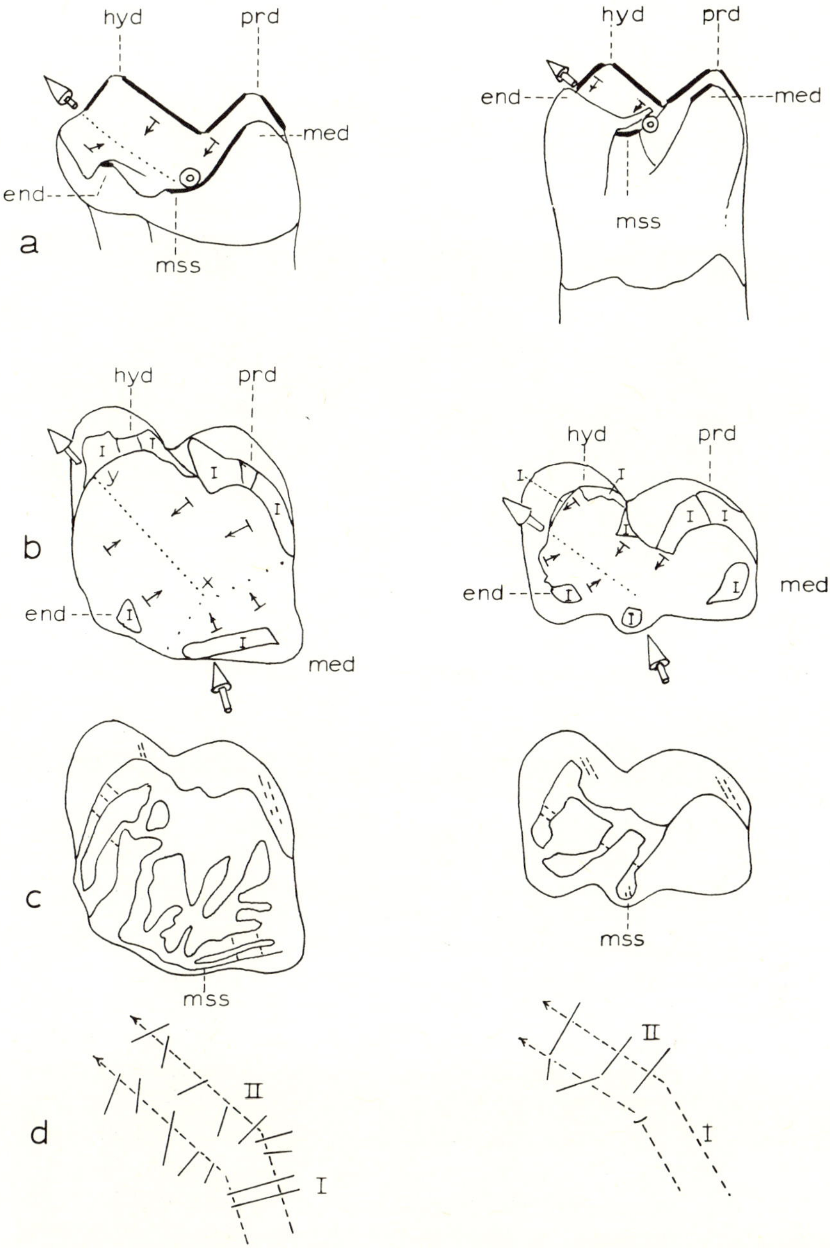

Figure 23.10. Motion of protocone during phase I and phase II across left M_2 in *Allomys* (left half of figure) and *Meniscomys* (right half of figure). *a*: anterolingual view, looking parallel to phase I facets (heavy lines) and motion; dotted line is center of diagonal valley and approximate path of apex of protocone during phase II. *b*: occlusal view of same tooth; *x-y* is valley center and phase II route of apex of protocone. I = phase I facets; small arrows identify direction of dip of facets in interior basin, details of crests omitted; 3-dimensional arrows show phase I (bottom) and phase II (upper left) directions. *c*: same tooth, with interior crest edges outlined, with striae indicating relative movement of protocone. *d*: Relationship of protocone path (dashed lines) and crest edges during phases I and II.

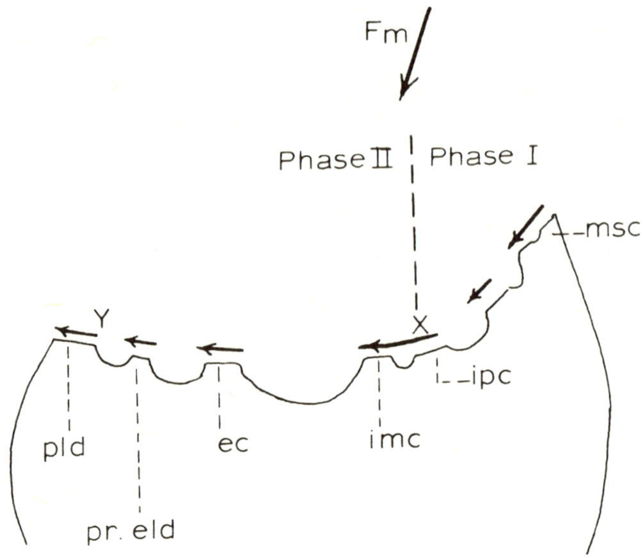

Figure 23.11. Posterolingual view of vertical section of left M_2 of *Allomys* sp. A, along phase I–II path of apex of protocone (bend in path straightened). Upper arrow (Fm) is approximate direction of a single masticatory force vector necessary to apply pressure to facets during phase I, and to later cause continued movement posterolabiad during phase II. Lower arrows show movement.

Key: msc = metastylid crest; ipc = tip of internal protoconid crest; imc = internal (anterior) mesoconid crest; ec = entoconid crest; pr. eld = process from ectolophid; pld = posterolophid. *X*, *Y* as in figure 23.10b.

of dip of these phase II facets are represented by arrows, it can be seen that these surfaces are tilted obliquely posteriad with respect to the relative direction of phase I motion of the upper molar (fig. 23.10b). The path of least resistance for the protocone of the upper molar as it strikes these facets runs obliquely posteriad. Striae on these facets confirm that this was the actual direction of motion of the protocone during phase II. In occlusal view (fig. 23.10b) the degree of change in direction of motion from phase I to phase II was measured as between 25 and 30 degrees in a little worn tooth of *Allomys*. If phase II in these rodents followed phase I in a single stroke (see comments at the end of this section), the only mechanism that is capable of producing such a pronounced change in such a small interval of space (approximately 0.2 mm) is deflection by the opposing tooth surface which in the case of the lower molars would be the inwardly facing surfaces of the protoconid and hypoconid. A sulcus or groove runs between points marked *X* and *Y* (fig. 23.10b) in the unworn lower molar. In the moderately worn tooth a worn groove occurs in the posterolophid at point *Y*, confirming that the line X–Y and the path of the apex of the protocone coincide. The point X represents the lowest point in the trigonid basin of the unworn molar and is the approximate point of most rapid change in direction of occlusal motion in both vertical (fig. 23.11) and horizontal planes.

In the upper molar, the same relationships between surface attitudes and directions of movement in phases I and II are evident. A central transverse valley (fig. 23.7a-d) runs obliquely anterolinguad from a deepest point just linguad of the ectoloph, and the axis of this alignment, when extended, passes anterior of the apex of the protocone. When one aligns the phase I facets of the ectoloph and thus views parallel to the direction of phase I motion of the hypoconid, it is apparent that the anterolabially facing facets of the metaconules deflect the hypoconid approximately 25 degrees from the phase I direction. A similar relationship holds for the interdental valley and the path of the protoconid.

In late stages of wear, the phase II motion of the lower molar has shifted to a more transverse, less anterior direction. The evidence for this is the shifting of the groove for the apex of the protocone to a more labial position on the posterolophid. Also, the horizontal angle between striae of the phase I facet on the anterior surface of the hypoconid and the phase II striae at the groove for the protocone is smaller than in little worn teeth. Because the angles of all facets become progressively lower in late stages of wear and produce eventually a single, flat occlusal surface, a lessening of the difference in direction of motion in phases I and II is expected if the deviation is caused by interaction of the facets. In other words, if the direction in phase II is controlled or limited by the attitude of the contact between the protocone and the hypoconid acting in oblique opposition to the masticatory vector, then reduction of this contact to a horizontal plane would remove such control or limitation, and would permit the motion to follow the path in the horizontal plane as determined by the masticatory vector alone.

In *Meniscomys* the relationships of facet angles, alignment of striae, and inferred directions of occlusal motion as indicated in little worn teeth (fig. 23.10, right column) are very similar to those described for *Allomys*. As in *Allomys*, a sulcus for the protocone runs in the direction of the phase II movement, and striae on the posterolophid of the little worn lower molar indicate that phase II movement corresponds to this axis. Sighting along the direction of the phase I movement shows that the phase II facets on the anterolabial side of the sulcus slant obliquely posteriad so that the protocone is deflected posteriad by contact with those surfaces (fig. 23.10, right a,b).

The most important difference between *Allomys* and *Meniscomys* is that this early stage of wear in *Meniscomys* does not persist long. In the typical intermediate stage of wear, the phase I facets form a small angle with the horizontal plane, and the striae of phases I and II are almost if not entirely parallel. In late stages of wear, motion of the lower tooth is horizontal and anterolingual, and two phases no longer exist.

The significant difference in occlusal motion in these genera is the dominance of essentially unidirectional movement during much of the life of the individual in *Meniscomys*, and dominance of bidirectional movement in *Allomys*.

The gradual but early loss of distinct phases in *Meniscomys* as the tooth flattens supports the conclusion that the distinct phases depend upon topographic relief. Butler (1973:23) suggested that phase II in Primates may be a separate mode of chewing, rather than a continuation of a single stroke that commenced as phase I. It seems possible that the aplodontids may have used phase I and phase II as separate modes of chewing when different types of food were processed. This would involve varying the activity pattern of the masticatory muscles. EMG data, which has now been published for the masticatory activity in a number of mammalian orders, are much more uniform throughout these groups than are the indicated profiles of the lower jaw movements (Hiiemae 1978:390), suggesting that changes in the direction and magnitude of the input masticatory vector occur but are not being detected. However, the apparently unidirectional occlusal motion in the many taxa of rodents with flat occlusal surfaces, and the gradual straightening of the angle between the directions of the two phases as the teeth wear flat in the more brachyodont taxa *Allomys* and *Meniscomys*, seem most simply explained by assuming only a single, sustained movement which is deflected after centric occlusion is reached. Otherwise it becomes necessary to invoke different models for juvenile and adult individuals.

Occlusal Motion and Crest Direction

The different patterns of crest arrangement in the aplodontid genera is related to occlusal motion, even though the pattern in *Allomys* seems unrelated to that in *Meniscomys*, whereas the directions of occlusal movement seem similar.

When the direction of occlusal movement of the protocone in a primitive species of *Allomys* during phases I and II is projected upon a diagram of the crest outlines of the lower molar (fig. 23.10d), it can be seen that the axes of the crests tend to be aligned transverse to the direction of motion, and to rotate with the change of the protocone path. At the start of phase I movement, the crests contacted are aligned in an almost anterior direction. At the transition between phases I and II, the crests contacted run anterolabiad. Most of the crests contacted during the remainder of phase II are directed more nearly labiad. Therefore, the apparently radial pattern is in major part related to the rotation of the path of the protocone.

In *Meniscomys*, because the mesostylid is positioned more posteriorly and no crests exist in the anterolingual quadrant of the tooth where the phase I position of the protocone occurs, almost all crests coming into contact with the protocone are phase II crests. Consequently, the few crests in *Meniscomys* are all aligned more or less parallel to one another and transverse to the phase II direction. The different crest pattern in *Meniscomys* is also related to the fact that the occlusal outline is smaller and the crests are heavier, allowing fewer crests, especially in the anterolingual region where phase I contacts would occur. But in both genera the crests tend to be aligned transverse to the direction of movement.

Crest Thickness and Randomness

The tendency in *Allomys* for the longer of the interior crests to variously shorten, curve and eventually anastomose as they became higher, suggests a response to increasing fragility. Alignment of crests transverse to the direction of chewing motion, and therefore transverse to the direction of stress, while most efficient in mastication, is also the direction of minimum resistance to strain and breakage. Therefore, as crests so aligned become higher without becoming thicker, a point is reached at which they must either cease becoming higher or sacrifice some degree of efficiency of direction by bending and anastomosing.

The alternative of acquiring thicker crests may not have been selected in *Allomys* because more extensive remodeling would have been required in the transition. Acquisition of thicker crests would not only require elimination of crests and thus reduction of masticatory efficiency, but also modification of the outline of the tooth, because stress concentration patterns would be greatly altered. The molar in *Allomys*, with its numerous slender crests, has a more uniform distribution of stress over the occlusal area than that of *Meniscomys*, and the major cusps are consequently not as thick. Bending and anastomosing of crests was, for *Allomys*, a much simpler evolutionary route to increasing the strength of the crests than acquiring thicker crests.

Crest height, Crown Height, and Diet

The elevation of crests in *Allomys* was a response to increased wear. Increased wear must be attributed to either quantitative or qualitative change in diet. The lophodont condition in both *Allomys* and *Meniscomys* is strongly suggestive of a diet containing fibrous plant parts as opposed to seed endosperm. The overall crown height in *Allomys* is somewhat greater than that in the seed eating sciurids and cricetines, and that of *Meniscomys*, is still greater. Fibrous foods, being tougher, require higher occlusal pressures and may be more abrasive. An increase in the percentage of fibrous food in the diet, or a change to a new and more fibrous type of food would therefore cause increased rate of wear. Either of these changes would also be accompanied by an increase in the quantity of food required and further increase the rate of wear because the more fibrous diets are lower in calories and easily digestible protein (Rensberger 1975:19–20).

Enamel is differentially thickened on the labial sides of the protoconid and hypoconid in both genera. This phenomenon is attributable to the tendency for food to be concentrated in front of enamel ridges on the downstroke side of the dentinal platform, and for selective pressure to locally extend the persistence of the enamel in the presence of abrasion. Differential thickening of the enamel is greater in *Meniscomys* than in *Allomys* (table 23.1), suggesting either more intensely abrasive food or greater food concentration in *Meniscomys*. The existence of a smooth, flat dentinal surface in front of the labial enamel during much of the mature life of *Meniscomys* (fig. 23.4d) indicates greater food concentration, and may account, at least in part, for this difference.

Although *Meniscomys* itself became extinct, the group to which *Meniscomys* belonged survived this interval and gave rise to two successful groups which remained important elements of the fauna in North America into the late Miocene. Both these surviving groups rapidly became extremely hypsodont forms, undoubtedly with very fibrous diets. *Allomys* was also widespread in the early Miocene and seemingly responded to

Table 23.1. Differences in thickness of labial enamel on the hypoconid in *Allomys* and *Meniscomys*.

Taxon	Mus. No.	TM_1	TM_2	LM_1	LM_2	RM_1	RM_2
Allomys sp. A	UCMP 105026	0.17	0.15	2.15	2.49	.079	.060
Allomys sp. A	UWBM 54800	0.18	0.18	2.19	2.28	.082	.079
Allomys sp. A	UCMP 86268	0.14	–	2.15	–	.065	–
Allomys sp. A	UWBM 54801	0.14	0.17	2.19	2.28	.064	.078
Allomys sp. A	UCMP 76990	0.15	0.19	–	2.58	–	.074
Allomys sp. D	UCMP 105038	0.13	0.17	2.11	2.19	.062	.078
Meniscomys sp. A	UWBM 31451	0.22	0.24	2.06	2.41	.107	.100
M. hippodus	UCMP 105102	0.18	0.22	1.68	1.94	.107	.113
M. hippodus	UCMP 105105	0.21	0.20	1.94	2.06	.108	.103
Meniscomys sp. C	UWBM 54900	0.26	0.26	2.41	2.54	.108	.102

Abbreviations: TM_1 = enamel thickness on M_1; TM_2 = enamel thickness on M_2; LM_1 = length of M_1; RM_1 = ratio of enamel thickness to length for M_1.

similar selective pressure, but much more slowly, and disappeared after the early Miocene (except for one anomalous tooth found in a later deposit).

Antecedent Morphologies

The structure of the lower molar in *Allomys* differs from that in *Meniscomys* in its greater interior area, which forms a broad, shallow basin. The basinlike shape is defined by both the occlusal surfaces of the interior crests as well as the bottoms of the valleys.

In species A, when the thin crests are worn away, the shape of the basin as defined by the bottoms of the valleys is essentially the same as the unworn shape defined by the crests. In contrast, the molar of *Meniscomys* never exhibits a basined structure, in either little worn or heavily worn stages. Instead, the adaptation in *Meniscomys* involves several heavy and persistent interior crests.

Both genera possess prominent phase I crests on the protoconid and hypoconid, however. The structural difference between these forms is that in both interior and peripheral structures, the lower molar in *Meniscomys* is dominated by crests, whereas the molar in *Allomys* in early forms combines an interior basin with peripheral crests. Because the trend in *Allomys* involved flattening of the interior basin by elevation of low crests, projection of this lineage back to its antecedent morphology predicts a smoothly basined molar bordered by phase I shearing crests.

This predicted ancestral condition is rather well developed in *Plesispermophilus*. The interior basin is large and relatively smooth, and is bounded by slender crests. The apices of the protoconid and hypoconid are set well toward the labial margin of the tooth, expanding the interior basin. The phase I facets therefore occupy most of the crests, and these structures form a narrow rim around the interior basin (Fig. 23.9d).

The morphology in *Plesispermophilus* differs from that of the ancestral rodents, the paramyines, in this development of the phase I crests. For example, the hypoconid and protoconid in the paramines are more bulbous and lack the anterolabial crests for phase I shearing against the paracone and metacone, and also lack the metastylid crest for phase I shearing against the protocone. Although phase I facets appear on the protocone and metacone in the paramyines, the facets occur on rounded surfaces, and sharp functional edges never develop. However, the interior basin is usually well developed in the paramyines (Wood 1962, fig. 18) and forms the basic functional mechanism. The best evidence for the specific function of the paramyine basin is found in the living seed-eating sciurid rodents. In the squirrels, especially the sciurines, the lower molar is strongly and smoothly basined, perhaps even more so than in the paramyines, and resembles the paramyine tooth more than does that of any other surviving group of rodents. It appears that the basin structure is an adaptation to masticating soft seed endosperm.

The lower molars of *Plesispermophilus*, as derivatives of paramyine teeth, represent a combination of the inherited basin structure for seed mastication, combined with newly added shearing crests *around* the basin where crest development doesn't alter the structure of the basin.

In other prosciurines there is similar evidence of basin structure combined with enhancement of phase I crests. The difference is that the North American prosciurines seem to have modified the basin by development of a strong entoconid crest and, in the group giving rise to *Meniscomys*, have added a strong mesostylid crest. In contrast, the ancestry of *Allomys* clung to the basin longer.

The evidence from the long succession of *Allomys* in the John Day Formation indicates that the group was responding to increased selective pressure for lophodonty. Early in this succession the thin, low interior crests added a degree of lophodonty active in phase II, yet preserved the basin structure, perhaps for a complex (seasonal?) diet of both seeds and more fibrous plant parts. Gradually, the basin structure was replaced by the elevation of interior crests, indicating a surrender of the old basin-lophodont combination in favor of lophodonty alone. The ultimate transformation of this reformed mechanism may have been restricted by its own structure. The later Miocene survivors of the early aplodontid populations, which exhibited extreme dental adaptation to fibrous foods, had dentitions derived from the *Meniscomys*, not the *Allomys* type of pattern. The compromises involved in the dual seed eating and fiber shearing apparatus in *Allomys* may have restricted the options available. In particular, the slender characteristic of the crests may have placed limits upon the adaptability of this tooth to high occlusal pressures involved in chewing highly fibrous foods, especially in the presence of a more optimal crest structure in *Meniscomys* and its relatives.

References

Butler, P. M. 1972. Some functional aspects of molar evolution. *Evol.* 26:474–483.

——1973. Molar wear facets of early Tertiary North American Primates. In M. R. Montagna and M. R. Zingeser, eds., *Symp. IVth Int. Congr. Primat.*, vol. 3, *Craniofacial Biology of Primates*, pp. 1–27.

Cope, E. D. 1873. Synopsis of new Vertebrata from the Tertiary of Colorado, obtained in the summer of 1873. Washington, D.C.: Govt. Printing Office.

——1879. On some characters of the Miocene fauna of Oregon. *Proc. Am. Phil. Soc.* 18:63–78.

——1881. Review of the Rodentia of the Miocene Period of North America. *Bull. U.S. Geol. Geogr. Surv. Terr.* 6:361–386.

Filhol, H. 1883. Description d'un genre nouveau de Rongeurs provenant des Phosphorites du Quercy. *Bull. Soc. Philomat. Paris*, ser. 7, 7:99–100.

Fisher, R.V. and J.M. Rensberger. 1972. Physical stratigraphy of the John Day Formation, central Oregon. *Univ. Calif. Publ. Geol. Sci.* 101:1–45.

Galbreath, E.C. 1953. A contribution to the Tertiary geology and paleontology of northeastern Colorado. Univ. Kansas Palaeont. Contr., *Vertebrata* 4:1–120.

Gingerich, P.D. 1974. Stratigraphic record of Early Eocene *Hyopsodus* and the geometry of mammalian phylogeny. *Nature* 248:107–109.

Hay, R.L. 1963. Stratigraphy and zeolitic diagenesis of the John Day Formation of Oregon. *Univ. Calif. Publ. Geol. Sci.* 42:199–262.

Hiiemäe, K.M. 1978. Mammalian mastication: a review of the activity of the jaw muscles and the movements they produce in chewing. In P. M. Butler and K. A. Joysey, eds., *Development, Function, and Evolution of Teeth,* pp. 359–398, London: Academic Press.

Marsh, O.C. 1877. Notice of some new vertebrate fossils. *Am. J. Sci.*, ser. 3, 14:249–256.

Matthew, W.D. 1903. The fauna of the *Titanotherium* beds at Pipestone Springs, Montana. *Bull. Am. Mus. Nat. Hist.* 19:197–226.

Matthew, W.D. and W. Granger. 1923. Nine new rodents from the Oligocene of Mongolia. *Am. Mus. Novitates* 102:1–10.

Mills, J.R.E. 1955. Ideal dental occlusion in the Primates. *Dent. Practnr. Bristol* 6:47–61.

Peck, D.L. 1961. John Day Formation near Ashwood, in north-central Oregon. *U.S. Geol. Surv. Prof. Pap.* 424-B, p. 153–156.

Rensberger, J.M. 1971. Entoptychine pocket gophers (Mammalia, Geomyoidea) of the early Miocene, John Day Formation, Oregon. *Univ. Calif. Publ. Geol. Sci.* 90:1–209.

——1973. An occlusion model for mastication and dental wear in herbivorous mammals. *J. Paleont.* 47:515–528.

——1975. Function in the cheek tooth evolution of some hypsodont geomyoid rodents. *J. Paleont.* 49:10–22.

——in preparation. Successions of meniscomyine and allomyine rodents (Aplodontoidea) in the Oligo-Miocene John Day Formation, Oregon. Manuscript.

Schlosser, M. 1884. Nachtrage und Berichtigungen zu: die Nager des europaischen Tertiars. *Palaeontogr.* 31:323–328.

Stehlin, H.G. and S. Schaub. 1951. Die Trigonodontie der simplicidentaten Nager. Schweiz. *Palaeont. Abhandl.* 67:1–385.

Viret, J. and M. Casoli. 1961. Sur un Rongeur de la famille américaine des Aplodontides découvert dans le Stampien supérieur de la Limagne bourbonnaise. *Eclogae Geol. Helv.* 54:541–545.

Wood, A.E. 1962. The early Tertiary rodents of the family Paramyidae. *Am. Phil. Soc. Trans.* n. ser., 52:1–261.

24. Evolutionary Trends of the Inner Enamel Anatomy of Deciduous Molars from Sangiran (Java, Indonesia)

CORNELIS A. W. KORENHOF

The evolutionary significance of the morphology of the enamel-dentin partition plane in medieval Javanese human teeth collected by G. H. R. von Koenigswald shortly before World War II was discussed in earlier papers (Korenhof 1960, 1961, 1963, 1978 a, b) (fig. 24.1). In these the upper and lower molars of the permanent dentition were subject of study.

Material and Methods

The Sangiran material contains permanent as well as deciduous teeth. The former are kept in the Dental School of the State University of Utrecht; the latter are now in the Senckenberg Museum in Frankfurt (Paleontological Department, Prof. Van Koenigswald).

The morphology of the inner surface of the separated enamel caps has been studied by means of so-called "endocasts" (figs. 24.6–24.7; 24.9–24.11; 24.15). The inner surface proved to be more complete than the outer surface, with more conservative anatomical characteristics, and it also shows details which are unknown for the outer surface of the human dentition, details which are well known in earlier hominoid teeth, particularly of Dryopithecinae and recent apes (figs. 24.2, 24.13). As explained before (Korenhof 1978a), this was to be expected on embryological grounds because the enamel-dentin partition plane is much more a genetic blueprint of the occlusal anatomy of teeth than the later-formed chewing surface (fig. 24.3). Because of the resemblance to the permanent

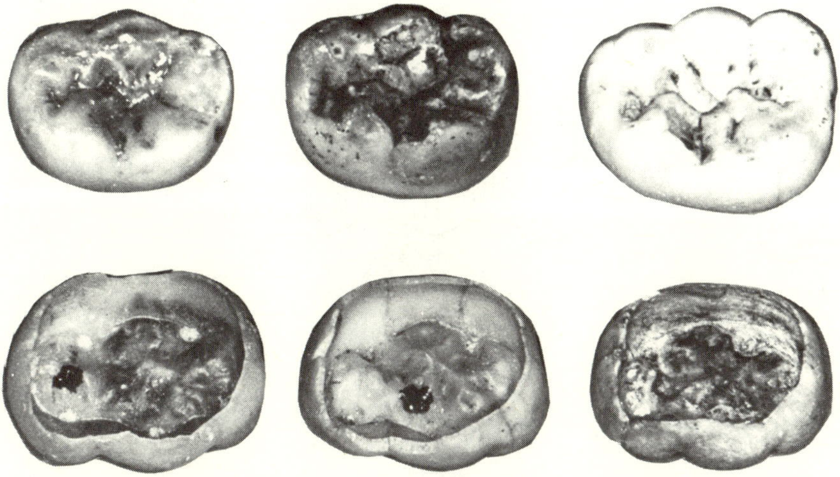

Figure 24.1. Second deciduous lower molars from Sangiran. Upper row: occlusal surface; lower row: inside enamel surface.

molars, the second deciduous molars from the lower and upper jaws have been studied, and the results will be compared with the results obtained in the earlier studies on the permanent molars.

Number of teeth for which endocasts have been prepared:

	Right	Left	Total
Lower second deciduous molar	31	21	52
Upper second dediduous molar	30	33	63

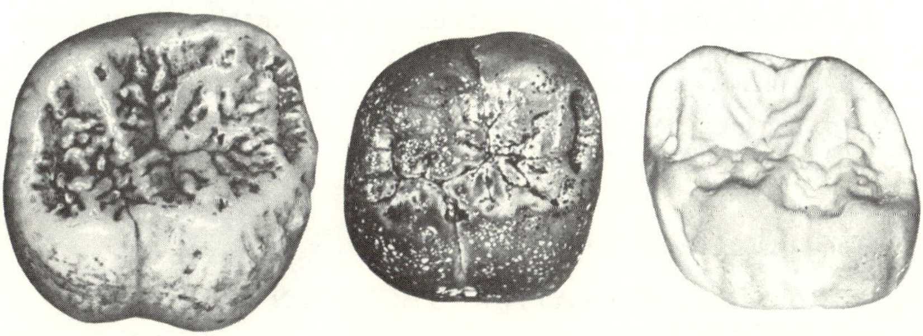

Figure 24.2. The anterior transverse crest and the oblique crest in fossil hominoids. From left to right: Pongo, Pleistocene, China; Pithecanthropus modjokertensis, Java: Dryopithecus rhenanus, Europe, cast. Collection Senckenberg Museum, Prof. G.H.R. von Koenigswald.

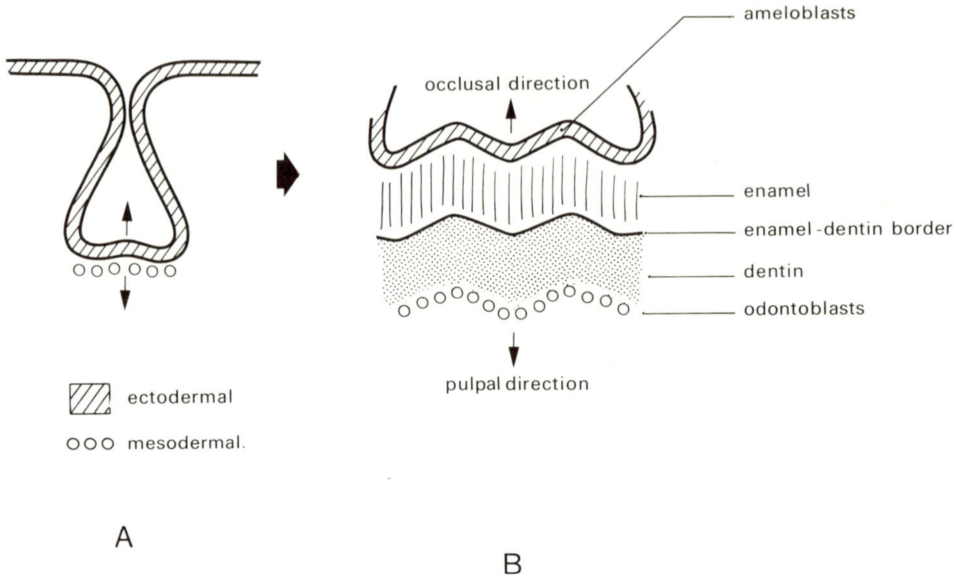

Figure 24.3. *A.* Schematical representation of early stage in the development of a tooth germ. *B.* Detail of A in a later stage. Ameloblasts and odontoblasts have moved in a occlusal and pulpal direction, respectively, after forming the enamel-dentin partition plane and succesive layers of their specific tissues.

In the endocasts of the *lower* molars we studied the following features: the frequency of occurrence of trigonid crest types in comparison with that in the permanent molars (figs. 24.4, 24.6, 24.7), and the frequency of occurrence of talonid crest types (figs. 24.8–24.10).

In the *upper* molars, we looked at the frequency of occurrence of two types of the crista obliqua, both in the endocasts and in the occlusal surface, and a comparison will be made with the frequency of occurrence in the permanent teeth (fig. 24.11). The same treatment was given to Carabelli's cusps and features (figs. 24.11, 24.15), and for the crista transversa anterior (fig. 24.14, 24.15).

The Lower Second Deciduous Molar

TRIGONID CREST PATTERNS BETWEEN PROTOCONID AND METACONID.

In many endocasts distinct trigonid crests or remnants thereof appear to run from the tip of the protoconid (or its distal ridge) to the tip of the metaconid, just as in the permanent molars (figs. 24.4, 24.6–24.10). The following six types can be distinguished (fig. 24.5):

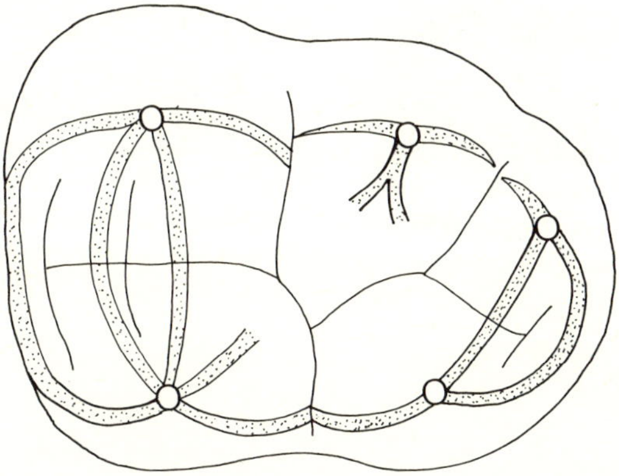

Figure 24.4. Remane's (1921) view of the presence of trigonid crests in Gorilla.

Type 1. Distally from the mesial marginal ridge, two complete crests between protoconid and metaconid (fig. 24.6, left lower, fig. 24.9 lower right).

Type 2. Distally from the mesial marginal ridge, two crests between protoconid and metaconid; the mesial crest is interrupted, the distal crest is complete (fig. 24.6, middle lower).

Type 3. Distally from the mesial marginal ridge, two crests; the mesial crest is complete, the distal crest is interrupted (fig. 24.6, right lower).

Type 4. Distally from the mesial marginal ridge, two crests between

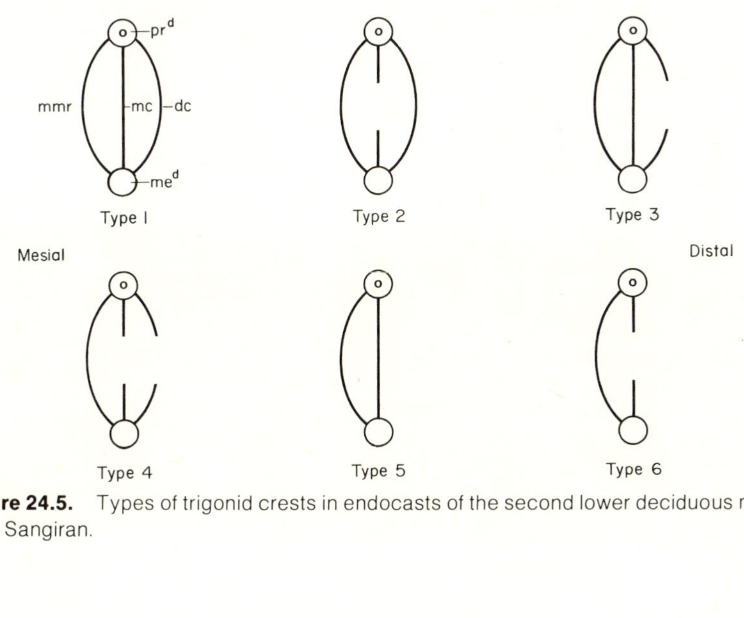

Figure 24.5. Types of trigonid crests in endocasts of the second lower deciduous molar from Sangiran.

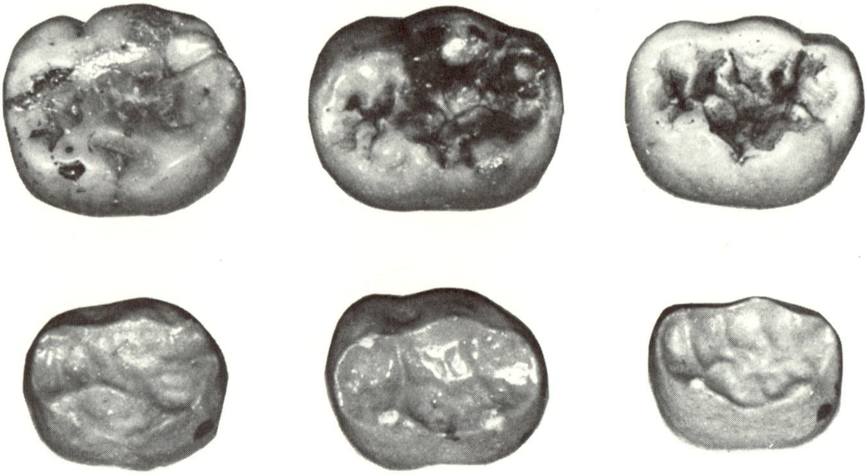

Figure 24.6. Types 1–3 trigonid crests in endocasts of second lower deciduous molars from Sangiran.

Lower row: endocasts.

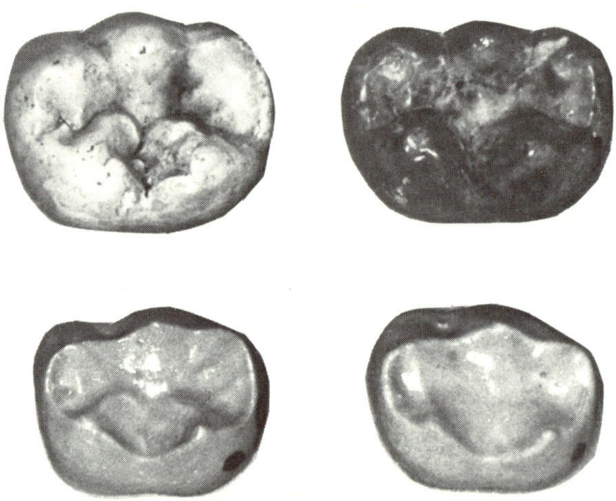

Figure 24.7. Types 4 and 5 trigonid crests in endocasts of second deciduous molars from Sangiran.

Lower row: endocasts.

Table 24.1. Frequency on occurrence of the six types of trigonid crest patterns in the second deciduous molar.

Type	Left	Right	L + R	%
1	5	12	**17**	**33**
2	3	5	**8**	**15**
3	7	6	**13**	**25**
4	1	3	4	8
5	5	4	9	11
6	–	1	1	2
Total	21	31	52	100

the protoconid and the metaconid; both are interrupted or incomplete over some distance (fig. 24.7, left lower).

Type 5. Distally from the mesial marginal ridge, one distinct and complete crest between the tip of the protoconid and the tip of the metaconid (fig. 24.7, right lower).

Type 6. Distally from the mesial marginal ridge, one interrupted or incomplete crest between the tip of the protoconid and the tip of the metaconid (fig. 24.10, middle lower).

Table 24.1 shows the frequency of occurrence of the six types of trigonid crest patterns.

Because of the variously wrinkled pattern of the occlusal enamel anatomy, the frequency of occurrence of the type numbers cannot be related to the occlusal groove pattern (Y, +, X) as for permanent molars. Therefore only the frequency of occurrence of the six types of trigonid crest pattern in the endocasts of the deciduous molars will be compared with that of the permanent molars. It appears that the low crest type numbers occur in a much higher percent than the high crest type numbers.

From table 24.2 it appears that in the deciduous molars all the endocasts

Table 24.2. Comparison of the frequency of occurrences of trigonid crests with the permanent molars.

	Permanent Molars		Deciduous Molars	
Type	Number	%	Number	%
1	18	3.6	17	32.7
2	37	7.4	8	15.3
3	7	1.4	13	25.0
4	60	12.0	4	7.8
5	70	13.9	9	17.3
6	196	39.0	1	1.9
0	114	22.7	–	–
Total	502	100.0	52	100.0

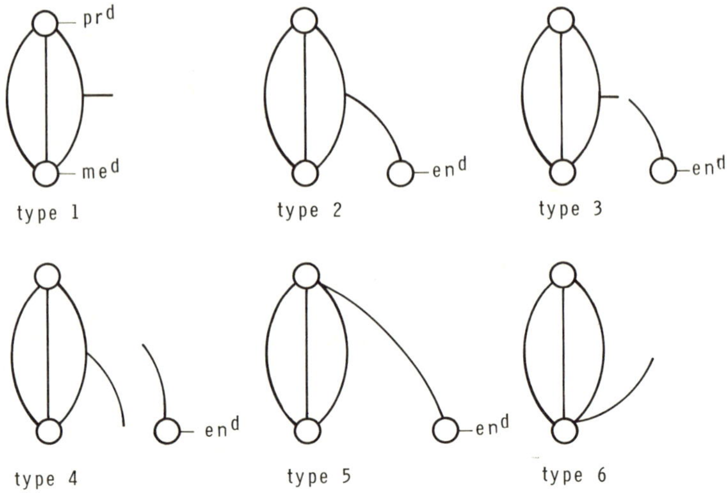

Figure 24.8. Types of talonid crests in endocasts of the second lower deciduous molar from Sangiran.

show traces of possible trigonid crests, whereas in the permanent molars this is only 77.3 percent. The lower crest type numbers are found in a higher percentage in the deciduous molars, whereas in the permanent teeth the higher type numbers are represented by a higher percentage.

If, as has been discussed earlier, it is assumed that the lower crest type numbers represent more conservative evolutionary conditions, it may be

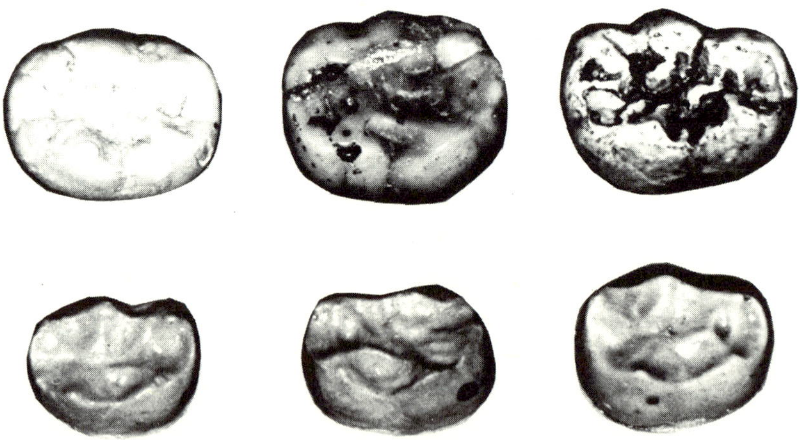

Figure 24.9. Types 1–3 talonid crests in endocasts of the second lower deciduous molar from Sangiran.

concluded that in this respect, the temporary teeth have more original anatomical characteristics.

THE CONNECTION OF TALONID CREST AND DISTAL TRIGONID CREST

The endocasts of the deciduous molars show a new and constant feature, which had not been observed in the permanent molars: a distinct crest extending from some part of the distal trigonid crest in a more or less distal direction. (figs. 24.9 and 24.10).

In most endocasts the attachment of this crest to the distal trigonid crest strongly resembles the original attachment of the earlier talonid crest; for this reason the crest was called "talonid crest."

The six distinguished types can be described as follows (fig. 24.8):

Type 1. From the middle of the distal trigonid crest a small crest runs for some length in distal direction, but is not connected to any other distinct landmark (fig. 24.9, lower left).

Type 2. From the middle of the distal trigonid crest an uninterrupted crest runs to the tip of the entoconid (fig. 24.9, lower middle).

Type 3. From the middle of the distal trigonid crest an interrupted crest runs to the tip of the entoconid (fig. 24.9, lower right).

Type 4. From the middle of the distal trigonid crest a small crest

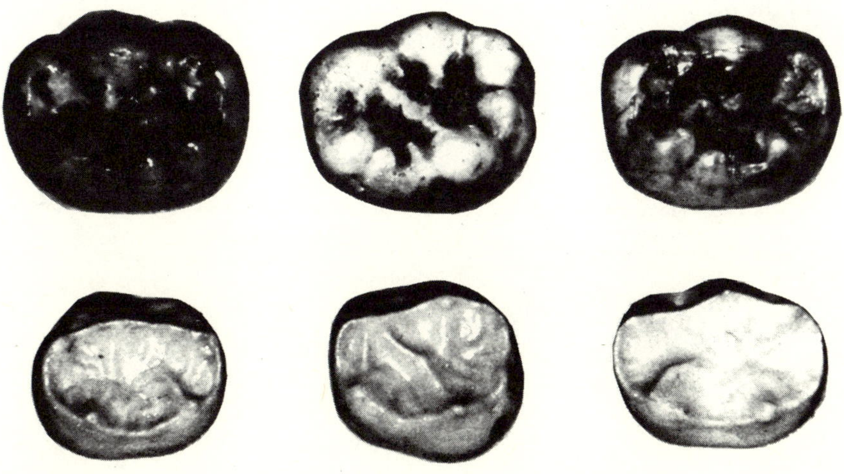

Figure 24.10. Types 4–6 talonid crests in endocasts of the second lower deciduous molar from Sangiran.

runs to the lingual side, which is reached slightly mesially from the tip of the entoconid (fig. 24.10, lower left).

Type 5. From the tip of the protoconid a crest runs to the tip of the entoconid (fig. 24.10, lower middle).

Type 6. From the tip of the metaconid a crest runs for some distance in distobuccal direction (fig. 24.10, lower right).

The first four types are considered as more conservative than types 5 and 6.

The frequency of occurrence of the six types of talonid crests and their distribution in the six types of trigonid crests is shown in table 24.3.

If these crests represent a reminiscence of the talonid attachment to the trigonid as in many other mammals, the deciduous molars in the endocasts may show conservative features, which are known neither from the occlusal anatomy nor from the permanent teeth, whether in endocasts or in occlusal surfaces.

This opinion is supported by the observation that the most conservative trigonid crest type is found in the highest frequency of the most conservative talonid crest type pattern. In general it can be stated that the three most conservative trigonid crest types contain most of the molars with the three most conservative talonid crest types.

The Upper Second Deciduous Molar

THE CRISTA OBLIQUA

In the permanent upper molars studied before, three types of the crista obliqua were distinguished (fig. 24.11):

Type A. A continuous crest between the tip of the protocone and the tip of the metacone, sometimes with a vertical notch at the

Table 24.3. Frequency of occurrence of talonid crests in deciduous molars with types of trigonid crests.

	Trigonid Crest Type						Total
Talonid Crest Type	1	2	3	4	5	6	Number
1	9	3	6	1	3	–	22
2	2	–	–	–	–	–	2
3	1	2	1	–	2	–	6
4	–	–	1	–	–	–	1
5	1	–	1	–	–	–	2
6	3	2	5	1	2	–	13
Total Number	16	7	14	2	7	–	46

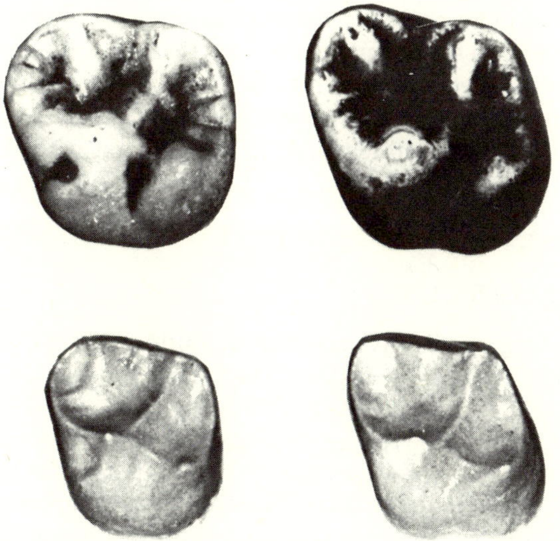

Figure 24.11. Types 1 and 2 crista obliqua in second upper deciduous molars from Sangiran.

Left upper: type 1; left lower: endocast of left upper molar also representing type 1 oblique crest; right upper: type 2; right lower: endocast of right upper molar representing type 1 crest.

place of the longitudinal main groove, but never fully cut by the latter.

Type B. A similar crest which, however, is distinctly interrupted by the longitudinal main groove.

Type C. A total absence of the crista obliqua. Contrary to the situation in Type B, the lingual ridge of the metacone and the distobuccal ridge of the protocone are not developed more strongly than those of the paracone.

The same types were distinguished in the study on the deciduous molars. In table 24.4 the frequency of occurrence of these types is shown both for the occlusal side and for the endocasts.

From the table it appears that the most conservative crest type on the

Table 24.4. Frequency of occurrence of the crista obliqua on the occlusal surface and in the endocasts of second deciduous upper molars.

Inside of the Enamel	Outside of the Enamel		
	Type A	Type B	Type C
Type A	54	9	–
Type B	–	–	–
Type C	–	–	–

outside is found in 86 percent, whereas the endocasts show the A type in 100% of the cases. In the permanent molars these percentages were 63 and 85.7 percent respectively, where the C type crests occurred in 22.5 percent and 7.8 percent, respectively. This type does not occur in deciduous teeth. In conclusion it may be stated that, in comparison with the permanent molars, the deciduous molars again show much more original characteristics. See also figure 24.14

CARABELLI'S CUSP

Following Jørgensen's division and definitions (1956), three types of Carabelli's structures can be distinguished (fig. 24.12):

Type a. This comprises an ". . . occlusally convex, well developed furrow (fissure or groove). The area of the lingual surface gingival to the furrow is strongly developed in the occlusal direction and rises, at least opposite the occlusal part of the furrow, to a higher occlusal level than the bottom of the furrow." This type corresponds to the true tuberculum anomale as defined by Carabelli (1842).

Type b. "The convex furrow is more or less strongly developed (the mesial half in almost every case is definitely more pronounced than is the distal). The area gingival to the furrow does not rise occlusally so far as the bottom of the occlusal part of the furrow."

Type c. "There is no trace of a convex furrow; but on the lingual surface of the protocone is found a vertical protoconule groove, which is shallow and has a gently rounded fundus."

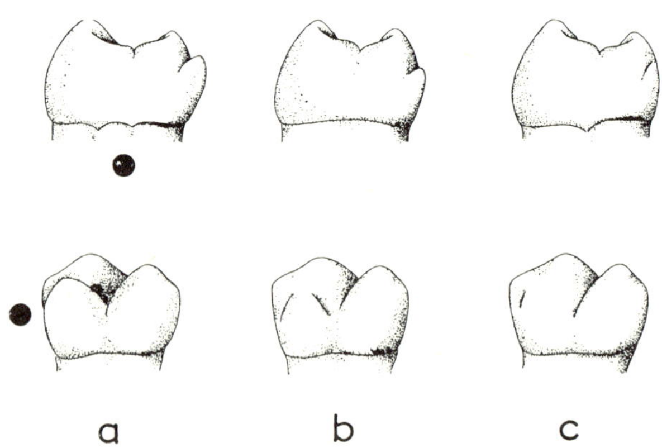

Figure 24.12. Carabelli structures types a–c.

Table 24.5. Frequency of occurrence of Carabelli's structures on the occlusal surface and in endocasts of the second deciduous molar.

	Outside of the Enamel				
Inside of the Enamel	Type A	Type B	Type C	No structure	Total
Cusp	1	6	7	–	14
Complete cingulum	–	–	5	4	9
Indications of cingulum	–	–	21	10	31
Pits	–	–	4	–	4
No structure	–	–	–	2	2
Total	1	6	37	16	60

The frequency of occurrence of Carabelli's structures on the enamel outer and inner surfaces and the correlation in types follows from table 24.5.

From the table it may be concluded that real Carabelli's cusps (types a and b) are found in only seven cases on the outside enamel anatomy, pits and small grooves in 37 cases (type c).

On the endocasts, however, complete or incomplete cingulums are present in 26 cases, i.e., in nearly half of the collection. See also figs. 24.11 and 24.15. In the permanent molars, the same applies to only 10 percent of the endocasts. Carabelli's cusps (types a and b) are also present in 10 percent of all endocasts in the permanent teeth (48 out of 475 molars); this is about the same percentage as in the temporary dentition.

The buccal side of the deciduous molars shows neither traces of supernumerary tubercles nor their derivatives neither on the outside nor on the inside surface.

THE CRISTA TRANSVERSA ANTERIOR

This anatomical landmark is a constant characteristic occurring in Primates, not only in recent ones (fig. 24.13), but also in Dryopithecinae, Pithecanthropus modjokertensis, and the Pleistocene Pongo from Java an China (fig. 24.2). The anterior transverse crest has been studied on both the outer and the inner enamel surface.

The five types earlier distinguished in the permanent molars can be defined as follows (fig. 24.14):

- Type 1. A lingual paracone ridge, connected with the mesial marginal ridge.
- Type 2. A distinct ridge, mesially from the lingual paracone ridge, not connected with the marginal ridge, and not extending in distal direction.
- Type 3. A complete crest from the tip of the paracone to the mesial marginal ridge (fig. 24.11, lower left).

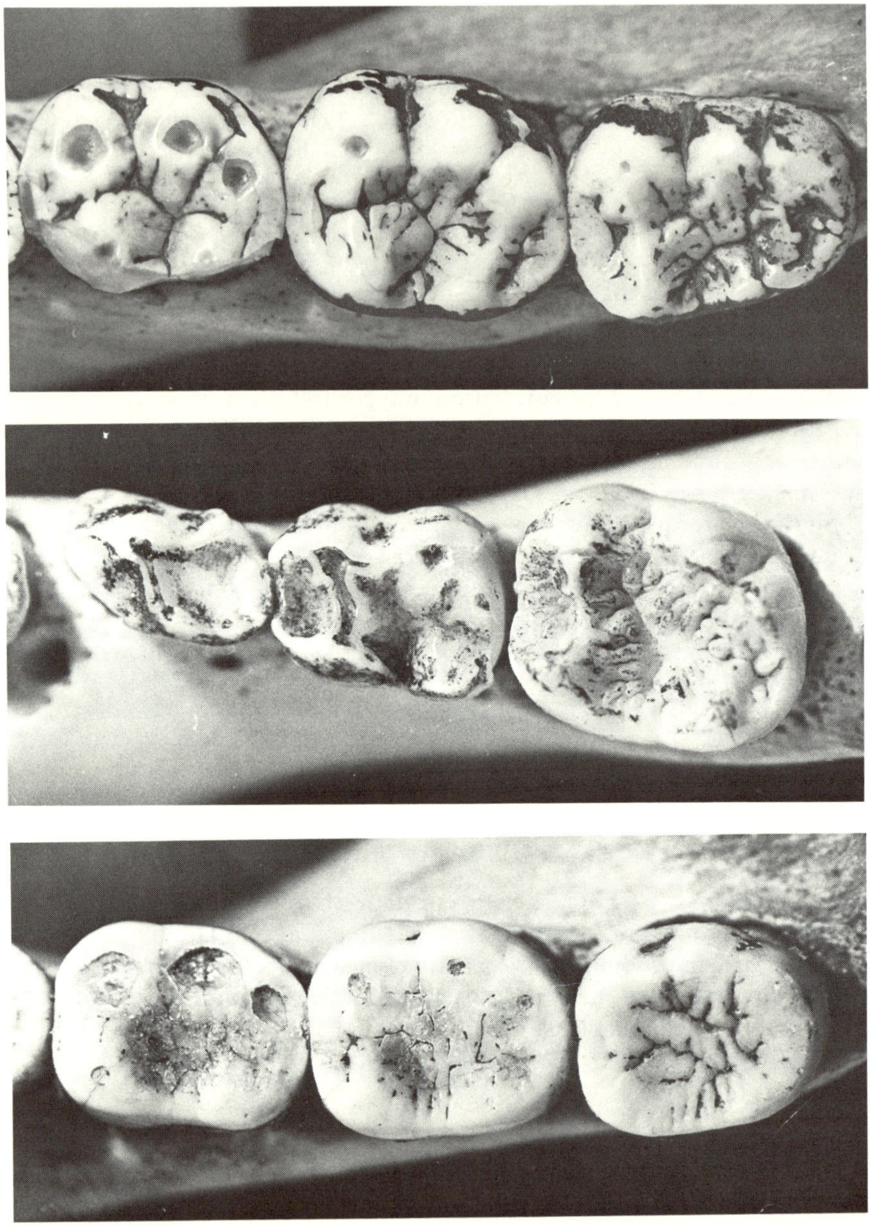

Figure 24.13. Lower molars of Gorilla (A), Pan (B) and Pongo (C) showing the anterior transverse crest.

Type	Definition	
1	Lingual paracone ridge connected with M. M. ridge	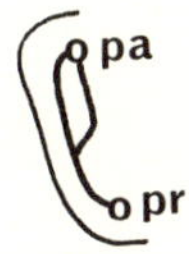
2	Distinct ridge, mesially from lingual paracone ridge, not connected with M. M. ridge	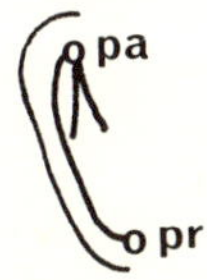
3	Complete crest from tip paracone - mes. marg. ridge	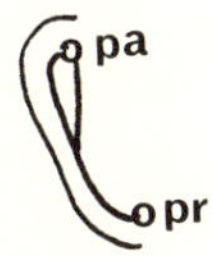
4	Uninterrupted connection from tip paracone - tip protocone	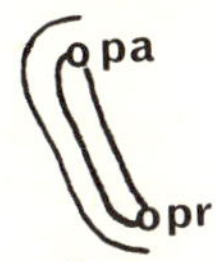
5	C - shaped crest between two localities on mes. marg. ridge	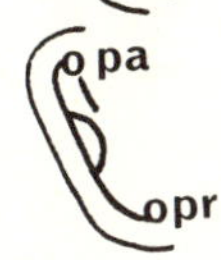

Figure 24.14. Types of the anterior transverse crest.

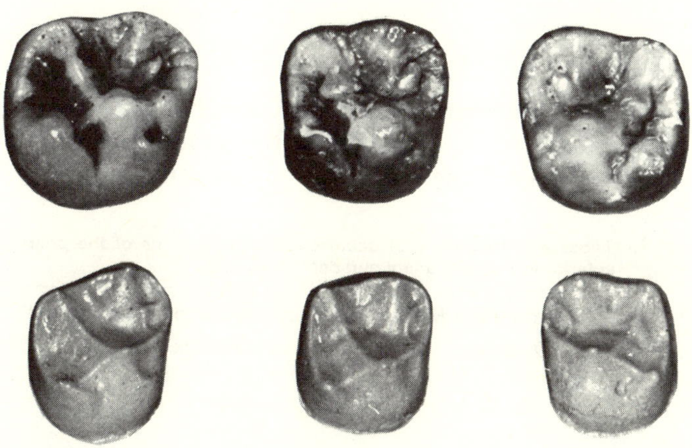

Figure 24.15. Type 5 examples in endocasts of upper second deciduous molars from Sangiran.

Upper row: occlusal aspects; lower row: endocasts.

Table 24.6. Frequency of occurrence of the crista transversa anterior on the occlusal surface and in endocasts of the second deciduous molar.

Type	Occlusal Surface	Endocasts	Total Number
1	–	1	1
2	2	8	10
3	–	–	–
4	–	–	–
5	3	8	11
Total Number	5	17	22

Type 4. An uninterrupted connection from the tip of the paracone to the tip of the protocone.

Type 5. A C-shaped crest between two points on the mesial marginal ridge (fig. 24.15, all three endocasts).

Table 24.6 shows the frequency of occurrence of the types for the outside enamel surfaces.

Though the numbers are too small to permit distinct conclusions it may be said that the enamel-dentin surface shows many more anterior transverse crests than the occlusal surface. In table 23.7, these findings are compared with those of the earlier studied permanent teeth.

From this table it may be concluded that the anterior transverse crests are found in a higher percentage in the deciduous teeth than in the permanent teeth, both on the outer and inner enamel surfaces.

Conclusions

The significance of the anatomy of the enamel-dentin border has been stressed and demonstrated by an examination of lower and upper second deciduous molars from Sangiran (Java). In comparison with the occlusal

Table 24.7. Types and frequency of occurrence of indications of the crista transversa anterior in the permanent and deciduous molars.

	Permanent Molars (448)		Deciduous Molars (63)	
Type	Outside	Inside	Outside	Inside
1	8	9	–	1
2	2	16	2	8
3	4	8	–	–
4	1 (?)	6	–	–
5	7	9	3	8
Total	1.5%	10.2%	8%	27%

anatomy, the endocasts (replicas of the enamel-dentin border plane) show more original conditions for the crista obliqua, Carabelli's trait or complex, and the crista transversa anterior. In comparison with the permanent dentition, in the milk teeth more original conditions can be concluded for the trigonid crest pattern, the crista obliqua and the crista transversa anterior. In the endocasts a new anatomical landmark has been discovered, the talonid crest, which is not present in the permanent teeth. This crest too, might be an indication of a more conservative construction of the temporary molars.

References

Carabelli, G. 1842. *Anatomie des Mundes.* Wien: Braumüller u. Seidel.

Jørgensen, K. D. 1956. The deciduous dentition. Thesis, University of Copenhagen, *Acta Odontologica Scandinavica.* Copenhagen: Bianco Lunos Bogtrykkeri A-S.

Korenhof, C. A. W. 1960. Morphogenetical aspects of the human upper molar. A comparative study of its enamel and dentine surfaces and their relationship to the crown pattern of fossil and recent primates. Utrecht: Acad. proefschrift, Uitg. mij. Neetlandia.

—— 1961. The enamel-dentin border: a new morphological factor in the study of the (human) molar pattern. *Proc. Koninkl., Nederl. Acad. Wetensch.*, series B. 64:639–664.

—— 1963. The enamel-dentin border: a new morphological factor in the study of the (human) molar pattern. *Nederl. Tijdschr. voor Tandh. Suppl.* 70:30–57.

—— 1978a Remnants of the trigonid crests in medieval molars of man of Java. In P. M. Butler and K. A. Joysey, eds., *Development, Function, and Evolution of Teeth,* pp. 157–170. New York: Academic Press.

—— 1978b De evolutie van het ondermolaarpatroon en overblijfselen van het trigonid bij de mens. *Nederl. Tijdschr. voor Tandh.*, 85:456–549 en 493–499.

Remane, A. 1921. Beiträge zur Morphologie des Anthropoidengebisses. *Archiv für Naturgeschichte* 87 (Abt. A):1–179.

25. Dental Reduction *Selection or Drift?*

PATRICIA SMITH

ABSTRACT

In each of four regions—Australia, South America, the Near East, and North Africa—a similar pattern of gradients in tooth size is repeated. Populations living in river valleys with a high proportion of fish and shellfish in the diet have larger teeth than their neighbors living in a drier environment, where seeds and nuts are an important food staple. Diachronic and synchronic comparisons of tooth size in Australian populations support the hypothesis that ecological factors have influenced the rate of dental evolution within *Homo sapiens*.

Dental reduction is a well-known characteristic of hominoid evolution (Le Gros Clark 1964, Bailit and Friedlander 1966; Smith 1977). In the earliest Hominidae, *Australopithecus africanus*, dated at 4 million years B.P. to 2.3 million years B.P. (Tobias 1978), the anterior teeth are small and the cheek teeth (premolars and molars), large and broad. In the line that gave rise to *Homo habilis*, the teeth gradually reduced in size, along a posterior-anterior gradient, both mesiodistally and buccolingually. This trend continued and in *Homo erectus*, the cheek teeth are 7 percent smaller than those of *Australopithecus africanus*, although the incisors and canines show no tendancy to reduction but rather a tendency to become larger. Indeed the upper lateral incisors in *H. erectus* from Choukoutien as measured by Weidenreich (1937), are larger than the maximum values

I would like to thank T. Brown for his help and encouragement with my research on Australina dentitions W. Lambert, for her assistance with data analysis and G. Townsend, P. Telfer, and L. Richards for their many helpful suggestions. I would also like to thank G. Pretty and W. Woods, for allowing me access to their collections.

The research was supported in part by grants from the Wenner Gren Foundation and Joint Research Fund of Alpha Omega and Hadassah.

Table 25.1. Overview of reduction in tooth size associated with changing diet and technology.

Years B.P.	Percent reduction in tooth size in maxilla	Sources of Food	Diet	Technology
>3,000	25	Agriculture/horticulture; animal/domestication.	Further increase in cereal consumption milk & milk products.	Metallurgy and pottery.
>10,000	21	Incipient agriculture; animal domestication?	Incrase in cereal consumption; possible use of milk & milk products.	Microliths projectile point.
>14,000	17	Increased dependancy on one or more resources.	Fish, birds, shellfish included in the diet; cereals probably ground into flour.	Blade tools, fishhooks, grindingstones, mortars, & pestles.
>40,000	13	Hunting and gathering, especially herd animals.	Meat & vegetable foods.	Awls, Leval ois tech, earth over.
>100,000	7	Hunting and gathering, including big-game hunting.	Meat & vegetable foods.	Handaxes fire-use
$(2.3\text{–}4)10^{6}$*	–	Hunting and gathering. Scavenging?	Meat & vegetable foods.	Chopping-tools & flakes.

SOURCE: Based on Robinson (1956), Isaac (1954), and Tobias (1978).
NOTE: Reduction calculated as percentage of summed lengths (MD), of maxillary teeth in each group/summed length of maxillary teeth in Australopithecus africanus.
*A. africanus

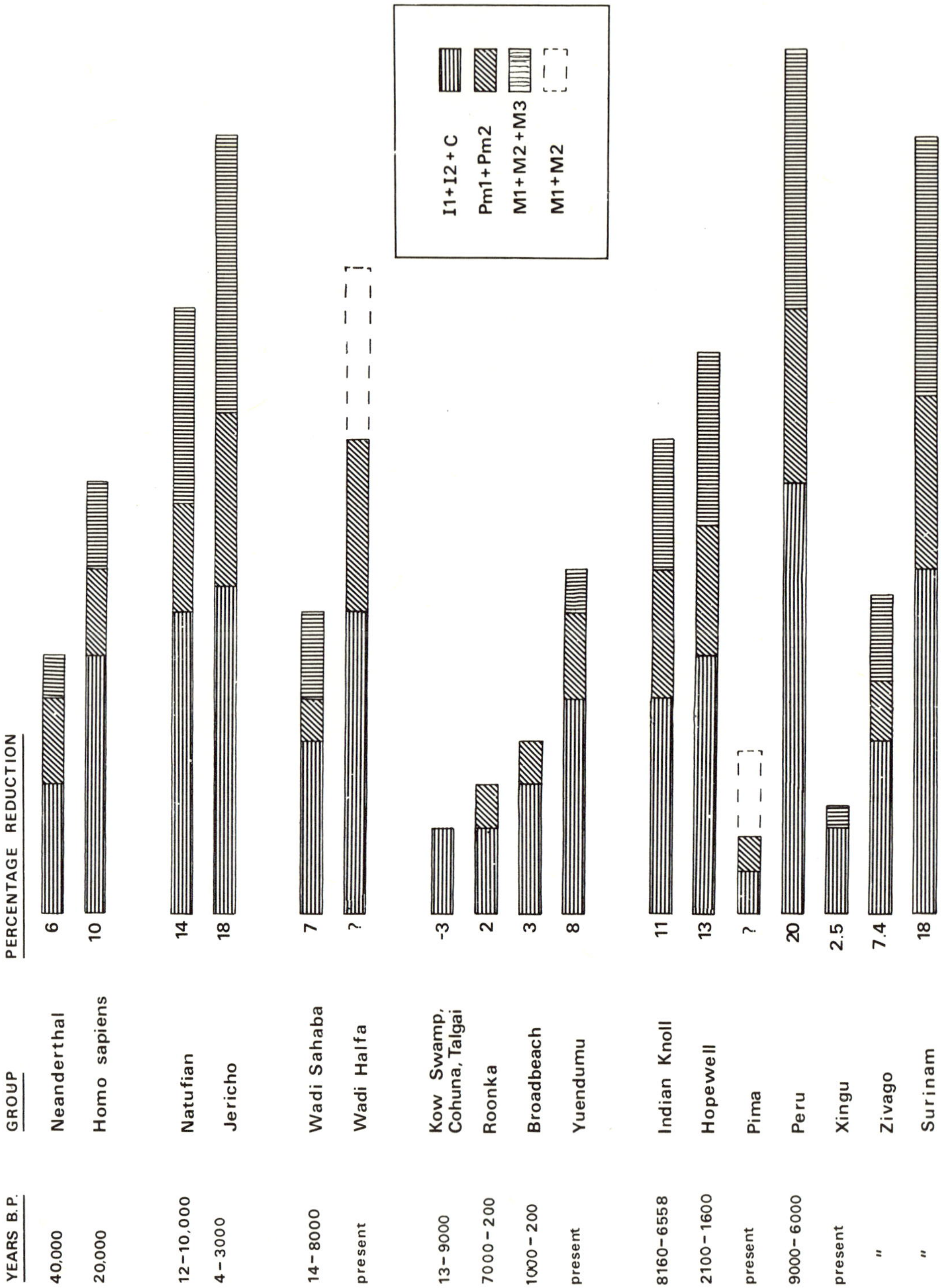
I1+I2+C
Pm1+Pm2
M1+M2+M3
M1+M2
YEARS B.P.
GROUP
PERCENTAGE REDUCTION
40,000 Neanderthal 6
20,000 Homo sapiens 10
12–10,000 Natufian 14
4–3000 Jericho 18
14–8000 Wadi Sahaba 7
present Wadi Halfa ?
13–9000 Kow Swamp, Cohuna, Talgai -3
7000–200 Roonka 2
1000–200 Broadbeach 3
present Yuendumu 8
8160–6558 Indian Knoll 11
2100–1600 Hopewell 13
present Pima ?
9000–6000 Peru 20
present Xingu 2.5
" Zivago 7.4
" Surinam 18

Note: All values calculated as percentage of values obtained for mesiodistal diameters in *Homo erectus*. Data for *H. erectus*, Neanderthals and *Homo spaiens* 20,000 B.P. from Smith 1977; for Natufian and Wadi Sahaba from Smith 1979; for Jericho and Wadi Halfa from unpublished data, Hebrew University. Data for Yuendumu taken from Townsend and Brown 1979; for Indian Knoll from Perzigian 1979; for Hopewell from Scinlli 1979; for Pima from Hanihara 1976; for Peru from Scott 1979; for Xingu, Zivago, and Surinam from de Smet and Brabant 1969.

Values for *Homo erectus*, Neanderthal, Natufian, Jericho, Wadi Sahaba and Peru are based on mixed samples, containing unknown proportions of males and females. Since teeth in females are some 3 to 9 percent smaller than teeth in males, this may have decreased the values relative to those of other populations, where all male values were used. However, even with this caution synchronic variations clearly exceed diochromic variation, and this applies to tooth groups affected as well as to total amount of reduction. Premolars and molars in males from the Kow Swamp-Cohuna-Talgai group, are 5 percent larger than those of the mixed *H. erectus* series, and molars in males from Roonka and Yuendumu are 1 percent larger. Pima premolars are equal in size to those of *H. erectus*.

Figure 25.1. Percentage reduction in maxillary teeth of different populations.

recorded for any Australopithecine, even taking into consideration the Laotolil and Hadar specimens, who are reported to have the largest incisors and canines of any Australopithecine (Johanson and White, 1979).

It is in Neanderthal populations that the anterior teeth first show a definite reduction, and the summed length of mesiodistal diameters of the maxillary teeth in Neanderthals is 6 percent smaller than that found in *H. erectus*. This trend is continued, and greatly accelerated in some *H. sapiens* populations (table 25.1), but by no means in all (fig. 25.1). What factors then may have affected the rate of dental evolution in hominids?

The dental changes outlined above, which are associated with reduction of the jaws, increased body size and cranial capacity, have been related to dietary changes associated with behavioral modifications and technological developments. The switchover to meat eating and tool use, characteristic of *Homo habilis* (Isaac 1971; Tobias 1978), was presumably initiated in ancestral *Australopithecus africanus* populations, in whom the masticatory system is small relative to *Australopithecus robustus* (Robinson 1956, 1961). Hunting activities became more sophisticated throughout the Pleistocene, with *H. erectus* populations efficient big game hunters and conversant with the use of fire at least in Europe and Asia (Howell 1960, 1967; Isaac 1974). The location of Neanderthal living sites demonstrates the ability of Neanderthal populations to cope with cold conditions, including those in which plant foods may have been seasonally scarce. Upper Paleolithic *Homo sapiens* sites show a tendency to change. Shell middens, fish and avian remains, mortars and pestles, and grinding stones, provide good evidence of a wide range of specialized adaptations (Freyer 1977; Smith, 1979).

Each of these changes is alleged to have reduced the functional load borne by the teeth, first through the adoption of a meat based diet, but second through tool use in food preparation, and finally cooking (Dahlberg 1965). Some workers have, however, stressed the importance of large teeth as tools in primitive populations. Specifically, claims have been made that only with the introduction of blade tools was there relaxation of selective pressures maintaining large anterior teeth in presapiens populations (Brace 1967, 1979a; Brose and Walpoff 1971). But this equation of tooth size with tool technology, rather than with the physical consistency of the diet, does not adequately explain the wide range of variation of tooth size, and specifically the size of the anterior teeth as found within each of the major racial groups of *H. sapiens*. Some but not all recent Australian aboriginal groups have teeth larger than those of Neanderthals, as do some living Amazonian Indian tribes. Even in Epipaleolithic populations from Egypt and Israel, regional variations in tooth size are considerable (fig. 25.1).

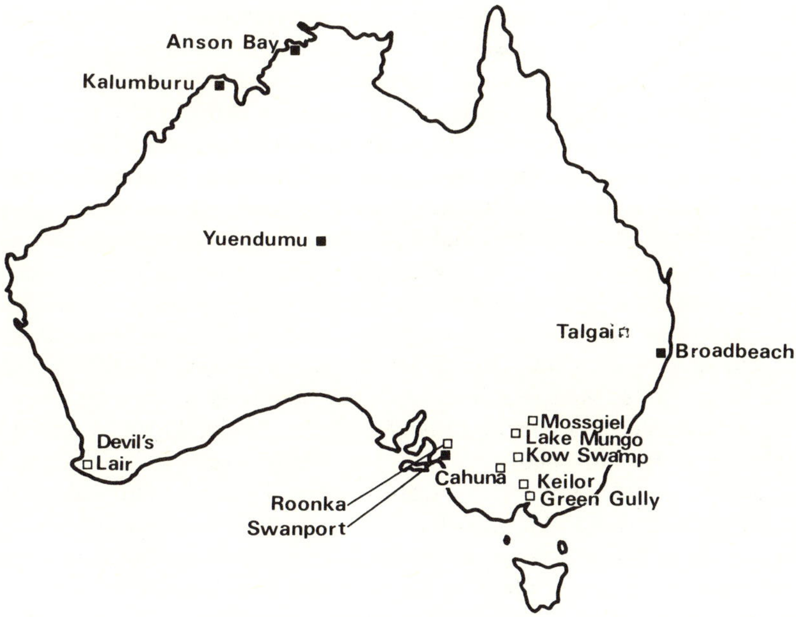

Figure 25.2. Location of sites sampled.

Since it is the most isolated region and therefore presumably the least affected by outside influences, let us examine the Australian sequence in some detail (fig. 25.2). During the Upper Pleistocene, Australia, New Guinea, Tasmania, and the intervening islands formed one large continent. The climate was colder and wetter than it is now, and there were numerous inland lakes and rivers (Bowler 1976). Archeological surveys have shown that the coastal zone and inland waterways have been inhabited for at least 30,000 years and the semi-arid interior for at least 10,000 years (Mulvaney 1975). Bowdler (1977) has suggested that the earliest settlers may have reached Australia by boat, were fully adapted to a littoral economy based on fishing and shellfish collecting, and were spread throughout the country along the coast and rivers. Toward the end of the Pleistocene, sea levels rose, cutting off New Guinea and Tasmania and many small islands. Within Australia, the climate became hotter and drier. Many lakes and rivers dried out, affecting food resources and triggering off new adaptations to cope with altered ecological conditions.

Lake Mungo, the earliest hominid site with reliable dates (Thorne 1976), offers convincing documentation of this hypothesis. Located in southeast Australia along a dried-out lake bed, numerous occupation levels have been identified. The earliest, with hominid remains, dates at around 30,000 B.P. Fish bones and mussel shells testify to utilization of lacustrine resources, while hearths and fire-cracked lumps of mud, probably heated and used for heating food in earth ovens, resemble those used at the time of European contact just over 100 years ago. A number of fragmentary human skeletal remains have been recovered, dated at 28,000–30,000 B.P. and 24–25,000 B.P., but their teeth are few and badly worn.

The earliest site with a reasonable number of specimens is Kow Swamp, dated at 13,500–9,500 B.P., followed by Roonka with three levels, dated at 7,000–4,000 B.P., 4,000–2,000 B.P. and 2,000 B.P. to contact. In addition there are over 12 specimens from Chowilla, dated at 6,000–4,000 B.P.; isolated specimens from Cohuna and Keilor at around 13,000 B.P.; one from Talgai, dated back to at least 13,000 B.P. and possibly even earlier, and other single specimens from Mossgiel, Green Gully, Lake Nitchie and Tartanga, all around 6,000 years old (Thorne 1976; Pretty 1977; Blackwood and Simpson 1973; Mulvaney 1975; Hale and Tindale 1930). Talgai was found in Queensland. All the other specimens were found in the Murray Valley region. The only other early fossil hominid teeth are the two Cossach specimens recently described by Larnach and Lofgren (1979) which are possibly up to 6,000 years old, and one tooth from the Devil's Cave, which is 8,000–12,000 years old, (Davies 1968). Both of these sites are in western Australia.

Table 25.2. Maxillary tooth diameters in male specimens from early sites in Australia.

Maxilla	Lake Nitchie	Mossgiel	Talgai	Keilor	Cahuna	Kow Swamp
I MD						10.6
BL						8.5
I MD						7.6
BL	7.6					7.1
C MD			9.5		8.3	–
BL	9.4		10.6		–	9.3
P MD					8.9	
BL					10.6	10.4
P MD				7.2	7.8	
BL				10.6	10.7	9.8
M MD	–		12.0	–	–	
BL			12.9	13.0	13.6	13.6
M MD	–		11.0	9.9	11.2	12.5
BL	13.9		13.6	13.8	14.2	14.2
M MD	10.6	9.2		9.4	11.3	11.2
BL	13.2	11.8		12.5	13.6	13.0

All values are averaged values of right and left. Individual specimens except for Kow Swamp where numbers range from 1 to 4. Note difference in values from those reported by Thorne (1976) where mixed samples were used.

Table 25.3. Mean values for tooth measurements in male aborigines from different regions in Australia.

Maxilla	Swanport			Yuendumu			Kalumburu			Anson Bay			Broadbeach			Chowilla			Roonka		
	No.	x	SD	No.	x	SD	No	x	SD	No	x	SD	No.	x	SD	No.	x	SD	No.	x	SD
I^1 MD	7	9.2	.3	206	9.4	.6	16	9.5	.7	4	9.4	.7	7	9.8	.8	4	10.0	.5	7	10.2	.5
BL	9	7.9	.4	205	7.9	.5	10	8.2	.3	6	8.0	.7	8	8.0	.6	4	8.1	.5	11	8.4	.7
I^2 MD	11	7.6	.4	203	7.5	.6	16	7.6	.9	3	8.0	.9	12	7.7	.6	6	7.9	.8	9	8.1	.6
BL	14	7.1	.5	194	6.9	.5	10	7.2	.5	6	7.0	.4	14	6.9	.5	6	7.1	.4	17	7.3	.5
C MD	15	8.3	.7	192	8.2	.5	10	8.2	.5	5	8.6	.3	14	8.3	.4	6	8.8	.3	17	8.6	.6
BL	16	9.2	.7	186	9.0	.6	9	9.1	.4	5	9.2	.6	15	9.0	.5	6	9.5	.3	18	9.6	.5
P^1 MD	19	7.6	.4	200	7.6	.4	12	7.8	.4	5	7.8	.8	20	7.8	.6	5	8.1	.5	17	8.0	.5
BL	20	10.4	.5	199	10.2	.5	10	10.5	.4	6	10.6	.6	18	10.4	.5	6	10.8	.4	17	10.8	.7
P^2 MD	18	7.3	.3	191	7.1	.4	9	7.2	.3	4	7.5	.6	19	7.4	.5	6	7.7	.5	17	7.6	.4
BL	20	10.4	.6	191	10.2	.5	10	10.6	.3	6	10.6	.8	20	10.4	.6	5	10.3	.8	18	10.7	.7
M^1 MD	23	11.4	.5	208	11.2	.6	16	11.2	.5	4	11.6	.8	9	11.9	.6	7	11.7	.4	16	11.6	1.0
BL	26	13.0	.7	208	12.5	.6	13	12.5	.6	6	13.4	.7	9	13.0	.5	6	13.4	.5	17	13.0	.8
M^2 MD	20	10.8	.6	158	10.8	.6	9	11.0	.1	5	10.5	.9	20	11.6	.9	7	11.2	.3	18	11.2	.5
BL	23	13.5	.5	154	12.8	.7	7	13.2	.4	6	13.3	.7	19	13.4	.7	6	13.7	.6	18	13.7	.8
M^3 MD	16	9.8	.6	57	10.0	.7	2	9.9	–	7	10.1	.7	22	10.1	1.0	4	10.1	.6	19	10.3	.8
BL	17	12.7	.7	48	12.3	.8	2	12.2	–	7	13.2	.7	22	12.4	1.0	5	13.5	.7	19	12.4	1.1

I measured all of these, with the exception of the Tartanga and western Australian specimens, and the values calculated for tooth diameters in males are shown in table 25.2. To obtain data on the more recent past, I also measured tooth size in recent specimens from the North (Anson Bay and Kalumburu), Murray River south of Roonka (Swanport) and Eastern

Coast (Broadbeach). The values obtained, together with those of Townsend (1976) for the central desert Walbiri are shown in table 25.3. Tooth size in the Kow Swamp specimens, and especially molar size, is larger than that of the 7,000–4,000-year-old group, but there are no discernable differences between these and the more recent groups represented by the later levels at Roonka and Broadbeach although there are differences between them and other recent populations. There is considerable homogeneity between the north, central, and south regions with teeth in all four groups significantly smaller than those of Broadbeach and Roonka. Yuendumu, the desert group, has the smallest teeth, with 21 of 32 diameters calculated as smaller than those of Swanport and Kalumburu.

Although I did not measure the teeth from Yuendumu myself, Townsend and I compared measurement techniques. On repeated measurements our average differences were of the order of ± 0.2 mm. so that our measurements have been compatible, while Richards and Telfer (1979) found an increased frequency of reduction of lower second molar cusp number in Yuendumu than in other Australian populations examined by them. This I consider independent evidence of smaller tooth size at Yuendumu.

Unfortunately, there are very few specimens from Australia earlier than the Kow Swamp group. At Lake Mungo, only one third molar could be measured. This was small, as were the roots and sockets of the remaining teeth, which were too badly worn or damaged to be measured. It seems, therefore, that the teeth were smaller than those of Kow Swamp, and this confirms to the pattern of skeletal differences reported by Thorne (1976) with Mungo Lake specimens more gracile. Outside of Australia, the only specimens from this age range in southeast Asia are the Niah Cave adolescent which may be up to 40,000 years old, and the mandible and possibly frontal bone from Tabon Cave in the Philippines, which is 24,000 years old. Howells (1972) and Macintosh (1978) consider the Tabon specimen to resemble early Australian populations and to be non-Mongoloid. Thorne (1977) agreed, and said that both the Tabon and Niah Cave specimens resembled the Lake Mungo specimens more than the Kow Swamp specimens.

The material at our disposal does not then favor the hypothesis of an extra robust founding population in Australia, but rather suggests that within the continent, selective forces favored the maintenance of robust skeletal characteristics and large teeth. The post-Pleistocene populations represented by the earliest level at Roonka show some reduction of tooth size and overall skeletal robusticity when compared with Kow Swamp, but over the past 7,000 years little further change seems to have taken place in the population in this immediate area or indeed in the populations of the eastern coast, as represented by Broadbeach. Tooth size and skeletal morphology were, however, modified in the populations to the north, in the central desert, and to the south, including the lower Murray region, with the desert tribes having the smallest teeth.

Brace (1979), found tooth size in some northeastern coastal groups, notably from Cape York and Cairns, to be smaller than those of the groups described in this study, and suggested that size differences in teeth within the continent of Australia showed a north-south cline, resulting from genetic admixture with small-toothed newcomers. The numerous exceptions to this cline he dismissed as reflecting relict populations that had been bypassed. Certainly genetic admixture would account for the small teeth of the Cape York and Cairns groups, which do show many resemblances to New Guinean populations and differ from other aboriginal groups (Macintosh and Larnach 1973). However the north-south cline he proposed does not apply to the data presented here, which show no differences between tooth size in the north (Anson Bay and Kalumburu) and the south (Swanport). Neither does it explain the disconformities shown by large teeth at Roonka and Broadbeach and small teeth in the Yuendumu, living to the east of these sites. Finally, tooth sizes in the Tasmanians (surely the most remote of aboriginal "relict" populations to use Brace's term (1979), are the same dimensions as those of Swanport (Adams 1940). Differences in tooth size appear rather to reflect differences in habitats utilized. In sub-Saharan Africa, a number of studies have demonstrated a high correlation between habitat and body build, with climate playing an important role (Passarello and Vecchi 1979).

Taken over the whole continent, tooth size in Australia seems to reflect long term localized adaptations. The ethno-archeological evidence indicates that diet and cooking methods have changed little in coastal regions and river valleys, where fish, shellfish, small game and plants were eaten, but that these do differ in the interior. Grinding stones present around 18,000 B.P. in the north, first appear in the southeast at around 13,000 B.P., when the climate was becoming drier, and may reflect changing strategies associated with shortage of traditional foods (Mulvaney 1975).

Desert tribes exploit a wider variety of foods than coastal groups, and this includes the collection of grass seeds which are winnowed, ground, mixed with water, and then baked in hot ashes. The resulting cake is abrasive, but needs moistening rather than chewing to be digested and contains carbohydrates in high concentration. Cereals also contain high concentrations of phytic acid which retard Ca^{++} absorption. Is it too fanciful to suggest that this combination of factors may favor selection for smaller teeth and reduced skeletal robusticity? Certainly in other parts of the world there is a correlation between diet, tooth size and body build.

Those populations with a heavy reliance on riverine resources—fish, shellfish, green plants and tubers—have larger teeth and are of sturdier build than their contemporaries living in more arid zones, where seeds and nuts are an important food staple. In Australia this applies to the differences found between the Murray Valley and Broadbeach populations and those of the semi-arid interior. The percentage for differences found between mesiodistal tooth diameters of various *Homo sapiens* populations and those of *Homo erectus* as shown in table 25.2 now fall into a consistent pattern.

Tooth size in river valley living South American populations (Xingu and Jivago) shows less reduction than is found in cereal eating Peruvians to the Southeast and Venezualans and Surinam Indians to the north. Scott's (1979) findings for tooth size in early Peruvians, show that these differences are long standing, and antedate organized farming activities in Peru. Certainly in North America, the differences in tooth reduction between late Archaic Indian Knoll populations and living Pima Indians, like those found in southwest Asia and Egypt between Epipaleolithic Nubians and Natufians, are to be ascribed to differences in food utilization rather than technology. As pointed out by Dahlberg (1965), and as the Australian record demonstrates, marked differences in physical consistency of foods utilized may occur between groups of similar technological status adapting to different ecological conditions.

In each region, reduction in tooth size has occurred in varying amounts, and in each region a characteristic tooth size profile is maintained despite differences in absolute dimensions of the teeth. This stability of tooth pattern has previously been commented on (Bailit and Friedlander 1966), but its significance, in relation to Brace's (1967) hypothesis relating reduction of anterior teeth to the development of blade tools, seems to have been largely ignored. If selective forces were acting specifically on one dental segment, this unit would be maintained at the expense of the other teeth and influence skeletal morphology. If, on the other hand, it is the total amount of tooth material that is being acted upon with a balance maintained between the economical advantage of smaller teeth and functional demands necessitating adequate tooth mass, then the rate of reduction will reflect this. The pattern of reduction, i.e., tooth group affected, will vary in relation to external factors affecting the shape of the jaws; factors which include accommodation of the brain, organs of special sense and balance, muscle function and adaptation to temperature and humidity of the air. This is what seems to have taken place. Skeletal differences in early hominids are more pronounced than dental differences; the amount of dental reduction is related to changes in diet and skeletal robusticity, but the pattern of reduction appears to vary with ancestry rather than with technology, and the differences extend back at least 40,000 years.

References

Adam, W. 1940. The jaws and dentition of the Tasmanian Aborigines. D. Dent. Sci. thesis, Melbourne.

Bailit, H. L. and J. S. Friedlander. 1966. Tooth size reduction: a hominid trend *Am. Anthrop.* 68:665–672.

Blackwood, R. and K. N. G. Simpson. 1973. Attitudes of Aboriginal Skeletons excavated in the Murray River region between Mildura and Ronmark, Australia. *Mem. Nat. Mus. Victoria* 34:99–150.

Bowdler, S. 1977. The coastal colonisation of Australia In. J. Allen, J. Golson, and R. Jones, eds., *Sunda and Sahul*, pp. 205–246. New York: Academic Press.

Bowler, J. M. 1976. Recent developments in reconstructing late Quaternary environments in Australia in R. L. Kirk and A. G. Thorne, eds., pp. 55–57. Canberra: Australian Institute of Aboriginal Studies.

Brace, C. L. 1967. Environment, Tooth form and size in the Pleistocene *J. Dent. Res.* 46:809–816.

—— 1979a. Krapina, 'Classic Neanderthals' and the evolution of the European face. *J. Hum. Evolution* 8:527–550.

—— 1979b. Australian tooth-size clines and the death of a stereotype. *Curr. Anthrop.* 21:141–153.

Brose, D. S. and M. H. Walpoff. 1971. Early Upper Paleolithic Man and Late Middle Paleolithic tools. *Am. Anthrop.* 73:1156–1194.

Dahlberg, A. A. 1965. Geographic distribution and origin of dentitions. *Int. Dent. J.* 15:348–355.

Davies, P. L. 1968. An 8–12,000 year old human tooth from Western Australia. *Oceania* 3:33–40.

de Smet, R. and H. Brabant. 1969. Observations Anthropologiques sur la Denture Indiens Jivagos. *Bull. Soc. Roy. Belge Anthrop. Prehist.* 80:97–123.

Freedman, L. and M. Lofgren. 1979. Human Skeletal remains from Cossack W. Australia. *J. Hum. Evolution* 8:283–299.

Freyer, D. W. 1977. Metric dental change in the European Upper Paleolithic and Mesolithic. *Am. J. Phys. Anthrop.* 46:109–120.

Hale, H. M. and N. B. Tindale. 1930. Notes on some human remains in the lower Murray Valley S. Australia. *Records S. Aust. Mus.* 4:145–218.

Hanihara, K. 1976. Statistical and comparative studies of Australian Aboriginal Dentition. *Bull. No. 11* Tokyo: University Museum, University of Tokyo.

Howell, F. C. 1960. European and northwest African Middle Pleistocene hominids. *Curr. Anthrop.* 1:195–232.

Howell, F.C. 1967. Recent advances in human evolutionary studies. *Quart. Rev. Biol.* 42:471–513.

Howells, W. W. 1972. Analysis of patterns of variation in crania of Recent Man. In R. Tuttle, ed., The *Functional and Evolutionary Biology of Primates*, pp. 123–151. Chicago: Aldine.

Isaac, G.L. 1971. The diet of early man: aspects of archeological evidence from Lower and Middle Pleistocene sites in Africa. *World Archaeol.* 2:278–299.

Isaac, G.L. 1974. Stratigraphy and patterns of cultural change in the Middle Pleistocene. Report of discussion on. *Curr. Anthrop.* 15:508–514.

Johanson, D.C. and T.D. White. 1979. A systemtic assesment of early African Hominids. *Science* 202:321–330.

Le Gros Clark, W.E. 1964. *The Fossil Evidence for Human Evolution.* Chicago: University of Chicago Press.

Macintosh, N.W.G. 1978. The Tabon Cave mandible. Arch. and Phys. Anthrop. *Oceania* 13:143–166.

Macintosh, N.W.G. and S.L. Larnach. 1973. A cranial study of the Aborigines of Queensland, with a contrast between Australian and New Guinea crania In R.L. Kirk, ed., *The Human Biology of Aborigines in Cape York.* pp. 1–12. Australian Aboriginal Studies 44. Canberra: Australian Institute of Aboriginal Studies.

Mulvaney, D. 1975. *The Prehistory of Australia.* London: Pelican Books.

Passarello, P. and F. Vecchi. 1979. Morphologic variability of human African populations South of Sahara. *J. Hum. Evolution* 8:467–474.

Perzigian, A.J. 1976. The Dentition of the Indian Knoll Skeletal population; odontometrics and cusp number. *Am. J. Phys. Anthrop.* 44:113–122.

Pretty, G. 1977. The cultural chronology of Roonka Flat—a preliminary consideration In R.V.S. Wright, ed., *Stone Tools as Cultural Markers: Change, Evolution and Complexity.* Canberra: Australian Institute of Aboriginal Studies.

Richards, L., and P. Telfer. 1979. Dental traits in Australian Aboriginals. *Arch. and Phys. Anthr. in Oceania* 14:184–194.

Robinson, J.T. 1956. The dentition of the Australopithecinae. *Mem. 9.* Pretoria: Transvaal Museum.

Robinson, J.T. 1961. The origin and adoptive radiation of the Australopithecines In G. Kurth, ed., *Evolution and Hominisation*, pp. 120–140 Stuttgart: G. Fischer.

Sciulli, P.W. 1979. Size and morphology of the permanent teeth in Prehistoric Ohio Valley Amerindians. *Am. J. Phys. Anthrop.* 50:615–628.

Scott, E.C. 1979. Increase of tooth size in prehistoric coastal Peru, 10,000 B.P.–1,000 B.P. *Am. J. Phys. Anthrop.* 50:251–258.

Smith, P. 1977. Selective Pressures and Dental Evolution in Hominids. *Am. J. Phys. Anthrop.* 47:453–458.

—— 1979. Regional variation in Epipaleolithic populations OSSA 6:243–250.

Thorne, A.G. 1976. Morphological constrasts in Pleistocene Australians In R.L. Kirk and A.G. Thorne, eds., *The Origin of the Australians*, pp. 95–112. Canberra: Australian Institute of Aboriginal Studies

—— 1977. Separation or Reconciliation? Biological clues to the development of Australian Society In J. Allen, J. Golson and R. Jones, eds., *Sunda and Sahul*, pp. 187–205. New York: Academic Press.

Tobias, P.V. 1978. The earliest Transvaal members of the genus Homo with another look at some problems of hominid taxonomy and systematics. *Z. Morph. Anthrop.* 69:225–265.

Townsend, G.C. 1976. Tooth size variability in Australian Aboriginals: a descriptive and genetic study Ph. D. dissertation, University of Adelaide.

Townsend, G.C. and T. Brown. 1979. Tooth size characteristics of Australian Aborigines. *Occasional Papers in Human Biology* 1:17–28, Canberra: Australian Institute of Aboriginal Studies.

Weidenreich, F. 1937. The dentition of Sinanthropus pekinensis. *Paleont. Sinica* (Series D, vol. 1.).

Index